Molecular Aspects of Innate and Adaptive Immunity

Dedication

This book is dedicated to all former members of the Medical Research Council Immunochemistry Unit, 1967–2008, and especially in remembrance of the staff members Rod Porter, Rachel Fruchter, Lawrence Mole, Alan Williams, Tony Gascoyne, Audrey Richards, and visiting scientists Marion Koshland, Metta Strand, Walter Palm and Gordon Ross.

Molecular Aspects of Innate and Adaptive Immunity

Edited by

Kenneth BM Reid and Robert B Sim

MRC Immunochemistry Unit, Department of Biochemistry, University of Oxford, Oxford, UK

RSC Publishing

ISBN: 978-0-85404-698-0

A catalogue record for this book is available from the British Library

Published by The Royal Society of Chemistry,
Thomas Graham House, Science Park, Milton Road,
Cambridge CB4 0WF, UK

Registered Charity Number 207890

For further information see our website at www.rsc.org

Preface

This volume provides a survey of topics in the area of innate and adaptive immunity, which have been researched within the MRC Immunochemistry Unit, Oxford University, over a period of 40 years. The Unit was formed in 1967, with Professor R. R. Porter (who was awarded a Nobel Prize in 1972 for his research on antibody structure) as its first Director, with Professor K. B. M. Reid being Director from 1986 onwards.

The extension of the research on antibody structure into a study of one of the body's major defence systems, the complement system (which can be triggered by many stimuli including antibody recognition of invading microbes) led to the molecular characterization of many of the proteins of this important cascade system. Particular attention was paid to the structure–function relationships and genetics of the early components of the system, especially components C1q, C1r, C1s, C2, factor B, factor D, the allotypes of C4, the covalent binding properties of activated C3 and C4, and also the proteins that regulate complement activation – such as C1inhibitor, factors H and I, C4b-binding protein and properdin. Studies of complement receptor specificity and structure were also undertaken, as well as a range of studies on complement evolution and phylogeny.

The domain organization and functions of human leukocyte integrins (two of which are complement receptors), particularly with respect to heterodimer formation and activation of integrins, analysis of their deficiency [leukocyte adhesion deficiency (LAD)] and the involvement of these molecules in the recognition and resolution of inflammation, was another long-term research topic.

The study of the molecular genetics of complement component C4 led on to the analysis of a range of disease-susceptibility genes within the human major histocompatibility complex (MHC) Class III region, and of associated novel genes within the human MHC. These included heat shock protein (HSP) 70, tumour necrosis factor (TNF) alpha and cytochrome (Cyt) P450 loci and their association with immune-related diseases, a topic that has been a major research interest within the Unit.

Molecular Aspects of Innate and Adaptive Immunity
Edited By Kenneth BM Reid and Robert B Sim

Published by the Royal Society of Chemistry, www.rsc.org

The characterization of membrane proteins on leukocytes was initiated in the early 1970s, and led to the concept that many of these molecules belong to an immunoglobulin super family. This research was initiated by Alan F. Williams in the Immunochemistry Unit and continued, from 1978 onwards, in the MRC Cellular Immunology Unit.

In more recent years, structural and functional studies of hyaluronan (HA)-binding proteins and receptors, involved in extracellular matrix remodelling and leukocyte migration during inflammation, formed a major research focus in the Unit. HA is a high molecular weight polysaccharide found in tissues of all vertebrates. The interaction of HA-binding proteins with HA has a central role in the formation and stability of extracellular matrix (*e.g.* in cartilage) and is also important in immune-cell trafficking. The research has concentrated on TSG-6 (the protein product of TNF-stimulated gene-6), an HA-binding protein which is only expressed in adult tissues during inflammation, and the ubiquitous HA-receptor CD44, which is involved in matrix assembly and leukocyte migration.

Another major interest within the Immunochemistry Unit has involved the family of collectin molecules – which contain distinct globular carbohydrate-binding domains linked to collagen-like regions – and play important roles in innate immunity in the lungs and bloodstream by immediate recognition and clearance of microbial pathogens. This was most recently extended to include the ficolins, which have structural and functional similarities to the collectins. The serum collectin mannan-binding lectin (MBL) and ficolins recognize arrays of neutral sugars or acetylated compounds on microorganisms as being foreign and, after binding to them, stimulate activation of MBL-associated serine proteases (MASPs), which result in activation of the serum complement system – thus recruiting one of the major defence systems in the body. Collectins found in lung surfactant (surfactant proteins A and D (SP-A and SP-D)) also recognize and clear microorganisms, but without involvement of complement activation. The collectins and ficolins, in addition to their interactions with microorganisms, can resolve inflammation by binding to and promoting clearance of apoptotic cells and DNA – and by modulating the manner in which allergens are processed by dendritic cells. Recombinant forms of the collectins and ficolins are considered as candidates for new therapeutics for the treatment of infection and inflammation.

Each chapter in the volume gives a brief historical background to a topic, with some emphasis on work carried out within the Immunochemistry Unit, and then provides a survey of recent advances in the field. The main theme running through most of the chapters is that of protein structure–function relationships – good examples being the descriptions of quaternary structures of large oligomeric proteins, of factor H and C1q binding to specific ligands, or of the chemistry of the mechanism of catalysis of the covalent binding of activated C3 and C4 proteins to nucleophilic groups on microbial surfaces.

Kenneth B. M. Reid
Robert B. Sim

Contents

Molecular Aspects of Innate and Adaptive Immunity
Edited By Kenneth BM Reid and Robert B Sim

Published by the Royal Society of Chemistry, www.rsc.org

Section 2 The Complement System

Chapter 3 The Evolution of Complement Systems

Alister W. Dodds

Chapter 4 Structure and Function of the C1 Complex: A Historical Perspective

Gérard J. Arlaud

Section 6 Hyaluronan-Binding Proteins in Inflammation

Section 1
Antibodies

CHAPTER 1

R. R. Porter and the Structure of Antibodies

LISA A. STEINER[a] AND JULIAN B. FLEISCHMAN[b]

[a] Department of Biology, Massachusetts Institute of Technology, Cambridge, MA 02139, USA; [b] Department of Molecular Microbiology, Washington University School of Medicine, St. Louis, MO 63110, USA

1.1 Introduction

In 1967 Rodney R. Porter was invited to succeed Sir Hans Krebs as Whitley Professor of Biochemistry at Oxford. Porter founded and headed the MRC Immunochemistry Unit there, which now celebrates its 40th anniversary.

Porter's background and his classic work in the field of immunochemistry had begun well before his arrival at Oxford. As a student of Fred Sanger at Cambridge he was introduced to the structural basis of protein function, a topic that caught his fancy and pointed him toward immunology. Porter was intrigued by Landsteiner's work on the specificity of antibodies and the biochemical basis for the 'antibody paradox', namely how a diverse group of proteins, such as antibodies, could have remarkably different specificities yet have apparently similar structures. In 1958 Porter was a cofounder of the Antibody Workshop,[1] a small international group of researchers that met regularly in a series of informal and often lively sessions over a period of seven years to discuss fundamental research problems in immunology. This chapter outlines Porter's fundamental contributions to our present knowledge of antibody structure and the biochemical basis of their specificity. In recognition of this work, Porter shared the Nobel Prize in Physiology or Medicine with Gerald Edelman in 1972.

Molecular Aspects of Innate and Adaptive Immunity
Edited By Kenneth BM Reid and Robert B Sim

Published by the Royal Society of Chemistry, www.rsc.org

The early demonstration of the passive transfer of immunity by serum had paved the way for establishing the molecular nature of antibodies. Methods still were needed to fractionate serum and purify its constituents, and it is, indeed, striking that the understanding of antibody structure closely followed advances in methods to purify and characterize proteins. In the 1930s a centre that developed such methodologies was the Svedberg laboratory in Uppsala, and it was there that the first steps were taken toward determining the component of serum that carried antibody activity. Ultracentrifugation indicated that antibodies sedimented either at 17–19 S or at 6–7 S, corresponding to molecular weights of about one million and 160 000, respectively.[2–4] With the technique of free boundary electrophoresis, Tiselius and Kabat[5] demonstrated that the antibody activity in a rabbit antiserum to ovalbumin was confined to the γ-globulin region (fraction migrating slowest toward the anode). Sometimes, however, antibodies migrated faster on electrophoresis. Furthermore, some proteins without antibody activity (*e.g.*, properdin, a protein of the alternative complement pathway) also migrated in the γ-globulin fraction. Subsequently it was realized that all antibodies, even those of different classes [*e.g.*, immunoglobulin G (IgG), IgM, IgA, see below], share many basic structural features, despite differences in size or in electrophoretic properties. Eventually the term 'immunoglobulin' was introduced to include the set of all proteins that share the essential structural features of antibodies.[6] The term immunoglobulin refers to the antibody as a protein, regardless of its antigen-binding activity.

1.2 The Papain Fragments

In the 1950s the pioneering work of Fred Sanger in developing methods to establish the amino acid sequence of insulin ushered in an era of rapid advances in the determination of the sequences of proteins.[7,8] Rodney Porter, a PhD student of Sanger, was interested in understanding the chemical basis for antibody activity. At that time the only proteins for which amino acid sequences had been determined (insulin, ribonuclease, lysozyme) were at least an order of magnitude smaller than antibody molecules. An even more formidable problem was that different preparations of antibody, known to bind to the same antigenic determinant, varied measurably in molecular properties (*e.g.*, as demonstrated by electrophoresis), and were impossible to fractionate into homogeneous constituents. Nevertheless, heterogeneity did not obscure the substantial similarity among all immunoglobulin molecules. The major distinction between antibodies was in their recognition of different antigens, not in their overall molecular structure. As Porter noted in his Nobel lecture,[9] "This combination of an apparently infinite range of antibody combining specificity associated with what appeared to be a nearly homogeneous group of proteins astonished me and indeed still does". Thus antibodies contrasted strikingly with enzymes, which typically differ substantially from one another in structure as well as in specificity.

Porter's plan was to simplify the sequencing problem by breaking the antibody molecule into fragments, hoping that one or more of the smaller pieces would retain specificity for antigen. He was influenced by the work of Landsteiner who had shown that in many cases only a small part of an antigen was able to bind to its antibody, suggesting that the combining site of the antibody also may be smaller than the whole antibody molecule.[10] The plan also assumed that the heterogeneity of an antibody preparation would not prevent the isolation of its constituent fragments.

Porter's approach was to digest the antibody with papain. His initial efforts and those of others had shown that treatment with proteolytic enzymes yielded fragments of lower molecular weight that still could bind to antigen, but these products had not been isolated or characterized in detail.[10–14] These early experiments were limited by the lack of pure enzymes and effective methods to fractionate mixtures of proteins. However, by the late 1950s these materials and techniques had improved significantly. Thus the availability of highly purified papain[15] meant that the specific digestion products would not be substantially contaminated by enzyme. In addition, the newly introduced carboxymethylcellulose ion-exchange resins[16] efficiently separated the digestion products.

In a renewed effort, Porter treated several rabbit antibodies, each specific for a different antigen, with crystalline papain; the resulting digests were fractionated on columns of carboxymethylcellulose. In each case, three fractions of approximately equal size were obtained. Porter named them fractions I, II and III (Figure 1.1), in order of their elution; these fractions together accounted for almost all of the original antibody molecule and were resistant to further digestion with papain.[17,18] None of the fractions precipitated with the corresponding

Figure 1.1 Rodney Porter describing his studies on the structure of IgG in the early 1960s.

antigen, but fractions I and II, which were very similar in size and amino acid composition, specifically inhibited the precipitation of the antigen by the homologous antiserum. This suggested that the original antibody was 'multivalent' and could cross-link molecules of antigen, but that fractions I and II were 'univalent', *i.e.* they still could specifically bind to antigen molecules but could not cross-link them. Fraction III, which could not specifically inhibit precipitation, was shown to contain structures that enabled the transmission of the antibody across the placenta,[19] the binding of antibody to guinea pig skin permitting anaphylactic reactions upon antigenic challenge,[20,21] and the activation of complement.[21,22]

Fraction III crystallized readily when dialyzed against buffers of neutral pH, an unexpected result since the starting antibody preparation did not crystallize. This suggested that this fragment of the antibody might be more structurally homogeneous than the original material. It was further inferred from these results that the IgG molecule may consist of three tightly folded globular segments that are resistant to further digestion by papain, whereas the polypeptide chain(s) connecting these segments are more exposed to proteolytic digestion.

In later years Porter enjoyed recalling that he first believed that the crystals, which appeared upon dialysis of the papain digest in the cold, consisted of cystine, the oxidation product of the cysteine that had been used to activate papain. Accordingly, for several months he discarded the crystals in the sink. As he noted in his Nobel lecture,[9] it was fortunate that his neighbour at the National Institute for Medical Research at that time was Olga Kennard, a crystallographer. When Porter finally asked her opinion about the crystals, she remarked that they looked like crystals of protein, not cystine. They then were identified as the fraction III obtained by separation of the papain digestion products of the antibody on carboxymethylcellulose.

Fractions I and II are similar. Their separation on carboxymethylcellulose was fortuitous and merely reflected chemical heterogeneity, a result of their overall charge and the column elution conditions. Fragments I and II later were renamed as fragment antigen-binding (Fab) and the crystallizable piece III was renamed as fragment crystallizable (Fc); see Ceppellini *et al.*[6] for a summary of immunoglobulin nomenclature. The striking differences between the Fab and Fc fragments were an early clue to the solution of the antibody paradox. The two Fab fragments on a single antibody molecule are identical to one another and contain the combining sites for antigen, but they differ from one antibody molecule to another. The Fc fragments are shared by all IgG molecules.

Another interesting proteolytic fragment of antibodies called Facb (for fragment antigen and complement binding) subsequently was discovered at the MRC Immunochemistry Unit at Oxford.[23,24] It consists of an IgG molecule from which only part of the Fc had been removed. The Facb fragment is bivalent and therefore can still precipitate and agglutinate antigens and it also retains complement-fixing activity.

1.3 The Four-chain Model

Another approach to analyzing the structure of antibodies was taken by Gerald Edelman while he was still a graduate student at the Rockefeller Institute. Edelman found that when IgG was reduced with a mercaptan in the presence of a dissociating solvent such as 6 M urea, its molecular weight dropped significantly, demonstrating that it consists of a number of polypeptide chains, cross-linked by disulfide bridges.[25,26] But these results disagreed with Porter's earlier end-group analyses that had shown approximately one free amino-terminal residue per antibody molecule, consistent with a single-chain protein.[27] To resolve the discrepancy and to obtain products that might retain some biological activity, Porter and colleagues modified the conditions of reduction and chain separation. A key step was the use of mild reduction conditions, which had been shown by Cecil and Wake to preferentially cleave interchain disulfide bonds.[28] The chains prepared in this way remained soluble after separation by gel filtration in 1 M propionic acid.[29]

By good fortune, Julian Fleischman, a postdoctoral fellow who joined the Porter laboratory at that time, brought antisera to the Fab and Fc fragments which he had helped prepare in Melvin Cohn's laboratory at Stanford. A simple double immunodiffusion experiment established the relation between the polypeptide chains and the Fab and Fc fragments. Porter first proposed the four-chain model for IgG at a meeting in New York. Some initial uncertainty about the molecular weights of the separated chains meant the possibility that the molecule consisted of only two or even three chains was also considered.[29,30] Re-determination of the molecular weights confirmed that there were two heavy and two light chains per molecule.[31] The now-familiar four-chain model was presented by Fleischman *et al.*[32]

As pointed out by Fleischman,[33] a critical finding for the final model was the observation of Alfred Nisonoff and colleagues that pepsin digestion of the IgG molecule at pH ~4.5 degrades the Fc fragment but leaves the two Fab-like fragments intact and linked to one another by a disulfide bond;[34] the resulting bivalent fragment is called $F(ab')_2$. Thus previous models, in which the combining sites were placed at the distal ends of a cigar-shaped molecule, were no longer tenable.

The four-chain IgG model, with minor modification mainly in the number and location of interchain disulfide bridges, has stood the test of time. It also formed the basic structure for the other immunoglobulin classes. In IgG there are two identical heavy chains and two identical light chains of molecular weight 50 000 and 23 000, respectively. Each Fab fragment consists of one entire light chain plus the amino-terminal half of one heavy chain; the Fc fragment consists of the remainder of both heavy chains. The presence of two Fab fragments, each of which binds antigen, was consistent with the bivalence of IgG antibodies. The model was reconciled with Porter's end-group analyses when it was realized that the heavy chains have a blocked amino-terminus and that the light chains are heterogeneous at the amino-terminus; thus the yield of a single end-group per molecule did not reflect the actual number of chains.

1.4 V and C Regions of Immunoglobulin Chains

Once the four-chain structure of IgG had been established, Betty Press (E. M. Press) and others in Porter's laboratory turned to amino acid sequence determination in immunoglobulins to help elucidate the structural basis of antibody specificity. The most suitable materials at hand were the myeloma immunoglobulins because of their abundance and homogeneity. Although the combining specificities of these immunoglobulins were unknown, their primary structures might suggest the ways in which conventional antibodies would combine with antigens. For their analysis they chose the Fd fragment, the segment of heavy chain in Fab.[35] This sequence work had begun at St Mary's and much of it continued in Porter's laboratory at the MRC Unit in Oxford.

Amino acid sequence studies of immunoglobulin light chains from two different myeloma immunoglobulins had revealed that they consisted of a variable (V) segment and a constant (C) segment, each comprising about half of the chain.[36,37] The amino acid sequences of the Fd fragments of two human IgG myeloma immunoglobulins showed that there were also V segments in heavy chains similar in length to those of the light chains.[35] This dramatically confirmed that antibodies contained sequences specific for particular antigens (the V regions) and sequences common to all immunoglobulins of a particular class (the C regions). The structural basis underlying the antibody paradox, first suggested in Porter's identification of the Fab and Fc fragments of the antibody molecule, had been revealed, and it resided in the primary structure, the amino acid sequences, of the antibody polypeptide chains themselves.

The presence of V and C regions in immunoglobulins was established largely by analysis of myeloma proteins. However, Porter also investigated actual antibodies with known specificity for antigen. Although rabbit antibodies were heterogeneous and therefore problematic for amino acid sequence determination, partial amino acid sequences of rabbit immunoglobulin heavy chains were obtained in Porter's laboratory and they led to similar conclusions.[38–40] Porter also was interested in the genetic basis for antibody diversity, and rabbit antibodies were potentially useful for such studies because they expressed inherited variants of antibody molecules called allotypes. In 1969 Fleischman had purified a single, relatively homogeneous rabbit antibody to streptococcal carbohydrate and in Porter's laboratory at the MRC Immunochemistry Unit at Oxford he determined partial amino acid sequences in the V region of its heavy chain.[41,42] The results, together with previous work,[40] suggested that allotype-related amino acid sequences were probably present in the V region, and thus V region diversity was likely to be related to structural genes carried in the germ line.

Following his sequence studies on immunoglobulins, Porter turned his attention to the structures and genetics of the complex complement proteins described in other chapters in this volume. Porter's landmark career in immunochemistry ended abruptly with his untimely death in a road accident in 1985, only a few weeks after his retirement from the Whitley Chair of Biochemistry. A special symposium had just been held at Oxford in his honour to commemorate that occasion, and the proceedings published as Biochemical

Society Volume 51 – *Genes & Proteins in Immunity*.[43] Porter had planned to continue his research as director of the MRC Immunochemistry Unit. His loss was deeply felt by his family and by all of his students and colleagues. His achievements in science stand as a tribute to his life and his work.

References

1. R. R. Porter, *Perspect. Biol. Med.*, 1985, **29**, S108.
2. M. Heidelberger, K. O. Pedersen and A. Tiselius, *Nature*, 1936, **138**, 165.
3. M. Heidelberger and K. O. Pedersen, *J. Exp. Med.*, 1937, **65**, 393.
4. E. A. Kabat, *J. Exp. Med.*, 1939, **69**, 103.
5. A. Tiselius and E. A. Kabat, *J. Exp. Med.*, 1939, **69**, 119.
6. R. Ceppellini, S. Dray, G. Edelman, J. Fahey, F. Franěk, E. Franklin, H. C. Goodman, P. Grabar, A. E. Gurvich, J. F. Heremans, H. Isliker, F. Karush, E. Press and Z. Trnka, *Bull. Wld. Hlth. Org.*, 1964, **30**, 447.
7. F. Sanger and H. Tuppy, *Biochem. J.*, 1951, **49**, 463.
8. F. Sanger and E. O. P. Thompson, *Biochem. J.*, 1953, **53**, 353.
9. R. R. Porter in *Les Prix Nobel en 1972*, The Nobel Foundation, Stockholm, 1973, p. 173.
10. R. R. Porter, *Biochem. J.*, 1950, **46**, 479.
11. I. A. Parfentjev, 1936, U.S. Patent No. 2,065,196.
12. M. L. Petermann and A. M. Pappenheimer, Jr., *J. Phys. Chem.*, 1941, **45**, 1.
13. J. H. Northrop, *J. Gen. Physiol.*, 1941–1942, **25**, 465.
14. M. L. Petermann, *J. Amer. Chem. Soc.*, 1946, **68**, 106.
15. J. R. Kimmel and E. L. Smith, *J. Biol. Chem.*, 1954, **207**, 515.
16. E. A. Peterson and H. A. Sober, *J. Amer. Chem. Soc.*, 1956, **78**, 751.
17. R. R. Porter, *Nature*, 1958, **182**, 670.
18. R. R. Porter, *Biochem. J.*, 1959, **73**, 119.
19. F. W. R. Brambell, W. A. Hemmings, C. L. Oakley and R. R. Porter, *Proc. R. Soc. Lond. B*, 1960, **151**, 478.
20. Z. Ovary and F. Karush, *J. Immunol.*, 1961, **86**, 146.
21. K. Ishizaka, T. Ishizaka and T. Sugahara, *J. Immunol.*, 1962, **88**, 690.
22. A. Taranta and E. C. Franklin, *Science*, 1961, **134**, 1981.
23. G. E. Connell and R. R. Porter, *Biochem. J.*, 1971, **124**, 53P.
24. M. Colomb and R. R. Porter, *Biochem. J.*, 1975, **145**, 177.
25. G. M. Edelman, *J. Amer. Chem. Soc.*, 1959, **81**, 3155.
26. G. M. Edelman and M. D. Poulik, *J. Exp. Med.*, 1961, **113**, 861.
27. R. R. Porter, *Biochem. J.*, 1950, **46**, 473.
28. R. Cecil and R. G. Wake, *Biochem. J.*, 1962, **82**, 401.
29. J. B. Fleischman, R. H. Pain and R. R. Porter, *Arch. Biochem. Biophys.*, 1962, **Suppl. 1**, 174.
30. R. R. Porter, in *Basic Problems in Neoplastic Disease*, ed., A. Gellhorn and E. Hirschberg, Columbia University Press, New York, 1962, p. 177.
31. R. H. Pain, *Biochem. J.*, 1963, **88**, 234.
32. J. B. Fleischman, R. R. Porter and E. M. Press, *Biochem. J.*, 1963, **88**, 220.
33. J. B. Fleischman, *Current Contents*, 1981, **24**, 17.

34. A. Nisonoff, F. C. Wissler, L. N. Lipman and D. L. Woernley, *Arch. Biochem. Biophys.*, 1960, **89**, 230.
35. E. M. Press and N. M. Hogg, *Biochem. J.*, 1970, **117**, 641.
36. N. Hilschmann and L. C. Craig, *Proc. Natl. Acad. Sci. U.S.A.*, 1965, **53**, 1403.
37. K. Titani and F. W. Putnam, *Science*, 1965, **147**, 1304.
38. J. J. Cebra, L. A. Steiner and R. R. Porter, *Biochem. J.*, 1968, **107**, 79.
39. R. G. Fruchter, S. A. Jackson, L. E. Mole and R. R. Porter, *Biochem. J.*, 1970, **116**, 249.
40. L. E. Mole, S. A. Jackson, R. R. Porter and J. M. Wilkinson, *Biochem. J.*, 1971, **124**, 301.
41. J. B. Fleischman, *Biochemistry*, 1971, **10**, 2753.
42. J. B. Fleischman, *Immunochemistry*, 1973, **10**, 401.
43. Biochem. Soc. Symp. 1986, Vol. **51** Genes and Proteins in Immunity. K.B.M. Reid and M.A. Kerr (eds).

CHAPTER 2

Chemical Engineering of Therapeutic Antibodies

GEORGE T. STEVENSON AND WENG LEONG

Tenovus Research Laboratory, Southampton University Hospitals, Southampton SO16 6YD, UK

2.1 Introduction

The studies described in this chapter originated with observations of Ig chain interactions, and of interactions of haptens with Ig molecules and their chains, undertaken in the Immunochemistry Unit at Oxford during 1967–1970.[1–5] They also owe much to earlier descriptions from Rupert Cecil's team at Oxford of protein sulfhydryl (SH) and disulfide (SS) bonds and their reactivities.[6–8] After setting up a team at Southampton in 1970 my colleagues and I turned our attention to the therapeutic potentials of antibodies and their derivatives, particularly with regard to their abilities to engage specific molecules on neoplastic cell surfaces and thereby recruit lethal effector mechanisms to the targeted cell.[9–12] A major strategy which emerged was the enhancement of antibody cytotoxicity by permutations of the functional modules of IgG antibodies. The new molecular geometries which resulted entailed a variety of manipulations of SS bonds.

The approach is described in this chapter. First, the modular structure of IgG and some aspects of the sulfur chemistry of proteins are considered.

2.1.1 Structure of IgG

IgG is the predominant class of antibody molecules in mammals. It consists of four peptide chains (two heavy and two light) folded and joined in a tripartite

Molecular Aspects of Innate and Adaptive Immunity
Edited By Kenneth BM Reid and Robert B Sim

Published by the Royal Society of Chemistry, www.rsc.org

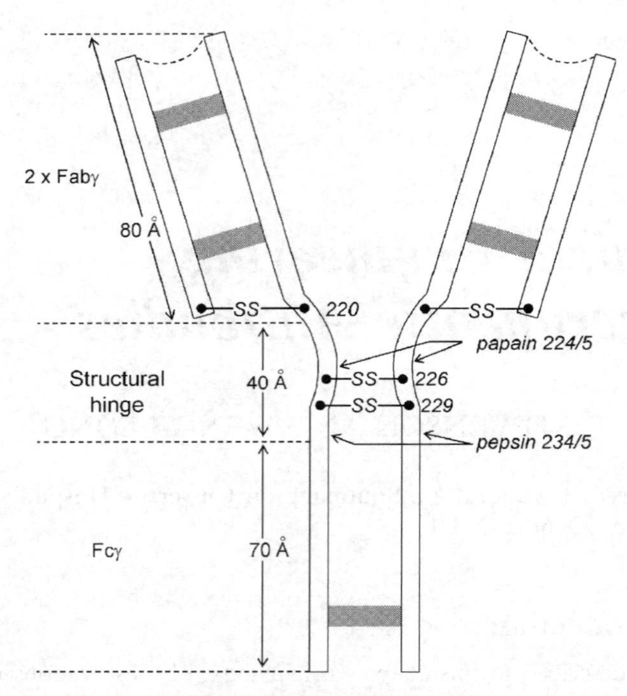

Figure 2.1 The interchain bonds and hinge region of human IgG1. Four chains that consist of two identical pairs are depicted, with their N-termini at the top of the diagram. The larger ('heavy') chains in this Ig class are called γ chains. The smaller ('light') chains can be either of two classes, κ and λ – which is present in a given antibody molecule and is not relevant in the present context. Antibody function is served by a series of 'complementarity-determining regions' distributed through the N-terminal half of each light chain and the N-terminal quarter of each γ chain; the two antibody sites are formally indicated by dashed lines. Only the γ chains serve effector functions. Cross-hatching indicates sets of non-covalent interchain bonds that link Ig domains: from top to bottom, V_L to V_H, C_L to $C_{\gamma}1$, $C_{\gamma}3$ to $C_{\gamma}3$.[52] About half-way along the γ chains a curved section depicts the *genetic hinge* (residues 216–230, Eu numbering), defined by having its own exon. Overlapping but not coincident with the genetic hinge is the *structural hinge*,[53] defined by two areas of marked mobility, residues 221–225 and 231–237. These lie above and below a rigid, SS-bonded, proline-rich hinge core. The structural hinge is very extended, occupying ~21% of the length of the molecule, but containing only 2.6% of the amino acid residues. The extended form renders it susceptible to endoproteases – the best known cleavages are indicated by arrows.

arrangement in which three compact modules are joined at an elongated and flexible hinge (Figure 2.1). Two identical modules, 'fragment antigen-binding' (Fab), consist each of one light chain joined to the N-terminal half of the heavy chain, with the antigen-binding site shared between the N-terminal domains of the two chains. The third module, 'fragment crystalline' (Fc) because it sometimes crystallizes, is responsible for antibody effector functions, such as recruiting complement and phagocytes to the site of antibody-coated targets.

It is instructive to recall the role of Rodney Porter, the first director of the Immunochemistry Unit, in elucidating the structure depicted in Figure 2.1. In 1958 he reported that partial digestion of rabbit IgG by papain yields the functionally intact fragments, of molecular weight about 50 000, now known as Fab and Fc.[13,14] This dissection was later found to be applicable to many varieties of mammalian IgG molecules, being due to cleavage of the γ-chain hinge just N-terminal to inter-γ SS bonds (Figure 2.1). In 1959 Edelman[15] reported that the molecular weight of IgG fell when it was reduced in the presence of urea, suggesting a multichain structure stabilized by non-covalent and SS bonds. However, the denatured and reduced chains thus produced could not be satisfactorily characterized. On learning of the studies by Cecil and Wake[7] of SS reactivities in a variety of proteins, Porter suspected that any *interchain* SS bonds in IgG could be reduced by thiol in the absence of denaturant, with the *intrachain* SS remaining shielded from the reducing agent by the protein's native folding. This proved to be the case. Sequential reduction of interchain SS bonds, removal of the thiol, and exposure to a denaturant now caused the IgG molecule to fall apart into its γ and light chains, in a form susceptible to purification and characterization. Collating this information with that yielded by enzymic digestion enabled Porter[16] to present, to a small conference at Columbia University in 1962, a four-chain model of IgG that resembled substantially the structure shown in Figure 2.1.

2.1.2 Availability of SH groups

As can be seen in Figure 2.1, reduction of interchain SS bonds does not cause the chains of IgG to dissociate, provided the interchain non-covalent bonds remain intact. However the reduction provides opportunity in the form of SH groups, the most reactive groups occurring naturally in proteins. That there is variation in the number of interchain SS among IgGs from different species, and among the various subclasses of IgG within a single species, must be allowed for.

We utilize the SH groups provided by reduction of IgG interchain bonds to link Ig modules in a variety of geometries. The groups all belong to *cysteine residues in the hinge area of IgG*, being either in the primary structure of the hinge or brought into the vicinity of the hinge by protein folding.

2.1.3 Enzymic Dissections

In addition to the limited digestion by papain, limited digestion by pepsin at a pH ~4.0 (~1.5 above optimal) has proved useful[17,18] (Figure 2.2). It provides Fab modules in a form that retains the upper two-thirds of the hinge region, still in dimeric form because of the retention of inter-γ SS bonds. The modules can now be separated simply by reducing these SS bonds, because the only inter-γ non-covalent bonds have been lost with the fragmented and discarded Fc.

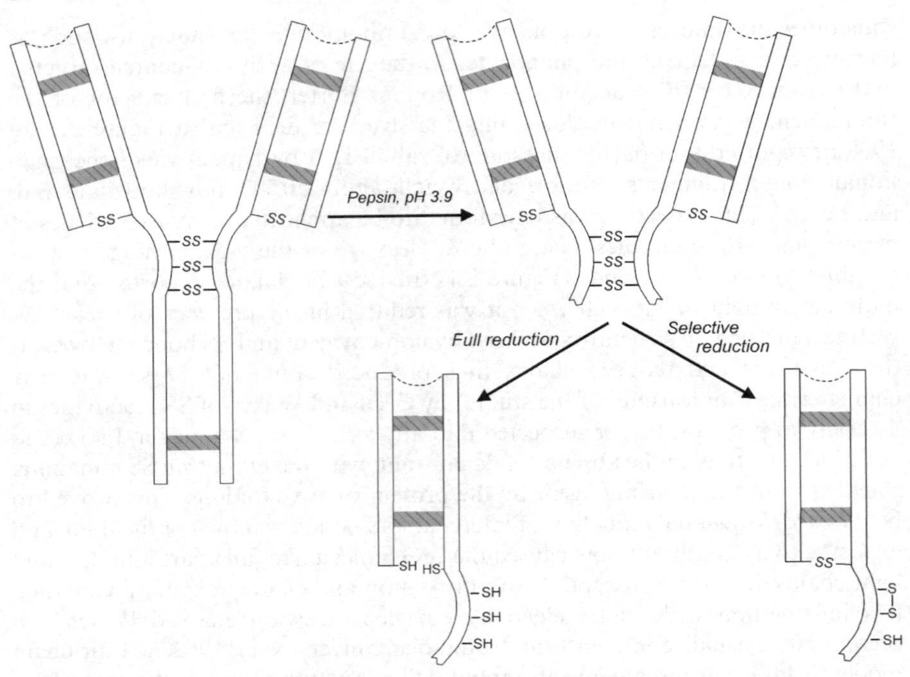

Figure 2.2 Cleavage of an IgG molecule, in this case mouse IgG1, by partial peptic digestion. The Fcγ is cut into fragments, which are usually discarded. Full reduction of the interchain SS bonds may be achieved by dithiothreitol (DTT) at pH ~8.0. Selective reduction requires two further SS interchanges after a full reduction (Section 2.3.2 and Figure 2.3).

To distinguish modules derived from IgG from those derived from other Ig classes (IgM, IgA, *etc.*) the Greek letter γ is appended to Fab or Fc. To denote the larger form of Fabγ yielded by pepsin, a prime is added. Thus digestion of IgG by papain yields Fabγ and Fcγ, while digestion by pepsin followed by reduction yields Fab′γ plus fragments of Fcγ.

2.2 Useful Chemistry of SH Groups and SS Bonds

Among the many reactions involved in the sulfur chemistry of proteins[6,19,20] two particularly concern us here: (1) SS interchange involving attack by a thiol (R–SH) on a protein SS bond, and (2) alkylation of SH groups by maleimide reagents.

During such manipulations one must be aware of the possible oxidation of SH groups of proteins or small thiols:

$$R\text{–}S^- + R'\text{–}S^- \rightarrow R\text{–}SS\text{–}R' + 2e^- \qquad (2.1)$$

This reaction has sometimes been utilized to link protein modules by SS bonds, and it has the advantage of simplicity. However, its rate is difficult to predict and the link is highly susceptible to reductive cleavage *in vivo*. Our approach to the reaction is to avoid it. To this end O_2 is removed from solutions by N_2-flushing, and ethylenediaminetetraacetic acid (EDTA) is added to buffers to sequester catalyzing metal ions (chiefly Fe^{3+} and Cu^{2+}). We have found that SH groups on reduced Ig molecules remain intact for months when stored refrigerated at pH 5 in N_2-flushed EDTA-containing buffers.

2.2.1 SS Interchange

The general equations for SS interchange, the reduction of a disulfide (P–SS–Q) by a thiol (R–SH), may be written:

$$\mathrm{P{-}SS{-}Q + R{-}SH \rightleftharpoons P{-}SS{-}R + Q{-}SH} \tag{2.2}$$

$$\mathrm{P{-}SS{-}R + R{-}SH \rightleftharpoons R{-}SS{-}R + P{-}SH} \tag{2.3}$$

The major reaction mechanism involves a nucleophilic attack by the thiol ion $R{-}S^-$, so the usual pH dependence of the reaction rate (rapid at $pH > 8$, slow at $pH < 7$) reflects the fact that most thiol SH groups exhibit a pK_a of about 9.5. Interchanges following Equations (2.2) and (2.3) can eventually yield P–SH, Q–SH and R–SH, each attacking P–SS–P, Q–SS–Q, R–SS–R, P–SS–Q, P–SS–R and Q–SS–R, with the equilibrium depending on 18 rate constants. However, a number of factors that affect the equilibrium, considered below, are readily understood and can greatly assist manipulations.[21]

(a) Clearly the conversion of P–SS–Q to P–SH and Q–SH in Equations (2.2) and (2.3) can be driven arbitrarily close to totality by a sufficient surplus of R–SH when this is not ruled out by other factors.

(b) Certain dithiols, with dithiothreitol (DTT) a well-known example, can drive reduction of a disulfide to near completion without being present in a large molar excess, because of their tendency to cyclize by forming an intramolecular SS bond:

$$\mathrm{P{-}SS{-}Q + threitol({-}SH)_2 \rightarrow P{-}SH + Q{-}SH + threitol({-}SS)} \tag{2.4}$$

(c) A similar steric effect on SS-interchange equilibria is seen with the interchain SS bonds in IgG molecules. It can be seen from Figure 2.3 that reduction of any of these bonds will leave SH groups persisting in close proximity because of the quaternary structure imposed by interchain non-covalent bonds. This will favour restitution of the SS bonds. In contrast after limited peptic digestion of IgG the dimeric product – conventionally designated $F(ab'\gamma)_2$ – is linked only by the two inter-γ SS

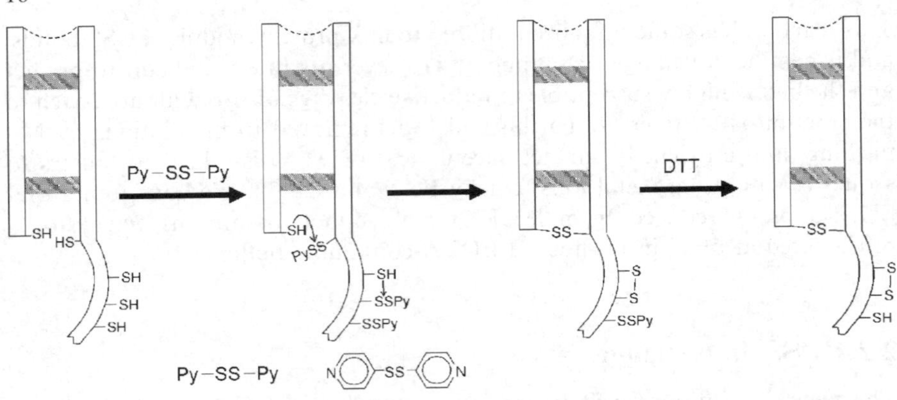

Figure 2.3 SS interchanges converting F(ab′γ)$_2$, from mouse IgG1 or IgG2a, to Fab′γ–SH. An initial reduction by DTT at pH 8.0 has cleaved the dimer to give Fab′γ(–SH)$_5$. The diagram shows the next reactions, all occurring at pH 5.0. (The exchange with Py–SS–Py could be carried out at a higher pH, but 5.0 has been chosen to slow the reaction and so give less trouble with concentration transients during mixing; the exchange with DTT must be at low pH.) The reaction with Py–SS–Py yields some Fab′γ–SS–Py groups sufficiently close to SH groups to give intramolecular SS bonds before further mixed pyridyl disulfides can form. The γ–light SS bond reappears and an intrachain SS loop appears in the hinge. (We do not know which two of the three hinge cysteines are most likely to form the SS loop, so depiction of the N-terminal pair is arbitrary.) The final attack by DTT breaks the electrophilic pyridyl-linked SS bond, and this bond only.

bonds and will dissociate into monomeric Fab′γ when they are broken, yielding an equilibrium for the SS interchange different from that in intact IgG. The inter-γ SS bonds in IgG have been labelled 'assisted', those in F(ab′γ)$_2$ 'unassisted', while the intrachain SS bonds (not shown in the diagrams) are 'shielded' because folding of the native molecule does not allow significant access of thiols at customary concentrations.[21]

DTT at ~2 mM and pH ~8 has long been a popular means of reducing interchain SS bonds in Ig molecules. At this concentration the cyclizing tendency of DTT is sufficient to overcome the assistance to interchain SS bonds provided by the protein's quaternary structure. Much greater concentrations are needed for full reduction of the bonds by a monothiol, such as 2-mercaptoethanol.

(d) Another example of essentially one-way SS interchanges, used in this case when one wishes to convert SH groups into SS bonds, is provided by compounds that exhibit SS bonds linked to the 2 or 4 carbon of a pyridyl ring.[20,22] We have used 4,4-dipyridyl disulfide (Py–SS–Py). Reaction with a thiol releases 4-pyridylthiol which, instead of remaining available for the reverse reaction, is almost entirely removed by a one-sided tautomeric equilibrium with the 4-thiopyridone:

$$\text{Py–SS–Py} + \text{R–SH} \rightleftharpoons \text{Py–SS–R} + \text{NC}_5\text{H}_4\text{–SH} \qquad (2.5)$$

$$NC_5H_4{-}SH \rightarrow HNC_5H_4{=}S \tag{2.6}$$

As a result the overall reaction, Equation (2.7), in many situations goes essentially to completion:

$$Py{-}SS{-}Py + 2R{-}SH \rightarrow R{-}SS{-}R + 2HNC_5H_4{=}S \tag{2.7}$$

A strong chromophoric signal from the thiopyridone $HNC_5H_4{=}S$ provides a useful spectrophotometric measure of the extent of reaction.[22]

(e) Vicinal charges sometimes profoundly affect the reactivities of SH groups and SS bonds, and are probably a major factor involved in the wide range of reactivities observed among cysteine residues in proteins.[19] These charges can markedly affect the pH dependence of SS interchange. The commonly seen pH dependence is exemplified by an attack of 2 mM DTT on human IgG. At pH 8.0 the four interchain SS bonds are reduced within 30 minutes at 30 °C, but at pH 5 <0.2 bond per molecule is cleaved, in accord with the usual pK_as of SH groups. However, when DTT attacks Py–SS–Py, or a mixed pyridyl disulfide formed on a cysteine residue of a protein, reduction by DTT at pH 5 still proceeds at a rapid rate because of the positive charge on the pyridyl nitrogen.[20,23] This has proved useful in our work (see Section 2.3).

2.2.2 Alkylation of SH Groups by Maleimides

Thiols add readily to the ring double bond of maleimides to give thioethers, and although the reacting species[6,19] is $R{-}S^-$, we find that the reaction is still conveniently rapid at pH 5.0. At this pH maleimides hydrolyze only slowly[24] and reactivity towards amino groups is negligible.

Reaction with maleimides is used either to block protein SH groups or to attach one end of a linking molecule to them. *N*-Ethylmaleimide (NEM) is the commonly used blocking agent. The *S*-succinimidylethyl group yielded by the reaction is shown in Figure 2.4.

Bismaleimide linkers were introduced by Moore and Ward[25] to link bovine plasma albumin to wool keratin, and used by Ishikawa's group[24] to link Ig molecules to enzymes for use in immunoassay. We modified this approach to link IgG modules together via cysteine residues in their hinge areas. Several bismaleimides are now available, some with problems because of low aqueous solubility. We standardized on *o*-phenylenedimaleimide (PDM) after comparing it with the somewhat less reactive alkyl-cored linker bis(3-maleimidopropionyl)-2-hydroxy-1,3-propanediamine. These yield distances between the linked S atoms of about 9 Å and 20 Å, respectively. The longer linker promised greater flexibility between the conjoined modules, but gave marginally less satisfactory results as judged by effector functions in the final constructs. The linkage between cysteine residues effected by PDM, *o*-phenylene–disuccinimidyl, is shown in Figure 2.4.

PDM is routinely dissolved in dimethylformamide (DMF) and used in reactions at ~1 mM in ~10% DMF. To link two Ig modules A and B, module

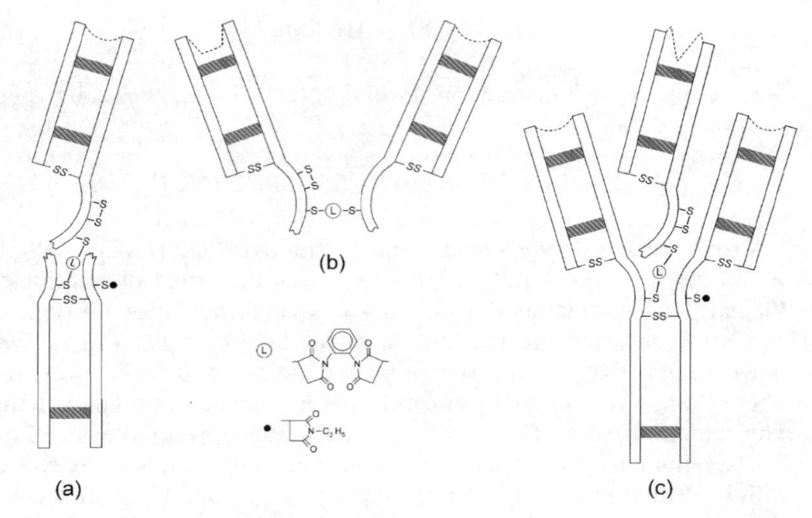

Figure 2.4 Some therapeutic constructs prepared from IgG modules. To simplify the nomenclature the primes and Greek letters used in naming the modules are omitted from the names of some of the final constructs. (A) FabFc, a univalent chimeric construct designed for rapidly modulating target molecules. Mouse Fab′γ is linked to human Fcγ. (B) F(ab′γ)$_2$, a bispecific antibody constructed from two mouse Fab′γ. One specificity is directed towards the target cell. The second specificity may also be directed towards a molecule on that cell, to increase specificity and affinity, or may be used to recruit an effector molecule or cell. (C) FabIgG, a bispecific molecule which retains its Fcγ module. Mouse Fab′γ is linked to the chimeric antibody rituximab, which has human IgG1 constant regions. The added Fab′γ arm is designed to recruit effector cells, and at the same time impede docking of Fcγ-receptors I, II and III on the Fcγ module. However, the Fcγ largely retains its affinity for the receptor FcRn, which prolongs metabolic survival.

A first reacts with PDM using relative concentrations that avoid homologous A–A cross-linking. One aims to obtain A–succinimidylphenylmaleimide, but it is important to realize that if A displays initially >1 SH group, the groups will be sufficiently close to be cross-linked intramolecularly by the PDM. This cross-linking occurs rapidly and cannot be prevented by raising the concentration of PDM. However, with an odd number of SH groups present one can rely on obtaining module A with a single maleimide group and its other SH cross-linked entirely.[26,27] Some approaches to the problem posed by an even number of SH groups per molecule are discussed in Sections 2.3.1 and 2.3.2.

2.3 Engineering IgG Modules

Our aims in engineering Ig molecules for therapeutic purposes, which vary with the situation in hand, can be to improve recognition of the cellular target, to improve effector recruitment and activity, to improve metabolic survival, and to reduce immunogenicity. These are among the broad aims of many chemical and genetic engineering programs.

Improved target recognition and affinity can be achieved by adding to an antibody a Fab module with an antibody site specific for an additional epitope on the target cell, thereby creating a bispecific antibody.

The second antibody activity in a bispecific antibody can also be used to bring to the surface of the target cell an effector cell, such as a macrophage or natural killer (NK) lymphocyte, with the aim of destroying the target. For this purpose the effector cell must be in an activated state, and frequently this activation must be achieved by choosing a molecular target on its surface capable of acting as a conduit for activation, in addition to being part of the coupling to the target cell.[28]

Human Fcγ modules are readily available in large amounts because of the large-scale preparation of human IgG from normal plasma for clinical purposes. Replacing the Fc on a mouse monoclonal IgG with human Fcγ improves the therapeutic potential of the antibody in three ways. It improves recruitment of complement and effector cells because it has better docking sites for these agencies; it prolongs the metabolic survival of the antibody through a better affinity for human FcRn, a receptor which safely sequesters much endocytosed IgG and returns it to extracellular fluid; and it reduces immunogenicity by removal of the major epitopes of a foreign IgG.

Recent extensions of antibody therapy to the treatment of autoimmune as well as neoplastic diseases have placed a premium on the removal of immunogenic epitopes from the antibody. Previously it was often acceptable to leave mouse Fabγ epitopes on the therapeutic construct, for example when treating a patient immunosuppressed by a lymphoid tumour and/or by chemotherapy. Now increasing numbers of antibodies are arriving in the clinic in which foreign Ig epitopes have been largely removed by the genetic engineering procedures of chimerization,[29] humanizing,[30] or raising in mice transgenic for human Ig.[31] Idiotypic epitopes will always be present on the antigen-combining site, but these are likely to be less troublesome in the absence of other immunogenic epitopes on the molecule.[32] The increasing availability of antibodies of predominantly human sequence from the pharmaceutical industry means that much chemical engineering can start from a highly characterized product of minimal immunogenicity.

2.3.1 Attaching the PDM Linker to Fcγ

The Fcγ module used in our work is derived from a brief (20 minute) digestion by papain of human IgG1. The module possesses two interchain SS bonds in its hinge, reduction of which by DTT yields four SH groups. All subsequent reactions[33] are carried out at pH 5. First, the four SH groups are reduced to three by reaction with a limiting amount of NEM so that an odd number of SH groups can be presented to the linker. Titrations have revealed that a molar ratio of NEM to protein of 0.75 yields an average 0.6 ± 0.1 alkylated group per molecule, a binomial distribution of which would give 52% of molecules with no alkylated groups, 37% with one, 10% with two, 1% with three and <0.1% with four. Subsequent yields have been consistent with this theoretical distribution. The Fcγ now undergoes SS interchange with Py–SS–Py, by which paired SH groups are restored to SS bonds and unpaired groups are left as 4-dithiopyridyls (Fcγ–SS–Py). An analogous reaction is shown in Figure 2.3. The protein is stored in this stable state until required.

When the linker is to be attached the Fcγ–SS–Py is allowed to react with DTT at pH 5. The mixed pyridyl disulfide is converted into a protein SH group and 4-thiopyridone, but the interchain SS bond is left intact. After removal of the DTT and thiopyridone, reaction with PDM yields ~33% of the Fcγ modules with a single maleimide group ready to be attached to any suitable partner displaying an SH group. The remainder of the Fcγ can be discarded as it comes from a readily available source.

The SH-displaying module to which the Fcγ-maleimide may be attached need not necessarily be an Ig module: any ligand homing to a targeted cell could be coupled to Fcγ in order to recruit complement and cellular effectors, and to benefit from the long metabolic survival conferred by Fcγ. The final construct consists of two modules attached by thioether bonds to either side of a bis-succinimidylphenylene core (Figure 2.4) derived from the linker. These tandem thioether bonds have proved stable *in vivo*, in animals and humans.[12]

2.3.2 Attaching the PDM Linker to Fab′γ

The Fab′γ with which we deal comes from mouse IgG monoclonal antibodies, some of which will have been genetically engineered to possess predominantly human IgG1 sequences.

Most monoclonal antibodies which retain mouse sequences belong to the IgG1 and IgG2a isotypes, both of which possess one γ–light and 3 inter-γ SS bonds (apart from a minor allotype of IgG2a). When the peptic $F(ab'\gamma)_2$ module is reduced by DTT one obtains monomeric Fab′γ modules each displaying five SH groups. There are two ways to attach PDM to these modules. First, and the simpler, after removing the DTT and reducing the pH to 5.0 one may allow the module to react directly with PDM. Four of the SH groups will be cross-linked while the fifth will display a succinimidylphenylmaleimide group ready for linking to a chosen partner.

The problem with the above method is that there is no control over which cysteine residue displays the maleimide group: it might be any of the four on the γ chain or the one on the light chain. In many cases this need not be of functional importance for the final construct. However, when it is of functional or regulatory importance two intermediate steps can be inserted to ensure that the maleimide group appears in one of the three clustered SH groups derived from inter-γ SS bonds. With $Fab'\gamma(\text{–SH})_5$ at pH 5, successive SS interchanges at the same pH with Py–SS–Py and DTT yield Fab′γ–SH with the γ–light SS bond reconstituted, two hinge-region SH groups converted to an intrachain SS loop, and the remaining hinge-region cysteine displaying an SH group to which PDM can be attached[34] (Figure 2.3). It is of interest that formation of the SS-bonded loop has also been observed when SH groups on mouse Fab′γ are allowed to undergo spontaneous oxidation.[35]

It is becoming increasingly likely that any monoclonal Fab′γ to be manipulated for a therapeutic construct will have been converted into predominantly human sequences, will have only two inter-γ SS bonds, and so after reduction will present four SH groups. The approach used with human Fcγ to obtain an odd number of SH groups, by alkylating one group under limiting conditions, is too wasteful for the expensive Fab′γ. The best solution at present is to remove one cysteine residue, either from the γ-chain hinge or from the C-terminus of the light chain, during the genetic chimerization procedures which supply the human sequences. Then after reduction one obtains Fab′γ(–SH)$_3$, to be dealt with by one of the two methods outlined for mouse Fab′γ.

2.3.3 Examples of Therapeutic Constructs

2.3.3.1 Univalent Antibodies

The first class of therapeutic constructs we made were univalent antibodies used to attack idiotypic epitopes on the surface Ig of neoplastic lymphocytes. One problem of surface Ig (usually IgM) as a therapeutic target is that it rapidly undergoes antigenic modulation, the process by which initial exposure to antibody rapidly renders the cell resistant to the cytotoxic action of the same antibody.[36] It is associated with cross-linking of surface antigen, which leads to patching and internalization,[37] with only minimal aggregation required to confer resistance to antibody-induced complement lysis.[38] It appears more effective *in vivo* than *in vitro*, with secondary cross-linking of the attacking antibody by cells bearing Fc-receptors probably being a major factor.[39,40] Sometimes modulation appears to mimic the consequences of cross-linking by the physiological ligand, and the questions of involvement of lipid rafts and tetraspanin-enriched microdomains now make the subject appear very complex. Nevertheless, the need to cross-link target molecules has been observed by many, which suggests to us that this potent cellular defence mechanism could be avoided by the use of univalent antibodies.

The first univalent antibody tried was polyclonal rabbit IgG antibody with one Fabγ arm removed by brief exposure to papain to yield a derivative called Fab/c. The asymmetric digestion is possible because the predominant allotype of rabbit IgG is glycosylated at a hinge threonine residue on one side only, since the oligosaccharide, once in place, blocks glycosylation of the contralateral γ chain. It is this naked chain which is the more sensitive to papain.[41] The cells of a guinea-pig B-lymphoblastic leukaemia were killed strikingly more successfully by Fab/c from anti-idiotype IgG than by the parent bivalent antibody.[42]

Mouse and human IgG are not susceptible to unilateral removal of Fab arms. In its place two types of univalent derivative were constructed. Fab′γ from mouse monoclonal antibody specific for an idiotypic epitope on the surface of a B-lymphoid tumour was linked to either human normal IgG or to human Fcγ.[26,43] In both cases the two modules were linked, hinge to hinge, by tandem thioether bonds connected to a linker core. The simpler derivative, FabFc, is closely analogous to the Fab/c developed originally, and is illustrated in Figure 2.4.

2.3.3.2 *Bispecific Antibodies*

Most of the bispecific antibodies constructed in our laboratory have used one Fab arm to target a neoplastic cell, and a second arm to recruit an effector cell, such as a macrophage, to the surface of the target more efficiently than is done by a normal IgG antibody (Figure 2.4). The final construct might or might not display Fcγ. Retaining the Fcγ means that complement is still available as an effector to attack target cells, and that the construct should have a prolonged metabolic survival. The problem with a retained Fc is that it might cause significant damage by directing a cytotoxic attack on the specially recruited effector cells, and promote also a dangerous release of cytokines from these cells by helping cross-linking their surfaces.[28]

Martin Glennie has led the work in our unit on multi-Fab constructs which lack Fcγ. As well as $F(ab'\gamma)_2$ which recruit effector cells,[44] bispecific $F(ab'\gamma)_2$ have been used to deliver the ribosome-inactivating toxin saporin to the surfaces of lymphoma cells.[45] In another variation the development of trispecific $F(ab'\gamma)_3$ led to a construct which attached to two molecules (CD2 and CD3) on the recruited effectors (resting cytotoxic T cells), thereby activating them as well as linking them to the neoplastic targets.[46]

The final construct shown in Figure 2.4 is FabIgG, in which Fab′γ from a mouse anti-FcRIII (anti-CD16) monoclonal antibody is linked to a widely used therapeutic antibody, rituximab,[47] with human IgG1 constant domains. Rituximab is specific for CD20, a non-modulating antigen widely displayed by the B-lymphocytic lineage, and was developed for the treatment of human B-cell lymphoma. The Fab arm to be added to rituximab is specific for Fcγ-receptor III, an activating molecule on certain cytotoxic cells, notably macrophages and NK lymphocytes. These effectors are firmly linked to FabIgG-coated targets. *In vitro* the titre of antibody-dependent cellular cytotoxicity with NK effectors[48] is enhanced 10–50 times in the mid-plateau region. At the same time the binding of the Fcγ module of the construct to Fcγ-receptors I and II on the monocytic cell line THP-1 is seen, by flow cytometry, to be impaired, presumably because of steric hindrance around the receptor docking sites on the lower hinge and upper Fcγ.[49] Among these receptors is the Fcγ-receptor IIB, which mediates an important inhibitory signal on the macrophage lineage.[50] It is hoped that with binding to the III receptor increased, and to the IIB receptor diminished, the cytotoxicity of macrophages for antibody-coated targets will be greatly enhanced. It is hoped also that the impaired binding to Fcγ-receptors by the Fcγ module of FabIgG will ameliorate the cytokine release often troublesome at the onset of antibody therapy.[51]

Acknowledgements

Many colleagues at Southampton, some of whose names appear in the list of references, have helped in this work. The principal financial support has come from Tenovus, the Cancer Research Campaign (now amalgamated into Cancer Research UK), and the Medical Research Council.

References

1. G. T. Stevenson and K. J. Dorrington, *Biochem. J.*, 1970, **118**, 703.
2. G. T. Stevenson, H. N. Eisen and R. H. Jones, *Biochem. J.*, 1970, **116**, 153.
3. G. T. Stevenson, *Biochem. J.*, 1973, **133**, 827.
4. C. W. K. Lam and G. T. Stevenson, *Nature*, 1973, **246**, 419.
5. G. T. Stevenson and L. E. Mole, *Biochem. J.*, 1974, **139**, 369.
6. R. Cecil and J. R. McPhee, *Adv. Protein Chem.*, 1959, **14**, 255.
7. R. Cecil and R. G. Wake, *Biochem. J.*, 1962, **82**, 401.
8. R. Cecil and G. T. Stevenson, *Biochem. J.*, 1965, **97**, 569.
9. G. T. Stevenson and F. K. Stevenson, *Nature*, 1975, **254**, 714.
10. F. K. Stevenson, E. V. Elliott and G. T. Stevenson, *Immunology*, 1977, **32**, 549.
11. T. J. Hamblin, A. K. Abdul-Ahad, J. Gordon, F. K. Stevenson and G. T. Stevenson, *Br. J. Cancer*, 1987, **42**, 495.
12. T. J. Hamblin, A. R. Cattan, M. J. Glennie, M. R. MacKenzie, F. K. Stevenson, H. F. Watts and G. T. Stevenson, *Blood*, 1987, **69**, 790.
13. R. R. Porter, *Nature*, 1958, **182**, 670.
14. R. R. Porter and E. M. Press, *Ann. Rev. Biochem.*, 1962, **31**, 625.
15. G. M. Edelman, *J. Am. Chem. Soc.*, 1959, **81**, 3155.
16. R. R. Porter, in: *Basic Problems in Neoplastic Disease*, ed., A. Gellhorn and E. Hirschberg, Columbia University Press, New York, 1962, p. 177.
17. A. Nisonoff, F. C. Wissler and L. N. Lipman, *Science*, 1960, **132**, 1770.
18. A. Nisonoff, G. Markus and F. C. Wissler, *Nature*, 1961, **189**, 293.
19. T.-Y. Liu, in: *The Proteins, 3rd edit.*, ed., H. Neurath, Academic Press, New York, 1977, p. 329.
20. K. Brocklehurst, *Internat. J. Biochem.*, 1979, **10**, 259.
21. G. T. Stevenson, *Chem. Immunol.*, 1997, **65**, 57.
22. D. R. Grassetti and J. F. Murray, *Arch. Biochem. Biophys.*, 1967, **119**, 41.
23. C. E. Grimshaw, R. L. Whistler and W. W. Cleland, *J. Am. Chem. Soc.*, 1979, **101**, 1521.
24. S. Yoshitake, Y. Hamaguchi and E. Ishikawa, *Scand. J. Immunol.*, 1979, **10**, 81.
25. J. E. Moore and W. H. Ward, *J. Am. Chem. Soc.*, 1956, **78**, 2414.
26. G. T. Stevenson, M. J. Glennie, F. E. Paul, F. K. Stevenson, H. F. Watts and P. Wyeth, *Bioscience Reports*, 1985, **5**, 991.
27. M. J. Glennie, H. M. McBride, A. T. Worth and G. T. Stevenson, *J. Immunol.*, 1987, **139**, 2367.
28. D. M. Segal, G. J. Weiner and L. M. Weiner, *Curr. Opin. Immunol.*, 1999, **11**, 558.

29. G. L. Boulianne, N. Hozumi and M. J. Shulman, *Nature*, 1984, **312**, 643.
30. M. Verhoeyen, C. Milstein and G. Winter, *Science*, 1988, **239**, 1534.
31. L. L. Green, M. C. Hardy, C. E. Maynard-Currie, H. Tsuda, D. M. Louie, M. J. Mendez, H. Abderrahim, M. Noguchi, D. H. Smith and Y. Zeng, *Nature Genet.*, 1994, **7**, 13.
32. R. G. Mage, *Nature*, 1988, **333**, 807.
33. G. T. Stevenson, V. A. Anderson, K. S. Kan and A. T. Worth, *J. Immunol.*, 1997, **158**, 2242.
34. K. S. Kan, V. A. Anderson, W. S. Leong, A. M. Smith, A. T. Worth and G. T. Stevenson, *J. Immunol.*, 2001, **166**, 1320.
35. M. E. Shott, K. A. Frazier, D. K. Pollock and K. M. Verbanac, *Bioconjugate Chem.*, 1993, **4**, 153.
36. L. J. Old and E. A. Boyse, *J. Nat. Cancer Inst.*, 1963, **31**, 977.
37. C. W. Stackpole, J. B. Jacobson and M. P. Lardis, *J. Exp. Med.*, 1974, **140**, 939.
38. J. Gordon and G. T. Stevenson, *Immunology*, 1981, **42**, 13.
39. R. W. Schroff, M. M. Farrell, R. A. Klein, H. C. Stevenson and N. L. Warner, *Blood*, 1985, **66**, 620.
40. A. C. Lane, S. Foroozan, M. J. Glennie, P. Kowalski-Saunders and G. T. Stevenson, *J. Immunol.*, 1991, **146**, 2461.
41. M. W. Fanger and D. G. Smyth, *Biochem. J.*, 1972, **127**, 767.
42. M. J. Glennie and G. T. Stevenson, *Nature*, 1982, **295**, 712.
43. G. T. Stevenson and M. J. Glennie, *Cancer Surveys*, 1985, **4**, 213.
44. J. Greenman, A. L. Tutt, A. J. T. George, K. A. F. Pulford, G. T. Stevenson and M. J. Glennie, *Mol. Immunol.*, 1991, **28**, 1243.
45. M. J. Glennie, D. M. Brennand, F. Bryden, H. M. McBride, F. Stirpe, A. T. Worth and G. T. Stevenson, *J. Immunol.*, 1988, **141**, 3662.
46. A. Tutt, G. T. Stevenson and M. J. Glennie, *J. Immunol.*, 1991, **147**, 60.
47. D. G. Maloney, A. J. Grillo-López, C. A. White, D. Bodkin, R. J. Schilder, J. A. Neidhart, N. Janakiraman, K. A. Foon, T. M. Liles, B. K. Dallaire, K. Wey, I. Royston, T. Davis and R. Levy, *Blood*, 1997, 90, 2188.
48. R. J. Dearman, F. K. Stevenson, M. Wrightman, T. J. Hamblin, M. J. Glennie and G. T. Stevenson, *Blood*, 1988, **72**, 1985.
49. P. Sondermann, J. Kaiser and U. Jacob, *J. Mol. Biol.*, 2001, **309**, 737.
50. F. Nimmerjahn and J. V. Ravetch, *Immunity*, 2006, **24**, 19.
51. G. T. Stevenson, *Leukemia Res.*, 2005, **29**, 239.
52. J. B. Natvig and M. W. Turner, in: *Clinical Aspects of Immunology*, P. J. Lachmann, D. K. Peters, F. S. Rosen, M. J. Walport, (eds), Blackwell Scientific Publications, Oxford, 1993, p. 149.
53. D. R. Burton, *Mol. Immunol.*, 1985, **22**, 161.

Section 2
The Complement System

CHAPTER 3

The Evolution of Complement Systems

ALISTER W. DODDS

MRC Immunochemistry Unit, Department of Biochemistry, University of Oxford, South Parks Road, Oxford OX1 3QU, UK

3.1 Phylogeny in the MRC Immunochemistry Unit

From his first publications on immunoglobulin structure, Rodney Porter's major research model was the rabbit. He determined the N-terminal amino acid sequences from a mixture of antibodies[1] and showed that papain digestion generated a fragment [later to be named fragment antigen-binding (Fab)] that could inhibit the precipitin reaction of whole antibody.[2] As the work on immunoglobulin structure progressed other animal species and humans were studied for comparative purposes. The first experiments on complement within the newly formed MRC Immunochemistry Unit utilized guinea pig serum as a source of complement, as at that time this was the standard model in complement research.[3] When Ken Reid (who worked on cod insulin for his PhD[4]) began to characterize C1q the experiments were carried out in parallel in humans and rabbits.[5] However, after these first experiments, the focus of the Unit's complement research was almost exclusively human. Alan Williams also came to the Unit from a comparative background, in chicken haematopoiesis,[6] and was responsible for the first work within the Unit on an exotic animal when, in collaboration with Jean Gagnon, he characterized a protein like Thy-1/Ly-6 from the squid.[7] Coincidentally, Jean's research before joining the Unit also centred on chickens.[8] In addition, Duncan Campbell joined the Unit after completing his PhD on the bovine complement system.[9] My interest in the phylogeny of the complement system began with studies on the binding specificities of C4 purified from a range of mammals.[10] Alex Law and I were trying

Molecular Aspects of Innate and Adaptive Immunity
Edited By Kenneth BM Reid and Robert B Sim

Published by the Royal Society of Chemistry, www.rsc.org

to identify the residues responsible for the differences in reactivity of the thioester of human C4-A and C4-B. The finding that sheep and cattle have a similar pair of isoforms[11] supported our hypothesis that a His residue in C4-B was responsible for the difference in reactivity. On the basis of this work, Keith Whaley asked me to write a chapter on 'complement in funny animals' for the second edition of his book *Complement in health and disease*.[12] In the 15 years since Tony Day and I wrote this review, I have collaborated on many projects on a range of invertebrate and vertebrate complement systems. It has been a pleasure.

3.2 Evolution of Immune Systems

Living organisms are nutrient-rich and protected environments that are attractive habitats for other organisms. Some internal parasites (including commensals, *etc.*) cause no harm or are even beneficial to their hosts. In living their parasitic lifestyle, pathogens are those organisms that cause harm directly by causing tissue damage, and secondarily by causing inflammation and also by weakening the host through overfeeding. All organisms from bacteria to higher plants and animals therefore have a range of defensive mechanisms to prevent colonization by viruses, bacteria, fungi, protozoans and multi-cellular parasites. More than 1400 organisms that are infectious to humans have been identified, of which over 500 have humans as their only known host.[13] It is reasonable to assume that other species have similar numbers of species-specific parasites; free-living organisms are therefore in the tiny minority. However, if we look at this the other way round, all immune systems are extremely successful and, for any given species, only a very small percentage of potential attackers have any success in gaining a foothold. A major role of immune systems is to protect against the many benign organisms that are harmless simply because they are dealt with so effectively.

No immune system is perfect; pathogens continually evolve new ways to overcome existing defence mechanisms and hosts retaliate by developing more complex immune systems. This arms race can be viewed as an extension of Van Valen's red queen hypothesis[14] on competition between species for resources, "It takes all the running you can do, to keep in the same place".[15] This constant striving to overcome ever more sophisticated pathogens has led to the evolution of a wide variety of different defence mechanisms, ranging from physical strengthening of cell walls and epithelia through numerous innate immune mechanisms to the adaptive immune mechanisms found in higher vertebrates. There is no evidence that those invertebrates and plants that depend mainly on innate mechanisms are any more susceptible to disease than are vertebrates with their much more complex acquired immune systems. There is, however, mounting evidence that rearranging immune receptors, not based on immunoglobulin domains, have arisen in some lines.[16] Comparative genomics has shown that in invertebrates, with only innate immune mechanisms, as well as in mammals, with their complex acquired immunity, there is more variation in number and polymorphism in immune genes than in any other group of genes. This suggests that immune genes are under greater evolutionary pressure than other parts of the genome.[17] Genes involved in immunity are estimated to comprise approximately 7% of the human genome.[18]

The evolution of immune systems and the evolution of pathogens are tightly interwoven. In an extremely instructive commentary, Hedrick[19] outlined the ways in which these processes can be viewed from an evolutionary–ecological point of view. The central points are that:

(a) It is impossible to evolve a perfect immune system because parasites can evolve faster than their hosts, and because the more complex immune systems become, the more sophisticated will be the mechanism employed by parasites to overcome them. Furthermore, immunity is costly. Too much expenditure on defence leads to fewer resources being available in other areas necessary for survival.

(b) To be successful, pathogens must balance virulence and infectivity. They must infect their host for long enough to reproduce and then be passed on. At the same time, it is not in their interest to harm or, worse, kill their host. Highly virulent, epidemic-causing pathogens, such as smallpox and influenza viruses, are often recently transferred animal infections that are poorly coevolved with their new host. There is strong selective pressure on pathogens to modulate their virulence.

Consequentially, the main driving force for the evolution of complexity and efficiency in immune systems is not competition between host and pathogen, but competition between host individuals to survive infections, while remaining fit for survival generally. Competition between pathogens to maximize infectivity in part by minimizing virulence ensures that only the least able hosts fail to survive.

To some extent, this is similar to the situation in predator–prey relationships. A zebra does not need to be able to outrun a lion, it only needs to be able to outrun the slowest zebra in the herd; and it is not in the interest of the lion to become such a powerful killing machine that it seriously depletes the zebra population. Camouflage, within an environment of long grass, is the major factor driving the evolution of stripes in animals. The zebra has hit upon a novel function for this feature using its stripes, not to hide, but to make it difficult for lions to pick an individual target in the herd. Wildebeest, the lion's other major food source, still use their less prominent stripes for camouflage but, unlike zebras, they have defensive horns. Different immune systems have developed new mechanisms to deal with specific threats and often put pre-existing systems and genes to new uses. As natural selection acts at the species level, each species and its ecosystem of parasites evolves independently of other species. New ways to deal with threats have been found in some lines, while others that have become redundant have been lost.

3.3 Evolution of the Complement System

3.3.1 The Complement Pathways

The three separate modes of complement activation are the classical, the alternative and the lectin pathways. Of these, the alternative pathway appears to be the simplest, and many believe that it is the ancestral pathway.[20–22] We have argued that the ancestral pathway was probably more akin to the lectin pathway than to the alternative pathway.[23,24]

The classical pathway is a specialized branch of the older lectin pathway in which the ancient lectin, C1q, has been recruited to recognize antigen-bound antibodies that first appeared at some time after the divergence of the jawless (lampreys and hagfish) from the jawed (sharks, bony fish and tetrapods) vertebrates.

The alternative pathway is an extremely unusual immunological mechanism, in that no recognition molecule is involved. A fluid phase tick-over leads to the generation of nascent C3b that can bind to any nearby surface. On self-surfaces, a range of soluble and membrane-bound inhibitors and inactivators inactivate the bound C3b. On non-self, the C3b is protected from inactivation and is able to form a C3 convertase that deposits more C3b around the site of initial non-specific binding. Eventually, from a single activation site, the whole target can become coated with opsonic C3. It seems highly improbable that the thioester protein (C3), activation enzyme (fD), amplification enzyme (fB) and control proteins could have originated simultaneously, yet if any one of these components had been missing the system would have been incapable of specifically attacking non-self.

A lectin pathway, in which the initiating enzyme is bound to the recognition molecule and opsonic molecules are deposited only within a small radius of the initiation site, is self-limiting. Such a system could have evolved in a stepwise fashion, with each step improving on the previous one. We have argued that only when a recognition-based pathway was well developed and multiple inhibitors were in place, could an alternative pathway-amplification mechanism have evolved to deposit opsonic thioester proteins distant from the site of recognition, while not attacking self. An alternative pathway that could activate spontaneously, without the participation of recognition molecules, was probably the final part of the complement system to evolve.[23]

When interpreting results that purport to demonstrate classical, lectin or alternative pathway activities in distant species, remember that there is a lot of crossover between pathways. For example, carbohydrates on immunoglobulins can activate the lectin pathway,[25] Ig and its fragments can activate the alternative pathway[26] and C1q can bind to a number of non-Ig targets and activate the classical pathway.[27] A number of bypass pathways have also been described that can operate in the absence of various components.[28] All of the above mechanisms have been identified by careful dissection of the human system; the assays used in comparative studies are often unable to distinguish between them.

Despite Shakespeare's contention that "That which we call a rose, by any other name would smell as sweet",[29] names alter our perceptions. It is unfortunate that when complement proteins have been discovered in lower animals, they have been assumed to be alternative pathway components and named C3 and fB. This has lead to the circular argument that because these species have C3 and fB, they must have an alternative pathway. If, as we contend, a recognition-based pathway is more ancient, the earliest thioester protein would have been deposited within a small radius of the initiation site, so may have been more C4-like than C3-like; C4 is also an opsonin. The amplification enzyme, also activated by an immobilised enzyme, would have been functionally more C2-like than fB-like, though its substrate (the same thioester protein) would have been deposited at a greater radius from the initiation site and would have been functionally more like C3 in the lectin pathway. In those species that arose before the duplication events that lead to higher vertebrate C3/C4/C5 and C2/fB, we therefore refer to the thioester proteins as C3/C4 and the amplification enzymes as C2/fB, even though their names in the original literature are C3 and fB.

3.3.2 Loss and Gain of Complement Components in Some Evolutionary Lines

Remember that extant species are not our ancestors, but our distant cousins. While the human complement system has increased in complexity, similar (but not the same) additions and changes may have occurred in other species. Even between mice and men there are a number of differences, such as humans having two C4 genes while mice (in an independent duplication event) have gained C4 and SLP; humans have separate genes for CR1 and CR2 while the mouse has a single alternatively spliced gene for these receptors. It would therefore be extremely surprising if other vertebrates and, to a greater extent, invertebrates have not added quite different components to the simpler complement systems that were present in our common ancestors. The current practice among researchers to look only for those components that have been characterized in humans and, on finding them, to assume that they are very similar in function to their human counterparts, but on not finding them concluding that a much simpler system is present, is perhaps a mistake. There is a tendency among comparative immunologists and workers in comparative genomics generally to assume that proteins with similar structures have identical functions. It is worth noting that a recent study of a range of enzymes suggested that at a level of 50% sequence identity only about 30% of the proteins examined were identical in function.[30]

Most complement component deficiencies are rare in the human population, though isolated families that lack almost every component have been described.[31] These rare cases often have severe medical problems. They have arisen from recent mutations and (in the absence of antibiotics and other modern drugs) disappear within a few generations through natural selection. However, two complement deficiencies are fairly common in humans, mannan-binding lectins (MBL) and C9. Deficiency of either of the two human C4 genes is also common, though the presence of two isotypes, which are only slightly divergent in function, ensures that this has little effect on fitness and survival.[32]

Mice and lower primates have two MBL genes (1 and 2), but in humans, gorillas, chimpanzees and rats one of the MBL genes (MBL-1) has become a pseudogene. Human MBL-2 is highly polymorphic, with many of the allotypes producing defective proteins.[33] It has been argued that this loss of function is under positive selection and that it is advantageous to have low levels of this protein, perhaps because high levels increase inflammation or aid infection.[34] However, it seems equally likely that higher apes and rats lack pathogens for which MBL is an important defence mechanism and that the loss of MBL-1 was neutral in terms of natural selection. The MBL-2 gene may be following MBL-1, in degenerating into a pseudogene. If this is so, mice and lower primates presumably still have pathogens for which MBL is a significant control mechanism. It may be that, sometime in the future, humans will face a new pathogenic threat for which MBL is an important defence factor and natural selection will rapidly eliminate the defective genes from the population, though the MBL-1 gene, having been lost, is unlikely to be repaired by chance mutation.

In the Japanese and Korean populations C9 deficiency is fairly common (0.1% of Japanese are homozygous deficient). While there seems to be a slight increase in meningococcal meningitis, most deficient individuals are healthy.[35] Again, this suggests that loss of C9 is not a major problem. It is likely that the deficiency mutation arose in an Asian group that was ancestral to both the modern Korean and Japanese populations. Alternatively, it is possible that C9 deficiency is ancient, but is more harmful in Europe and Africa and so has been removed from these populations through natural selection. It has been noted that the chicken genome seems to lack a C9 gene, although this has been identified in fish and amphibians; this supports the hypothesis that components present in an ancestor can be lost in future generations.[22]

In the case of human C4, simultaneous deficiency at both loci is very rare, which suggests that C4 is essential but that to a large extent possession of one isotype is compatible with survival. The mouse also has two C4 loci, but in this case one (SLP) has lost its C4 function. Duplication of complement C4 genes is common in mammals and seems to have occurred independently in mice, primates, and sheep and cattle. Multiple copies of many complement genes have been identified in bony fish.[36,37] This has allowed the copies to diverge in sequence and possibly function. The complement systems of bony fish may therefore be considerably more sophisticated than those of mammals.

These observations suggest that in modern humans, MBL and C9 are not currently essential. In opposition to this idea is that these two proteins have been conserved in all vertebrate groups (though not necessarily all species, since only a small sample has been studied). It is conceivable that in the absence of natural selection acting against the loss of MBL and C9 these could disappear from the human gene pool. Two possible consequences could arise from this situation:

(a) A highly virulent pathogen could appear in future generations, defence against which is dependent on MBL or C9 and our species could become extinct.

(b) Other arms of the immune system could fill the gaps in our defences formerly filled by MBL or C9 and future generations of humans and future species evolved from humans would have no C9 or MBL.

Obviously, something similar to (a) has not happened in our line of descent, although something of this nature could have been the cause of extinction of some lines. An example of the loss of a complement component in the line of descent leading to humans, as in (b), is that of a fourth class of complement thioester protein. A protein that is slightly more C4-like than C3-like in primary structure has been found in fish[38] and birds (XP_417086). This must have been present in our common ancestor, but seems to have been lost from the human and mouse genomes.

3.4 Phylogeny of the Complement System

Since the dawn of research on the complement system, there has been an interest in comparative studies.[39] Up until the early 1970s most of the work involved the demonstration of haemolytic activity in different species.[40] During the 1970s and 1980s complement-related proteins were purified and characterized from a range of vertebrate species.[12] Polymerase chain reaction (PCR), based on conserved sequences, has led to a massive expansion in the number of complement-related sequences, which has been added to by comparative genomic studies.[22] Invertebrate complement-related sequences were first identified by Courtney Smith, in a sea urchin (*Strongylocentrotus purpuratus*).[41] The first demonstration of a functional complement system that utilized proteins related to vertebrate complement components came from the labs of Nonaka and Fujita, who demonstrated an opsonic lectin pathway in a sea squirt (*Halocynthia roretzi*).[42]

Figure 3.1 shows a simplified view of current thinking on the phylogenetic relationships between the species in which complement-related sequences and activities have been described.[43] Complement-like proteins have now been identified in all multicellular phyla studied, with the exception of sponges, but the sponge genome project is still in its infancy. The non-mammalian species that have been studied in most detail, at both the functional and molecular levels, include two bony fish (rainbow trout, *Oncorhynchus mykiss*,[37] and common carp, *Cyprinus carpio*[44]), a cartilaginous fish (nurse shark, *Ginglymostoma cirratum*[45]), a jawless fish (lamprey, *Lampetra japonica*[46]), an ascidian (sea squirt, *H. roretzi*[47]) and an echinoderm (purple sea urchin, *S. purpuratus*[48]), all of which fall within the deuterostome group of animals. The most extensively characterized non-deuterostome complement system is that of an arthropod (horseshoe crab, *Carcinoscorpius rotundicauda*[49]). Large amounts of sequence data have been generated from a range of genome projects.

Complement-related functions and sequences that have been found, to date, are summarized in Figure 3.2. Only those functions and pathways for which there is molecular evidence have been included. In the molecular section an

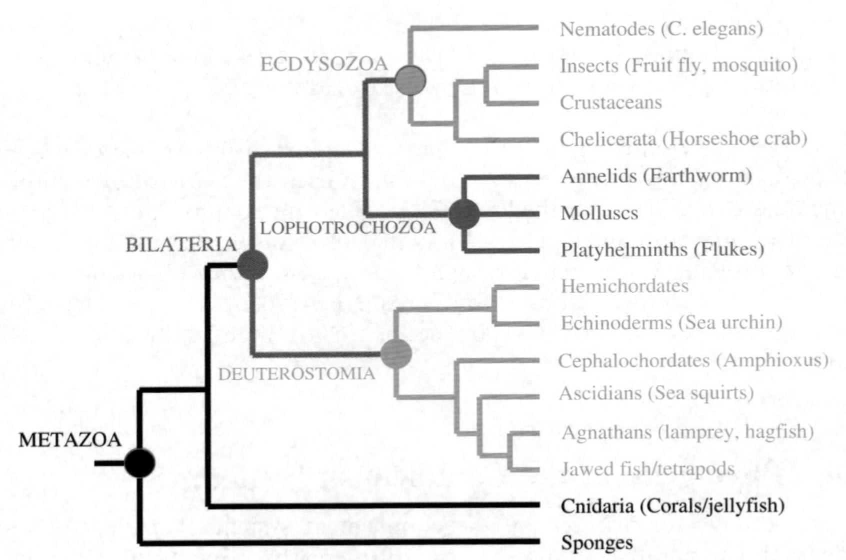

Figure 3.1 Animal relationships within the metazoa. Only the major groups in which complement-related proteins or activities have been described are shown. Adapted from Halanych (2004)[43] and Schubert *et al.* (2006).[72]

X indicates that a sequence that resembles a mammalian complement component has been determined or that a protein has been purified and partially characterized. In many cases the proteins represented by these sequences have not been demonstrated to have complement-related functional activity. Shading without an X indicates that phylogenetic relationships suggest that a gene should be present. However, this may not always be true. The horseshoe crab has C3/C4-like and C2/fB-like proteins, while *Drosophila* and mosquito do not. This suggests that these genes were present in an ancestral arthropod, but have been lost by (some) insects.

3.4.1 Recognition Molecules; Ficolins, MBL and C1q

The classical and lectin pathways require recognition of foreign material as the initial step in the activation process. In the mammalian lectin pathway this occurs directly via two distinct classes of recognition molecules, MBLs and ficolins. The carbohydrate recognition domain of MBL is a C-type lectin that, in humans, has affinity for a range of monosaccharides, including mannose, *N*-acetylglucosamine (GlcNAc), fucose and glucose.[50] In ficolins, the recognition regions are fibrinogen-like domains that have specificity for GlcNAc, *N*-acetylgalactosamine (GalNAc) and also, in some cases, sialic acid and lipopolysaccharide (LPS).[51] C1q, the recognition molecule of the classical pathway, has specificity for the fragment crystallizable (Fc) region of aggregated

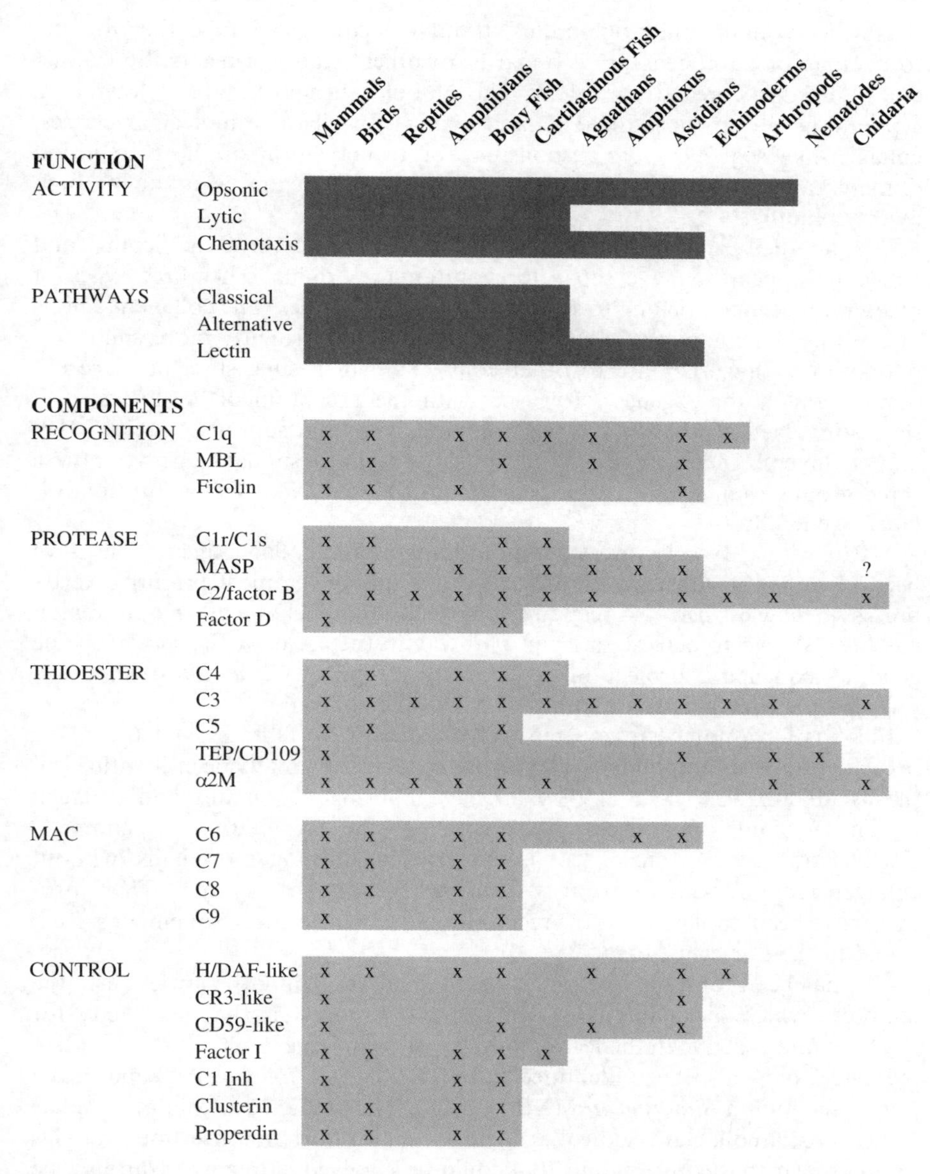

Figure 3.2 Complement-like functions and components described in metazoans. In the component section an X indicates that DNA sequence and/or protein has been described, which resembles a complement component. Pale shading indicates that phylogenetic relationships suggest that components should be present, but have not yet been described. This may not always be true, as the drosophila, mosquito and *C. elegans* genomes seem to contain no complement components. In the function and pathways section, dark shading indicates that there is evidence that certain pathways and functions are present. Sometimes this evidence is difficult to interpret and for this reason we have not represented more precise quantitation.

or surface-bound immunoglobulins, though it can also bind either directly to a range of pathogens or secondarily to other pathogen-recognition molecules, such as C-reactive protein. The globular heads of C1q belong to a separate family to those of MBL and ficolins. In all three molecular classes, collagen-like sequences are responsible for trimerization of the recognition domains and the formation of higher oligomers that are essential for high avidity binding.

Despite their similarity in overall molecular structure, C1q, ficolins and collectins appear to have separate evolutionary origins. The three types of recognition domain belong to separate structural families. The collagen regions of the three mammalian C1q chains A, B and C are slightly more similar to one another than they are to other collagens, which suggests a fairly recent duplication of these genes. However, with the exception of a short stretch that is involved in MASPs/C1r–C1s binding,[52] comparison of C1q with MBL and ficolin collagenous regions seems to show no more similarity between these three groups than would be expected from the Gly–Xaa–Yaa nature of collagens generally.

MBLs with C-type lectin-recognition domains and collagenous regions have been identified in most vertebrate classes, including the most primitive vertebrates, the jawless fish.[53] A related C-type lectin that lacks a collagenous region has been shown to activate a lectin pathway in the ascidian, *H. roretzi*.[54] The genome-sequencing project of a second sea squirt, *Ciona intestinalis*, has revealed the presence of a number of MBL-like collectins.[55]

Ficolins have not been as extensively studied in vertebrates, but they have been identified in amphibians and birds. Four ficolins have been identified in the ascidian *H. roretzi* and they all have fibrinogen domains and collagen regions.[56] Similar sequences are present in the *C. intestinalis* genome.[55] Tachylectins are a group of lectins that contain fibrinogen domains, without collagen regions, which were first described in horseshoe crabs.[57] These proteins have been implicated as the recognition molecules of a complement system found in the horseshoe crab.[49]

C1q has been found in all vertebrates studied, including the jawless fish, the lamprey, which lacks immunoglobulins.[46] Lamprey C1q has specificity for GlcNAc and is active in the lectin pathway of complement activation. C1q-like sequences have also been identified in the ascidian *C. intestinalis*[55] and in the purple sea urchin *S. purpuratus* (XP_782700). The predicted urchin protein has C1q heads, a collagen region that includes a kink-forming insertion and Cys residues near to the N-terminus that could be involved in oligomerization. The origins of the classical pathway therefore predate the appearance of the vertebrate adaptive immune system.[24] At the present time, the proposed C1q of the sea urchin is the earliest evidence for a lectin pathway that involves a collagen-containing lectin. However, as more genomic data become available it seems not unlikely that recognition molecules related to C1q, MBL or ficolin will be found in other phyla. When interpreting data from genomic sequencing remember that many non-complement proteins include C1q-like domains and some of these also have collagen sequences.[58]

3.4.2 Initiation Enzymes; Factor D, MASPs, C1r and C1s

Mammalian factor D (fD) may have dual physiological roles. As well as being the initiation enzyme of the alternative pathway it has been proposed to be active in lipid metabolism under its alternative name of adipsin,[59] though fD knockout mice seem to have no metabolic problems.[60] fD circulates as an active enzyme. Its only physiological plasma substrate appears to be factor B (fB) in complex with activated C3 (C3b or $C3H_2O$). It is unusual among serine proteases in that it has very low activity against small synthetic substrates. It has been proposed that the circulating enzyme has an inactive, zymogen-like, conformation, but that the substrate, C3bB, induces a conformational change to the active state.[61] fD has only a distant evolutionary relationship to the serine protease domains of MASP family proteins.[62] To date, only one non-mammalian protein that is structurally and functionally related to fD has been described from a bony fish, the common carp.[63] However, because fD consists only of a serine protease domain and is closely related to the granzymes of cytotoxic T-lymphocytes, it is difficult to identify in genomic data without supporting functional evidence.[22] It is also possible that other, structurally different, enzymes could fill the functional role of activating fB in complex with C3b in lower vertebrates and invertebrates.

The MASP family has been extensively studied across species. In humans, five members have been identified (C1r, C1s, MASP-1, MASP-2 and MASP-3) plus MAp19, an alternatively spliced variant of MASP-2 that comprises the N-terminal region of the non-catalytic chain. The roles of MASP-1, MASP-3 and MAp19 remain unclear, although all are found in complex with MBL and ficolin. C1r and C1s act as a tetrameric complex (two C1r + two C1s) and bind preferentially to C1q, while MASP-2 is active as a dimer and binds to both MBL and ficolins.

Bony fish have molecules that are more closely related to C1r and C1s than to MASP-2. Two molecules have been identified in the carp, though these appear to be the result of a recent duplication and are not equivalent to C1r and C1s,[64] and one from the rainbow trout.[65] All three molecules seem to be slightly more C1r-like than C1s-like. A C1s-like activity has been partially purified from trout serum and this enzyme cleaves trout C4, but not trout C3.[66] Genomic sequencing has revealed that both *Xenopus* (AAI08809, NP_001090130) and chickens (NP_001025948, XP_416518) have separate C1s-like and C1r-like genes. Based on the current data it seems likely that bony fish have a single molecular type that can autoactivate, like C1r, but with the complement-activating activity of C1s, as is found in mammalian MASP-2. Some time after the divergence of the bony fish and the tetrapods, a gene duplication led to the appearance of separate C1r and C1s proteins. The slightly greater similarity between the fish C1r/Cls and higher vertebrate C1r could be the result of the need to conserve structural features that are involved in autoactivation. However, there is little functional data outside that for the mammals.

MASP-like proteins have been identified in all vertebrate classes studied and also in a number of chordate invertebrates. In mammals, MASP-2 is the major activation protease for the lectin pathway. Like C1s, it can cleave C4 and C2 to form the C3 convertase of the lectin pathway; like C1r, it can autoactivate when in complex with MBL and ficolin bound to targets. MASP cDNA and gDNA sequences have been obtained from lamprey,[67] amphioxus[68] and ascidians.[55,69] They have not yet been found in echinoderms, but the probable presence of a C1q-like, C3/C4-like and C2/fB-like proteins suggest that they may be found. A MASP-like sequence has apparently been identified in the starlet sea anemone *Nematostella vectensis* genome (XM_001623051.1), but this sequence is suspiciously similar to amphioxus MASP and could have been derived from a chordate in the anemone's food.

The evolutionary history of MASPs is rather convoluted and has been discussed in a number of recent reviews.[47,70,71] Known MASP genes can be split into three groups. They all have heavy chain sequences that are split into a number of exons. The coding regions for the serine protease domains are of two sorts. In the first form the coding sequence is split into a number of exons and the active site Ser is coded by an AGY triplet, while in the second type the entire domain is coded by a single exon and the Ser is coded by a TCN triplet. The third group has both types of serine protease in tandem and these can be alternatively spliced to a single heavy chain. The suggested models propose that the AGY-type serine protease is ancestral, as this is found in ascidians. This was followed by the third type with an alternatively spliced gene, as has been found in amphioxus. The alternatively spliced gene has been retained by higher vertebrates, where it codes for MASP-1 and MASP-3. Genes with only an intronless AGY coding sequence are found in lampreys and higher vertebrates and code for MASP-2, C1r and C1s. It is proposed that these genes are derived from the alternatively spliced type by the loss of the TCN coding region. However, remember that ascidians, amphioxus and lamprey are not our ancestors, but our distant cousins. Making assumptions about what are ancestral characters and what are derived characters may be unwise. In this particular case, there may also be a phylogenetic problem, as recent studies suggest that vertebrates may be more closely related to ascidians than to amphioxus.[72]

Apart from the (possibly erroneous) sea anemone gene, MASP-like sequences have not yet been described in animals lower than the chordates and collagen-containing lectins have not been found outside the deuterostome group. However, it is likely that the horseshoe crab has a complement activation mechanism like that of a lectin pathway.[49] Presumably other types of recognition molecule and activation enzymes are present in non-deuterostome species. Possible recognition molecules are the ficolin-related tachylectins[57] and the pentraxins,[73] although many other types of lectin have been described in diverse species[74] and pathogen-associated molecular patterns (PAMPs), other than carbohydrates, could also be involved. Similarly, MASPs may be specialized for binding to collagen-containing lectins and, in species that utilize other lectin types, different interaction domains may be found in the heavy chain. Factor C, the initiating enzyme of the horseshoe crab clotting system,[57] is closely related to the MASPs.[62]

3.4.3 Thioester Proteins: C3, C4, C5, α2 M, CD109 and Insect Thioester Proteins

On the basis of primary sequence, thioester-containing proteins can be split into two families. Family 1 includes the α2 macroglobulin (α2 M) protease inhibitors, the insect thioester proteins (TEPs) and CD109, a cell surface protein. Family 2 comprises higher vertebrate complement components C3, C4 and C5 and the complement components of the lower vertebrates and invertebrates. The structure of α2 M is not yet available, but comparison of the recently derived structures of human C3[75] and TEP1 from *Anopheles gambiae*[76] confirms that both are similar in structure. Each has a core of eight domains, with a structure similar to the fibronectin-type-3 fold, that have been called macroglobulin (MG) domains. Inserted between MG7 and MG8 is a CUB domain split by a thioester domain. In both families the MG6 domain contains an insert. In TEP1 the insert contains a disordered region that is probably the equivalent of the α2 M bait region plus a link peptide that may be involved in stabilizing the native conformation of the molecule. In C3, the insert contains a link peptide, the α–β chain processing site, the anaphylotoxin domain and the activation site. C3 and other complement components also contain an extra C-terminal domain that has a structure unrelated to that of the MG domains.[77]

Protease inhibitors related to α2 M have been found in all vertebrates and many invertebrates. The most extensively characterized invertebrate protein is that of the Atlantic horseshoe crab (*Limulus polyphemus)*, but related sequences have been found in crustaceans, insects, molluscs and a tick.[78] Most members of this family are either dimeric or tetrameric, although monomeric examples have been described. They have a region that contains cleavage sites for many classes of protease, called the bait region. Proteolysis of any bond within this region leads to a large change in conformation that entraps the protease by a venus flytrap-like mechanism. Concurrently, the thioester becomes highly reactive and can bind covalently to the attacking protease, immobilizing it further.

The insect TEPs have been extensively studied in the fruit fly (*D. melanogaster*)[79] and in a mosquito (*A. gambiae*).[80] Related sequences are present in other insects, the nematode *Caenorhabditis elegans* and molluscs. The function of most insect TEPs has not been investigated, but TEP1 of the mosquito has opsonic activity[81] and thus seems to be acting in a complement-like fashion. The structure of mosquito TEP1[76] suggests that rather than having a specific activation site, which could be activated by a MASP or C2/fB-like protease, it has an α2 M-like bait region that may have cleavage sites for a range of proteases. There is no evidence from the genome sequence of the mosquito or *Drosophila* for a C2/fB-like amplification enzyme. It may be that the mosquito has reinvented a simple 'complement system' that is based on thioester protein and is activated by pathogen-derived proteases, depositing opsonic TEP directly onto the pathogen surface. As we have proposed previously[12,23] a mechanism of this type was possibly the origin of the complement system. Specificity can be introduced to such a system, without the need for recognition molecules, by varying the bait region sequence so that cleavage sites for self-proteases are removed while sites for the proteases of target pathogens are introduced.

CD109 is a GPI-anchored protein that is expressed on mammalian epithelial cells, platelets and activated T-cells.[82] A recent proposal is that it is involved in the TGF-β receptor system.[83] Sequence comparisons indicate that, like α2 M and the insect TEPs, it has a bait region rather than a specific activation site, but its primary structure is slightly more similar to the insect molecules than to α2 M. A CD109-like molecule has been identified in the ascidian *C. intestinalis*.[84]

The complement components of higher vertebrates, C3, C4 and C5, arose early in vertebrate evolution, probably as the result of two genome duplications.[85] Bony fish and higher vertebrates have all three proteins[37,44] and sharks have at least C3 and C4.[86] Lamprey, hagfish and invertebrates, however, each have only one type of complement-related thioester protein (though in some species more than one divergent gene copy is present). These complement thioester proteins have properties that make them functional and structural hybrids of some of the features of higher vertebrate C3, C4 and C5. For example, the primary structure of the lamprey protein is slightly more similar to C3 than to C4 but, like C4, it has three chains and is activated by MASP; it is likely that the activation peptide has anaphylotoxin activity, a function that is carried predominantly by C5 in higher vertebrates. C3/C4-like genes have been identified in jawless fish,[87,88] ascidians,[55,89] amphioxus,[90] echinoderms,[91] arthropods[49] and cnidarians.[92,93] Functional data show that the proteins from ascidians (*H. roretzi*)[89] and arthropods (*C. rotundicauda*)[49] are opsonic and, in both cases, activation appears to be via a lectin pathway-like mechanism.

Cobra venom factor (CVF) is an example of the recycling of a complement component to perform a new function. CVF has been known to immunologists for many years as a reagent that can decomplement the sera of many species. The protein is the product of a recently duplicated C3 gene, which undergoes processing to a C3c-like form. The CVF binds recipient fB that is activated by fD to give a fluid phase C3 convertase that is resistant to inactivation by the prey control proteins and so can consume all of the C3. The venom apparatus of snakes is thought to have evolved from the salivary glands and, like mammalian saliva, snake venoms contain digestive enzymes as well as toxins.[94] Snakes are incapable of chewing and therefore must swallow their prey whole. It is possible that by decomplementing the prey, the CVF aids digestion by allowing internal putrefaction via the prey's gut organisms.[12]

3.4.4 Amplification Enzymes; C2 and Factor B

Mammals, clawed frog (*X. laevis* ABB85337/NP_001081234) and a number of bony fish, carp (*C. carpio* BAA34707/BAA78416), trout (*O. mykiss* BAB19788/AAC83699), green spotted puffer (*Tetraodon nigroviridis* CAG06450/CAD21938) and zebra fish (*Danio rerio* AAI39710/NP_571413) have genes that code for separate proteins with C2-like and fB-like primary structures, though there is no evidence that the proteins have separate roles in the alternative and classical or lectin pathways, except in mammals. Two species of shark, *G. cirratum*[95] and *Triakis scylium* (BAF62177, BAB63203), each have two copies of a C2/fB gene that appear to have duplicated after the divergence of the sharks from the bony fish, in an event distinct from the duplication that led to separate C2 and fB genes in bony fish and tetrapods. The primary structures of both types of shark protein are slightly more similar to mammalian C2 than to fB. There is functional evidence that *G. cirratum* has a C2-like activity that is active in a classical pathway.[96] Sequences related to C2/fB have been identified in jawless fish,[97] ascidians,[55,98] echinoderms,[99] horseshoe crab[49] and a sea anemone (XM_001635440.1). The vertebrate sequences are characterized as having three complement-control protein (CCP) domains, a Von Willebrand factor type A domain and a serine protease domain. The invertebrate proteins are similar, but have two extra CCP domains at the N-terminus. There is no functional information on any of the invertebrate proteins.

3.4.5 Lytic Proteins; C6, C7, C8 and C9

The proteins of the mammalian membrane attack complex (MAC) are coded by six genes, C6, C7, C8α, C8β, C8γ and C9. Examples of sequences related to each of these have been identified in a range of bony fish and all six are present in the rainbow trout (*O. mykiss* CAF22026, AAG30011, CAH65481, Q90×85, CAF22027, CAJ01692). Purified MACs from the common carp contain proteins similar in molecular mass to their human counterparts.[100] Sharks have lytic complement.[45] The lesions formed in erythrocytes are similar in morphology to those formed by mammalian complement, and mammalian C8 and C9 appear to be compatible with unidentified shark components in forming a MAC.[101] Many invertebrates and lower vertebrates have lytic mechanisms that in some ways resemble complement.[40] However, it was shown by Nonaka that in the most extensively studied of these, the lamprey, the process was probably not complement related.[87] Despite this, sequences similar to C6 in domain structure have been found in amphioxus[90] and an ascidian (*C. intestinalis*).[55] There are no functional data on MAC formation in these species.

3.4.6 Control Proteins and Receptors

The identification of Complement Control Proteins based entirely on sequence similarity is extremely difficult for two reasons:

(a) Many are known to be multifunctional (*e.g.* C1 inhibitor is active against non-complement enzymes such as kallikrein and clotting factors; CR3 and CR4 recognize many ligands other than C3 fragments) and, therefore, their presence in another species, without functional data, is not conclusive evidence that they are part of a complement system.

(b) Many belong to protein families that have many functions unrelated to complement (*e.g.* proteins with multiple CCP domains are widely distributed and in mammals are found in proteins such as clotting factor XIII; properdin-like molecules with multiple thrombospondin repeats have been found in malaria parasites[102]).

Despite this, any species that has a complement system must have regulatory proteins to ensure that the system does not become completely depleted on activation and to protect from indiscriminate self-destruction.

The complement system of the ascidian *H. roretzi* is opsonic and the C3/C4 receptor has been shown to be an integrin related to CR3 and CR4.[103,104] Antibodies raised against the α-subunit could bind to a subset of haemocytes that inhibit C3-dependent phagocytosis.[103] Proteins related to the regulation of complement activation (RCA) locus proteins fH, C4bBP, DAF, MCP, CR1 and CR2 are found throughout the animal kingdom and also in pathogens, which use them for defence against complement attack. However, apart from those of viruses and a bony fish (barred sand bass, *Parablax neblifer*),[105,106] there are few functional data outside that for the mammals. C1 inhibitor-like sequences are present in bony fish.[65] Factor I has a distinctive domain structure and has been found in *Xenopus*,[107] bony fish[108] and sharks.[109] Properdin-related structures have been found in bony fish (CAJ34423, XP_684282, CAF90479).

3.4.7 Possible Origins of Complement

The earliest 'complement system' may have had two independent evolutionary roots:

(a) A thioester protein, which resembled the opsonic TEP1 of the mosquito and may have been activated directly by proteases produced by pathogens, together with a receptor.
(b) A lectin or other PAMP-recognition molecule, probably also with its own receptor.

These two could have become integrated by the introduction an initiating enzyme, with MASP-like function, that could bind to the PAMP-recognition molecule and activate the thioester protein on pathogens that did not produce the appropriate proteases. The next stage could have been the recruitment to the system of a C2-like amplification protease that would have increased the amount of thioester protein deposited around each PAMP-recognition site. This secondary deposition would have been limited by the short half-life of the reactive thioester protein and the requirement that the 'C2' be activated by a surface-bound initiation enzyme. Addition of extra functions, components and control proteins could then have proceeded independently in different lines as a result of different pathogenic pressures. It is probable that collagen-oligomerized recognition molecules and MASPs arose in our deuterostome ancestors, replacing earlier recognition–activation complexes. At some stage a soluble fD-like component was introduced that could activate the 'C2' distant from the PAMP-recognition site, to give it fB-like function. No functional or molecular evidence suggests when this first alternative pathway amplification mechanism arose, but it is probably present in all jawed vertebrates and could be much older.[23]

3.5 Summary

Complement is one of many innate immune mechanisms utilized by animals. Evidence suggests that at least two complement-like proteins are present in sea anemones and corals, which indicates that this system is one of the most ancient and enduring and that it first emerged close to the appearance of multicellular animals. Other than possibly mistaken nomenclature there is no evidence for an alternative pathway activation mechanism in any species lower than the jawed vertebrates. Lower chordates, the jawless fish and sea squirts have been shown to have opsonic complement systems activated via a lectin pathway. The arthropod, horseshoe crab, also has an apparently recognition-based activation mechanism, though the lectin or other PAMP-recognition molecules involved have not yet been identified. The selective pressure imposed by co-evolving parasites will have led to the development of diverse complement systems in the different evolutionary lines that have led to our extant cousins. The sea squirt opsonic complement system, for example, appears to be rather simple, consisting of lectins, MASPs, thioester protein, amplification convertase and a receptor. However, this reflects the complement system of our last common ancestor and represents the components that are common to the ascidian and mammal systems. There is no reason to assume that while many extra components and functions have been added to the mammalian system, ascidian complement has fossilized. Other components and functions may have been added and it is also possible that either line has lost components that were present in the common ancestor. In the absence of time-travelling specimen hunters, we will never know the function of the primordial immunoglobulin molecules that arose in our jawless or early jawed ancestors in the seas of the Silurian period, about 450 000 000 years ago. Almost certainly they acted to enhance the recognition capabilities of a pre-existing innate immune mechanism. It is not impossible, therefore, that those first immunoglobulins could have been 'complement components'. Perhaps the history of complement phylogeny in the Unit is older than we imagine.

Acknowledgements

I thank all of my collaborators on 'funny animals'; notably Paul Levine (St Louis, MO, USA and Woods Hole, MA, USA), Sylvia Smith (Miami, FL), Misao Matsushita (Hiratsuka, Japan), Teizo Fujita (Fukushima, Japan), Miki Nakao (Kyushu, Japan), Sigrun Lange (Reykjavik, Iceland) and Alex Law

(Oxford). Ken Reid has been very supportive, allowing me to spend time working in many pleasant locations (fresh body fluid samples being absolutely vital in the case of most species).

References

1. R. R. Porter, *Biochem. J*, 1950, **46**, 473–478.
2. R. R. Porter, *Biochem. J*, 1950, **46**, 479–484.
3. N. E. Hyslop, Jr., R. R. Dourmashkin, N. M. Green and R. R. Porter, *J. Exp. Med.*, 1970, **131**, 783–802.
4. K. B. M. Reid, P. T. Grant and A. Youngson, *Biochem. J.*, 1968, **110**, 289–296.
5. K. B. M. Reid, D. M. Lowe and R. R. Porter, *Biochem. J.*, 1972, **130**, 749–763.
6. A. F. Williams, *J. Cell Sci.*,1972, **11**, 771–776.
7. A. F. Williams, A. G. Tse and J. Gagnon, *Immunogenetics*, 1988, **27**, 265–272.
8. J. Gagnon, R. D. Palmiter and K. A. Walsh, *J. Biol. Chem.*, 1978, **253**, 7464–7468.
9. R. D. Campbell, N. A. Booth and J. E. Fothergill, *Biochem. J.*, 1979, **183**, 579–588.
10. A. W. Dodds and S. K. Law, *Biochem. J.*, 1990, **265**, 495–502.
11. X. D. Ren, A. W. Dodds and S. K. Law, *Immunogenetics*, 1993, **37**, 120–128.
12. A. W. Dodds and A. J. Day, in: *Complement in health and disease*, eds, K. Whaley, M. Loos and J. M. Weiler, Kluwer, London, 1993, pp. 39–88.
13. L. H. Taylor, S. M. Latham and M. E. Woolhouse, *Philos. Trans. R. Soc. Lond. B. Biol. Sci.*, 2001, **356**, 983–989.
14. V. Van Valen, *Evolutionary Theory*, 1973, **1**, 1–30.
15. L. Carroll, *Through the looking-glass, and what Alice found there*, 1871.
16. G. W. Litman, L. J. Dishaw, J. P. Cannon, R. N. Haire and J. P. Rast, *Curr. Opin. Immunol.*, 2007, **19**, 526–534.
17. J. Trowsdale and P. Parham, *Eur. J. Immunol.*, 2004, **34**, 7–17.
18. J. Kelley, B. de Bono and J. Trowsdale, *Genomics*, 2005, **85**, 503.
19. S. M. Hedrick, *Immunity*, 2004, **21**, 607–615.
20. P. Lachmann, *Behring Inst. Mitt.*, 1979, **63**, 25–37.
21. L. C. Smith, *Adv. Exp. Med. Biol.*, 2001, **484**, 363–372.
22. M. Nonaka and A. Kimura, *Immunogenetics*, 2006, **58**, 701–713.
23. A. W. Dodds, *Immunobiology*, 2002, **205**, 340–354.
24. A. W. Dodds and M. Matsushita, *Immunobiology*, 2007, **212**, 233–243.
25. R. Malhotra, M. R. Wormald, P. M. Rudd, P. B. Fischer, R. A. Dwek and R. B. Sim, *Nat. Med.*, 1995, **1**, 237–243.
26. K. J. Gadd and K. B. M. Reid, *Biochem. J.*, 1981, **195**, 471–480.
27. N. R. Cooper, *Adv. Immunol.*, 1985, **37**, 151–216.

28. S. E. Degn, S. Thiel and J. C. Jensenius, *Immunobiology*, 2007, **212**, 301–311.
29. W. Shakespeare, *Romeo and Juliet, Act II, Scene 2*, 1597.
30. B. Rost, *J. Mol. Biol.*, 2002, **318**, 595–608.
31. H. R. Colten and F. S. Rosen, *Annu. Rev. Immunol.*, 1992, **10**, 809–834.
32. C. Y. Yu, E. K. Chung, Y. Yang, C. A. Blanchong, N. Jacobsen, K. Saxena, Z. Yang, W. Miller, L. Varga and G. Fust, *Prog. Nucleic Acid Res. Mol. Biol.*, 2003, **75**, 217–292.
33. M. W. Turner and R. M. Hamvas, *Rev. Immunogenet.*, 2000, **2**, 305–322.
34. J. Seyfarth, P. Garred and H. O. Madsen, *Hum. Mol. Genet.*, 2005, **14**, 2859–2869.
35. T. Horiuchi, H. Nishizaka, T. Kojima, T. Sawabe, Y. Niho, P. M. Schneider, S. Inaba, K. Sakai, K. Hayashi, C. Hashimura and Y. Fukumori, *J. Immunol.*, 1998, **160**, 1509–1513.
36. M. Nakao, J. Mutsuro, M. Nakahara, Y. Kato and T. Yano, *Dev. Comp. Immunol.*, 2003, **27**, 749–762.
37. H. Boshra, J. Li and J. O. Sunyer, *Fish Shellfish Immunol.*, 2006, **20**, 239–262.
38. J. Mutsuro, N. Tanaka, Y. Kato, A. W. Dodds, T. Yano and M. Nakao, *J. Immunol.*, 2005, **175**, 4508–4517.
39. H. T. Marshall, *J. Exp. Med.*, 1904, **6**, 309–332.
40. I. Gigli and K. F. Austen, *Annu. Rev. Microbiol.*, 1971, **25**, 309–332.
41. L. C. Smith, L. Chang, R. J. Britten and E. H. Davidson, *J. Immunol.*, 1996, **156**, 593–602.
42. M. Nonaka and K. Azumi, *Dev. Comp. Immunol.*, 1999, **23**, 421–427.
43. K. Halanych, *Annu. Rev. Ecol. Evol. Syst.*, 2004, **35**, 229–256.
44. M. Nakao, Y. Kato-Unoki, M. Nakahara, J. Mutsuro and T. Somamoto, *Adv. Exp. Med. Biol.*, 2006, **586**, 121–138.
45. S. L. Smith, *Immunol. Rev.*, 1998, **166**, 67–78.
46. M. Matsushita, A. Matsushita, Y. Endo, M. Nakata, N. Kojima, T. Mizuochi and T. Fujita, *Proc. Natl. Acad. Sci. USA*, 2004, **101**, 10127–10131.
47. T. Fujita, M. Matsushita and Y. Endo, *Immunol. Rev.*, 2004, **198**, 185–202.
48. L. C. Smith, L. A. Clow and D. P. Terwilliger, *Immunol. Rev.*, 2001, **180**, 16–34.
49. Y. Zhu, S. Thangamani, B. Ho and J. L. Ding, *EMBO. J.*, 2005, **24**, 382–394.
50. U. L. Holmskov, *APMIS Suppl.*, 2000, **100**, 1–59.
51. Y. Endo, M. Matsushita and T. Fujita, *Immunobiology*, 2007, **212**, 371–379.
52. U. V. Girija, A. W. Dodds, S. Roscher, K. B. M. Reid and R. Wallis, *J. Immunol.*, 2007, **179**, 455–462.
53. M. Takahashi, D. Iwaki, A. Matsushita, M. Nakata, M. Matsushita, Y. Endo and T. Fujita, *J. Immunol.*, 2006, **176**, 4861–4868.

54. H. Sekine, A. Kenjo, K. Azumi, G. Ohi, M. Takahashi, R. Kasukawa, N. Ichikawa, M. Nakata, T. Mizuochi, M. Matsushita, Y. Endo and T. Fujita, *J. Immunol.*, 2001, **167**, 4504–4510.
55. K. Azumi, R. De Santis, A. De Tomaso, I. Rigoutsos, F. Yoshizaki, M. R. Pinto, R. Marino, K. Shida, M. Ikeda, M. Arai, Y. Inoue, T. Shimizu, N. Satoh, D. S. Rokhsar, L. Du Pasquier, M. Kasahara, M. Satake and M. Nonaka, *Immunogenetics*, 2003, **55**, 570–581.
56. A. Kenjo, M. Takahashi, M. Matsushita, Y. Endo, M. Nakata, T. Mizuochi and T. Fujita, *J. Biol. Chem.*, 2001, **276**, 19959–19965.
57. S. Iwanaga, *Curr. Opin. Immunol.*, 2002, **14**, 87–95.
58. R. Ghai, P. Waters, L. T. Roumenina, M. Gadjeva, M. S. Kojouharova, K. B. M. Reid, R. B. Sim and U. Kishore, *Immunobiology*, 2007, **212**, 253–266.
59. R. T. White, D. Damm, N. Hancock, B. S. Rosen, B. B. Lowell, P. Usher, J. S. Flier and B. M. Spiegelman, *J. Biol. Chem.*, 1992, **267**, 9210–9213.
60. Y. Xu, M. Ma, G. C. Ippolito, H. W. Schroeder, Jr., M. C. Carroll and J. E. Volanakis, *Proc. Natl. Acad. Sci. USA*, 2001, **98**, 14577–14582.
61. J. E. Volanakis and S. V. Narayana, *Protein Sci.*, 1996, **5**, 553–564.
62. M. M. Krem, T. Rose and E. Di Cera, *Trends Cardiovasc. Med.*, 2000, **10**, 171–176.
63. T. Yano and M. Nakao, *Mol. Immunol.*, 1994, **31**, 337–342.
64. M. Nakao, K. Osaka, Y. Kato, K. Fujiki and T. Yano, *Immunogenetics*, 2001, **52**, 255–263.
65. T. Wang and C. J. Secombes, *Immunogenetics*, 2003, **55**, 615–628.
66. H. Boshra, A. E. Gelman and J. O. Sunyer, *J. Immunol.*, 2004, **173**, 349–359.
67. Y. Endo, M. Takahashi, M. Nakao, H. Saiga, H. Sekine, M. Matsushita, M. Nonaka and T. Fujita, *J. Immunol.*, 1998, **161**, 4924–4930.
68. Y. Endo, M. Nonaka, H. Saiga, Y. Kakinuma, A. Matsushita, M. Takahashi, M. Matsushita and T. Fujita, *J. Immunol.*, 2003, **170**, 4701–4707.
69. X. Ji, K. Azumi, M. Sasaki and M. Nonaka, *Proc. Natl. Acad. Sci. USA*, 1997, **94**, 6340–6345.
70. T. Fujita, *Nat. Rev. Immunol.*, 2002, **2**, 346–353.
71. M. Nonaka and F. Yoshizaki, *Immunol. Rev.*, 2004, **198**, 203–215.
72. M. Schubert, H. Escriva, J. Xavier-Neto and V. Laudet, *Trends. Ecol. Evol.*, 2006, **21**, 269–277.
73. P. B. Armstrong, S. Misquith, S. Srimal, R. Melchior and J. P. Quigley, *Biol-Bull*, 1994, **187**, 227–228.
74. G. R. Vasta, H. Ahmed and E. W. Odom, *Curr. Opin. Struct. Biol.*, 2004, **14**, 617–630.
75. B. J. Janssen, E. G. Huizinga, H. C. Raaijmakers, A. Roos, M. R. Daha, K. Nilsson-Ekdahl, B. Nilsson and P. Gros, *Nature*, 2005, **437**, 505–511.
76. R. H. Baxter, C. I. Chang, Y. Chelliah, S. Blandin, E. A. Levashina and J. Deisenhofer, *Proc. Natl. Acad. Sci. USA*, 2007, **104**, 11615–11620.

77. J. Bramham, C. T. Thai, D. C. Soares, D. Uhrin, R. T. Ogata and P. N. Barlow, *J. Biol. Chem.*, 2005, **280**, 10636–10645.
78. P. B. Armstrong, *Immunobiology*, 2006, **211**, 263–281.
79. M. Lagueux, E. Perrodou, E. A. Levashina, M. Capovilla and J. A. Hoffmann, *Proc. Natl. Acad. Sci. USA*, 2000, **97**, 11427–11432.
80. E. A. Levashina, L. F. Moita, S. Blandin, G. Vriend, M. Lagueux and F. C. Kafatos, *Cell*, 2001, **104**, 709–718.
81. E. A. Levashina, *Insect Biochem. Mol. Biol.*, 2004, **34**, 673–678.
82. K. R. Solomon, P. Sharma, M. Chan, P. T. Morrison and R. W. Finberg, *Gene*, 2004, **327**, 171–183.
83. K. W. Finnson, B. Y. Tam, K. Liu, A. Marcoux, P. Lepage, S. Roy, A. A. Bizet and A. Philip, *Faseb. J.*, 2006, **20**, 1525–1527.
84. J. A. Hammond, M. Nakao and V. J. Smith, *Mol. Immunol.*, 2005, **42**, 683–694.
85. M. Kasahara, *Curr. Opin. Immunol.*, 2007, **19**, 547–552.
86. M. Nonaka and S. L. Smith, *Fish Shellfish Immunol.*, 2000, **10**, 215–228.
87. M. Nonaka, T. Fujii, T. Kaidoh, S. Natsuume-Sakai, N. Yamaguchi and M. Takahashi, *J. Immunol.*, 1984, **133**, 3242–3249.
88. H. Ishiguro, K. Kobayashi, M. Suzuki, K. Titani, S. Tomonaga and Y. Kurosawa, *EMBO. J.*, 1992, **11**, 829–837.
89. M. Nonaka, K. Azumi, X. Ji, C. Namikawa-Yamada, M. Sasaki, H. Saiga, A. W. Dodds, H. Sekine, M. K. Homma, M. Matsushita, Y. Endo and T. Fujita, *J. Immunol.*, 1999, **162**, 387–391.
90. M. M. Suzuki, N. Satoh and M. Nonaka, *J. Mol. Evol.*, 2002, **54**, 671–679.
91. W. Z. Al-Sharif, J. O. Sunyer, J. D. Lambris and L. C. Smith, *J. Immunol.*, 1998, **160**, 2983–2997.
92. L. J. Dishaw, S. L. Smith and C. H. Bigger, *Immunogenetics*, 2005, **57**, 535–548.
93. D. J. Miller, G. Hemmrich, E. E. Ball, D. C. Hayward, K. Khalturin, N. Funayama, K. Agata and T. C. Bosch, *Genome. Biol.*, 2007, **8**, R59.
94. B. G. Fry, N. Vidal, J. A. Norman, F. J. Vonk, H. Scheib, S. F. Ramjan, S. Kuruppu, K. Fung, S. B. Hedges, M. K. Richardson, W. C. Hodgson, V. Ignjatovic, R. Summerhayes and E. Kochva, *Nature*, 2006, **439**, 584–588.
95. D. H. Shin, B. Webb, M. Nakao and S. L. Smith, *Dev. Comp. Immunol.*, 2007, **31**, 1168–1182.
96. S. Hyder Smith and J. A. Jensen, *Dev. Comp. Immunol.*, 1986, **10**, 191–206.
97. M. Nonaka, M. Takahashi and M. Sasaki, *J. Immunol.*, 1994, **152**, 2263–2269.
98. X. Ji, C. Namikawa-Yamada, M. Nakanishi, M. Sasaki and M. Nonaka, *Unpublished*, 2001.
99. L. C. Smith, C. S. Shih and S. G. Dachenhausen, *J. Immunol.*, 1998, **161**, 6784–6793.
100. M. Nakao, T. Uemura and T. Yano, *Mol. Immunol.*, 1996, **33**, 933–937.

101. J. A. Jensen, E. Festa, D. S. Smith and M. Cayer, *Science*, 1981, **214**, 566–569.
102. D. Goundis and K. B. M. Reid, *Nature*, 1988, **335**, 82–85.
103. S. Miyazawa, K. Azumi and M. Nonaka, *J. Immunol.*, 2001, **166**, 1710–1715.
104. S. Miyazawa and M. Nonaka, *Immunogenetics*, 2004, **55**, 836–844.
105. T. Kaidoh and I. Gigli, *Journal of Immunology*, 1989, **142**, 1605–1613.
106. A. Dahmen, T. Kaidoh, P. F. Zipfel and I. Gigli, *Biochemical J.*, 1994, **301**, 391–397.
107. L. M. Kunnath-Muglia, G. H. Chang, R. B. Sim, A. J. Day and R. A. Ezekowitz, *Mol. Immunol.*, 1993, **30**, 1249–1256.
108. M. Nakao, S. Hisamatsu, M. Nakahara, Y. Kato, S. L. Smith and T. Yano, *Immunogenetics*, 2003, **54**, 801–806.
109. T. Terado, M. I. Nonaka, M. Nonaka and H. Kimura, *Dev. Comp. Immunol.*, 2002, **26**, 403–413.

CHAPTER 4

Structure and Function of the C1 Complex: A Historical Perspective

GÉRARD J. ARLAUD

Laboratoire d'Enzymologie Moléculaire, Institut de Biologie Structurale, 41 rue Jules Horowitz, 38027 Grenoble Cedex 1, France

4.1 From the Antibody to the C1 Complex of Complement

Research on complement at the MRC Immunochemistry Unit was initiated in the early 1970s and was a natural extension of the long-lasting investigations carried out by R. Porter and co-workers, establishing the major features of the primary structure of immunoglobulins and their organization in domains.[1] The transition from antibody to complement research that took place at this period is probably best illustrated by the description and characterization of fragment antigen and complement binding (Facb), a truncated rabbit IgG molecule obtained by limited proteolysis with plasmin, which was shown to retain both the ability to bind antigen and to activate the C1 complex of complement.[2,3] C1 had been previously shown to consist of three entities, C1q, C1r and C1s,[4] and Reid focussed his attention on C1q, the subunit responsible for binding of the C1 complex to the fragment crystallizable (Fc) region of immunoglobulin (IgG). This was the beginning of an impressive series of investigations, initiated in 1972 with the description of a method to isolate human and rabbit C1q,[5] which led to a model of the C1q architecture[6] and ultimately to the resolution

Molecular Aspects of Innate and Adaptive Immunity
Edited By Kenneth BM Reid and Robert B Sim

Published by the Royal Society of Chemistry, www.rsc.org

of its complete primary structure.[7] Pioneering studies on the C1 complex and its proteases C1r and C1s were conducted in parallel by Sim in collaboration with Porter, Gigli, Dodds, Kerr and Reid. These delineated the biochemical and functional properties of C1r and C1s[8] and characterized the mechanism of C1 activation by antibody–antigen complexes.[9] In collaboration with Arlaud and Colomb, Sim characterized the reaction of C1-inhibitor with the active C1 complex,[10] and later, with Jackson, he investigated the mechanisms of auto-immune C1-inhibitor deficiency.[11] In the 1980s, Arlaud and Gagnon determined the primary structure of the catalytic chain of C1r[12] and, ultimately, that of the whole C1r molecule. While progressively turning his attention to proteins other than C1q, such as properdin and collectins, Reid became actively involved in a series of studies on the C1q receptors, mainly with Erdei, Ghebrehiwet and Peerschke.[13,14] During the past decade, with U. Kishore and others, he focussed on the modular organization and binding specificity of the C1q recognition domain.[15]

4.2 Biochemistry of C1q and Elucidation of its Primary Structure

Early work by Reid allowed him to determine a partial amino acid sequence for the A chain of human C1q,[16] and revealed that this contained a repeating sequence Gly–X–Y, where X was often proline and Y was often hydroxyproline. This provided the first direct evidence that a substantial part of the C1q molecule consisted of a collagen-like region, in keeping with previous analyses that indicated a high glycine content.[5] As expected from this observation, each of the three (A, B, C) C1q chains was shown to comprise an N-terminal collagen-like region susceptible to collagenase treatment, and a C-terminal, non-collagen-like region susceptible to pepsin digestion.[5,17] This division of the C1q molecule into two very distinct types of structures, allowing it to be split by selective proteolysis, is a key feature of this protein, a feature extensively used later on by several authors to decipher its interaction properties. Elucidation of the primary structure of C1q was based initially on protein sequencing, which determined the amino acid sequences of the collagen-like regions of the A, B and C chains[18] and then of the entire A and B chains.[19] cDNA clones encoding the three C1q chains were later isolated from a monocyte library, and found to be aligned in the order A-C-B on a 24-kb DNA stretch located on chromosome 1 p. This allowed the DNA sequence of the C chain to be determined, thereby yielding the entire derived amino acid sequence for human C1q.[7] A major finding from the primary structure was that the repeating Gly–X–Y sequences of the A and B chains are interrupted approximately half-way along each collagen-like region, with the insertion of a threonine in the A chain and the substitution of an alanine for a glycine in the C chain.[20] As discussed later, this feature is currently considered to have major structural and functional implications. Structure-prediction studies suggested that the C-terminal globular regions of C1q may adopt a predominantly β-type structure with little α-helical structure,[19] and sequence comparisons revealed strong similarity to the C-terminal regions of types VIII and X collagens, suggesting structural relationships between these molecules.[7] Both hypotheses proved to be correct when the structure of the C1q globular domain was solved later on by X-ray crystallography.[21]

The information provided by the primary structure of C1q, the division of the molecule into collagen-like and globular regions, along with electron microscopy studies showing six peripheral globular domains connected to a fibril-like central portion,[22–24] led Reid and Porter to propose a model of the C1q molecule.[6] In agreement with data obtained by Yonemasu and Stroud,[25] C1q was proposed to be assembled from 18 polypeptide chains of three different types (6A, 6B and 6C chains; Figure 4.1). Each chain has a short N-terminal region involved in the formation of A–B and C–C inter-chain disulfide bonds. This region is followed by a collagen-like sequence that gives rise to the formation of six heterotrimeric (A, B, C) collagen-like triple helices, which first associate as a common 'stalk' and then, because of the above-mentioned interruptions in the Gly–X–Y motif, diverge at about half-way to form six individual 'stems'. Each stem ultimately merges into a globular 'head' domain resulting from the heterotrimeric association of the C-terminal non-collagen-like region of one A, one B and one C chain. Based on its amino acid and carbohydrate content, the molecular weight of C1q was estimated at 459 300, a value that at that time was significantly larger than those reported previously, but is indeed very close to the mass of 460 793 $g\,mol^{-1}$ determined recently by mass spectrometry analysis.[26]

That the globular head domains of C1q were responsible for the ligand-recognition function of the molecule was inferred from their ability to inhibit binding of intact C1q to antibody-coated erythrocytes,[27] and could be visualized by electron microscopy of the interaction between C1q and cross-linked IgG dimers.[28] The collagen-like portion of the molecule left undigested after prolonged treatment with pepsin was shown to inhibit reconstruction of C1 from C1q, C1r and C1s,[29] which provided indirect evidence that it contained the site(s) responsible for interaction with C1r and C1s. More direct evidence of this location was obtained later by ultracentrifugation analyses,[30] whereas electron microscopy of the chemically cross-linked C1 complex revealed that the compact mass corresponding to C1r and C1s was located in the region of the six connecting collagen-like fibrils, *i.e.* between the C1q heads and the central stalk.[31] Using both electron microscopy[32] and fluorescence polarization techniques,[33] Schumaker and co-workers obtained evidence for a segmental flexibility in the C1q molecule, which they attributed to wagging motions of the individual stem-head segments. This strongly suggested that the interruptions observed at about halfway in the repeating collagen-like sequences not only create a bend in the triple helices, but also generate flexibility in this area of the molecule.

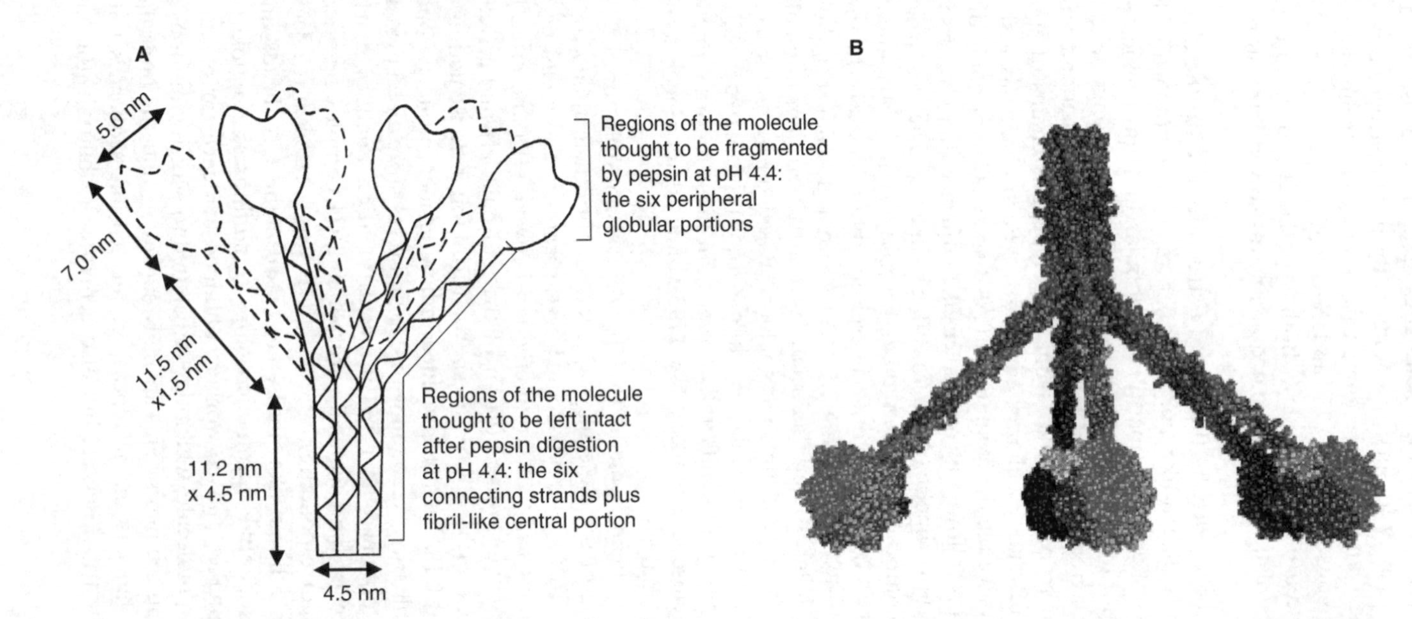

Figure 4.1 Two structural models of human C1q. (A) Model proposed in 1976 by Reid and Porter,[6] based on amino acid sequence, limited proteolysis and electron microscopy studies. (B) A more recent, three-dimensional version of model A, featuring the experimental X-ray structure of the C1q globular domain[21] and a homology model of the collagen-like region of the molecule, derived from information provided by the latter structure. Adapted from Reid and Porter[6] and Gaboriaud *et al.*[84]

4.3 Biochemistry of C1r and C1s and the C1 Activation Mechanism

4.3.1 Biochemistry of C1r and C1s

It had been recognized earlier that C1r and C1s were proteases responsible for the activation and activity of C1, respectively.[4,34] A method for the isolation from human serum of the C1 complex in both its proenzyme and activated forms was worked out by Gigli, Porter and Sim.[35] Concurrently with other groups (reviewed by Cooper[36]), this allowed C1r and C1s to be purified in good yield, making possible the first detailed comparison of their chemical and enzymatic properties.[8,37] C1r and C1s were found to have comparable molecular weights in dissociating solvents and similar amino acid compositions, but significantly different monosaccharide contents. Each proenzyme protein was shown to comprise a single polypeptide chain, whereas the activated forms each contained two disulfide-linked chains, termed A and B, derived from the N- and C-terminal parts of the proteins, respectively. The C1r A chain proved refractory to amino acid sequencing, but the N-terminal sequence of 29 residues of the C1s A chain was determined, showing no homology with proteins such as plasminogen or prothrombin. The 20 N-terminal residues of both B chains were identified, showing 60% identity between C1r and C1s, and strong homology with other known chymotrypsin-like serine proteases (SPs). Consistent with the observed sensitivity of C1r and C1s to di-isopropylfluorophosphate,[35,37] and with earlier findings,[38] this demonstrated that both proteins were typical SPs, their active site being located on the smaller B chain. When assayed for esterolytic activity, C1s cleaved several synthetic amino acid esters, but C1r did not hydrolyze any of the substrates tested except C1s, indicative of the highly restricted specificity of this protease.

4.3.2 Insights into the C1 Activation Mechanism

C1q, C1r and C1s were shown to account wholly for the haemolytic activity of human C1, and their approximate molar proportions in serum were found to be C1q:C1r:C1s = 1:2:2.[35] This provided the first indirect indication that C1 is assembled from the interaction between C1q and a tetrameric complex comprising two molecules, each of C1r and C1s, as demonstrated subsequently by others.[39,40] The activation of C1r and C1s in the C1 complex when bound to antigen–antibody aggregates was investigated in detail by A. Dodds and coworkers.[9] Using different C1q–C1r–C1s combinations in which C1r and C1s were proenzymic, or activated or blocked with di-isopropylfluorophosphate, it was demonstrated that C1r was responsible for its own activation and then, in turn, mediated C1s activation. C1s clearly played no proteolytic role in the activation process, but its presence in the C1 complex was nevertheless shown to be a prerequisite for C1r activation. The observation that C1r activation was inhibited reversibly by *p*-nitrophenyl-*p*′-guanidinobenzoate led Dodds and co-workers to hypothesize that C1r activation involved the formation of an intermediary form, C1r*, in which the peptide chain is unsplit, but a conformational change leads to the formation of a proteolytically active catalytic site.[9] As discussed later, this hypothesis is central to the C1 activation mechanism, and has been the subject of a number of investigations.

4.3.3 The Primary Structure of C1r

Further investigations on the amino acid sequence of human C1r were conducted later on by Arlaud and Gagnon, who initially solved about two-thirds of the amino acid sequence of its catalytic B chain,[41] and then reported its complete primary structure.[12] The sequence showed strong homology with other mammalian SPs, including the presence of the conserved histidine, aspartic acid and serine residues that formed the 'charge–relay' system. In contrast, C1r was found to lack the 'histidine-loop' disulfide bridge present in all other known mammalian SPs, a characteristic also shared by C1s.[42] Similarly, a partial amino acid sequence of the N-terminal A chain of C1r[43] and then its complete primary structure,[44] were determined by protein sequencing, yielding the complete amino acid sequence of C1r. Additional studies established that C1r activation involves the cleavage of a single Arg–Ile bond.[45] The cDNA coding for human C1r was isolated at the same period and sequenced,[46,47] confirming the protein sequence data. Similarly, the complete primary structure of human C1s was established from protein and DNA sequencing.[48–51] Comparative analysis of the primary structures of the A chains of C1r and C1s revealed that these were highly homologous and shared the same type of modular structure (Figure 4.2) with, from the C-terminus, a CUB module (first recognized in C1r/C1s, the sea urchin protein Uegf and the human bone morphogenetic protein-1), an epidermal growth factor (EGF)-like module, a second CUB module and a tandem repeat of complement-control protein (CCP) modules, which also occurred in complement regulatory proteins. As discussed later, this particular modular structure is a key feature of C1r and C1s and has multiple implications for their interaction and catalytic properties.

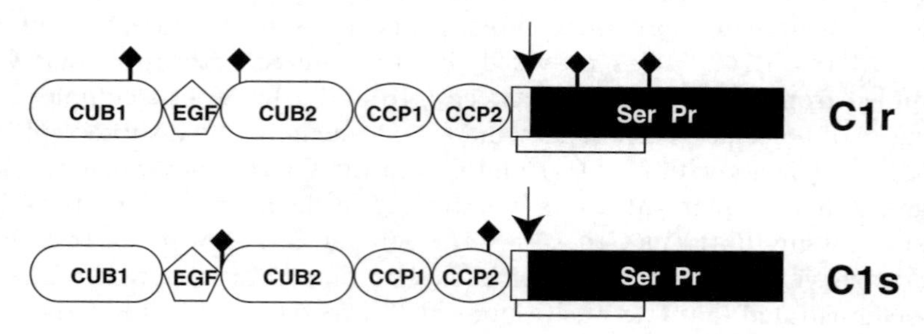

Figure 4.2 Modular structures of C1r and C1s. The arrows indicate the Arg–Ile bonds cleaved upon activation. The only disulfide bridge shown is that connecting the activation peptide to the serine protease (Ser Pr) domain. Closed diamonds represent N-linked oligosaccharides.

4.4 From the Domain Structure of C1r and C1s to a Model of C1 Architecture

4.4.1 The N-terminal Interaction Domains

The information provided by the primary structures of C1r and C1s, together with further studies based on limited proteolysis, electron microscopy and expression of recombinant fragments led to the identification of specialized functional regions in C1r and C1s. It was shown initially that both C1r and C1s comprised two domains, a smaller interaction domain derived from the N-terminal half of the A chain, responsible for Ca^{2+}-dependent C1r–C1s interaction, and a larger catalytic domain, comprising the C-terminal part of the A chain and the SP domain.[52] A model of the C1s–C1r–C1r–C1s tetramer was proposed, in which C1r forms a core, its distal interaction domains interacting with the corresponding domains of C1s. Fragments C1rα and C1sα, encompassing the N-terminal CUB_1 and EGF modules plus a short segment from the following CUB_2 module, were obtained by limited proteolysis.[53–55] Differential scanning calorimetry revealed that C1rα displayed a low-temperature transition that was shifted upwards in the presence of Ca^{2+} ions.[53] Similar data were obtained using the recombinant CUB_1–EGF segment of C1s (Thielens and Kardos, unpublished data), providing evidence for a Ca^{2+}-dependent stabilization of the CUB_1–EGF module pair in each protein. C1rα was shown to retain the ability of intact C1r to bind C1s in the presence of Ca^{2+} ions and, conversely, C1sα formed Ca^{2+}-dependent C1sα–C1rα heterodimers. The EGF module of C1r was synthesized chemically and analyzed by nuclear magnetic resonance (NMR) spectroscopy, providing direct evidence of its ability to bind Ca^{2+}, although with an affinity about 300-fold lower than that determined for the larger C1rα fragment.[56] Analysis of the chemical shift variations induced by Ca^{2+} were consistent with a binding mediated by residues homologous to those identified in other EGF modules belonging to the Ca^{2+}-binding subset.[57] Nevertheless, it became clear that residues located outside the EGF module either directly contributed to the Ca^{2+}-binding site or stabilized its conformation. This hypothesis was confirmed by a comparative analysis of the interaction properties of the CUB_1–EGF, CUB_1 and EGF segments of C1r by surface plasmon resonance spectroscopy.[58] The CUB_1–EGF segment readily bound to immobilized C1s in the presence of Ca^{2+}. In contrast, the individual CUB_1 and EGF modules each failed to do so. Based on these observations, it was proposed that Ca^{2+} binds primarily to ligands contributed by the EGF module, allowing formation of a compact CUB_1–EGF assembly that provided the conformation appropriate for mediating C1r–C1s interaction within the C1s–C1r–C1r–C1s tetramer.

Other studies provided evidence that the N-terminal regions of C1r and C1s also mediate interaction between the C1s–C1r–C1r–C1s tetramer and C1q. Thus, evidence for a low-affinity interaction between C1q and C1r was provided by ultracentrifugation analyses[59] and, indirectly, by the observation that C1q increases the activation rate of C1r in the presence of calcium.[60] Other experiments showed that treatment of the C1s–C1r–C1r–C1s tetramer with a carbodiimide prevents its interaction with C1q, because of modification of acidic residues of C1r.[61] Nevertheless, fragment C1sα[55] or the shorter C1s CUB_1–EGF segment[62] were shown to be required to promote efficient interaction of C1r with C1q and generate the formation of a complex with a stability comparable to that of intact C1. Current data are therefore consistent with the occurrence of a major interaction site in C1r, probably located within the N-terminal CUB_1–EGF segment, and acting in synergy with a site in the homologous domain of C1s.

4.4.2 The C-terminal Catalytic Domains

The C-terminal catalytic domain of C1s, comprising the CCP_1 and CCP_2 modules plus the SP domain, was obtained by limited proteolysis with plasmin and retained the ability to mediate cleavage of C4 and C2, the protein substrates of C1.[52,63] Insights into the role of individual domains of this region were made possible by expression of recombinant segments that lacked either the CCP_1 module (CCP_2–SP) or both CCP modules (SP).[64] Both could be activated by C1r and were in turn able to cleave C2 with similar efficiencies. In contrast, compared to intact C1s and to the whole CCP_1–CCP_2–SP segment, the C4-cleaving activity of CCP_2–SP was dramatically reduced, and that of SP was abolished. This indicated that, whereas C2 cleavage only involves structural determinants located in the SP domain, efficient C4 cleavage requires substrate-binding sites contributed by the CCP modules, in keeping with earlier studies.[65] Differential scanning calorimetry provided evidence for three independently folded domains, the CCP_1 and CCP_2 modules and the SP domain.[66] Further structural insights came from chemical cross-linking and homology modelling studies, which indicated that CCP_2 closely interacts with the SP domain on a side opposite to the active site and the susceptible Arg–Ile bond cleaved upon activation.[67]

The catalytic region of C1r, comprising the SP domain and the preceding CCP_1 and CCP_2 modules, was shown to associate as a non-covalent homodimer that forms the core of the C1s–C1r–C1r–C1s tetramer.[52,63] The corresponding fragment was obtained by limited proteolysis in its proenzyme and active forms, and found in both cases to retain the catalytic properties of the corresponding native C1r species.[68,69] Chemical cross-linking and homology modelling studies performed on the active form indicated a strong interaction between the CCP_2 module and the SP domain in each monomer and provided evidence for a cross-link between the N-terminal end of the CCP1 module of one monomer and the SP domain of its counterpart.[70] Based on these data, a three-dimensional model of the active C1r catalytic region was proposed, featuring a head-to-tail interaction of the monomers, with their active sites facing opposite directions towards the outside of the dimer. Further insights into the function of individual modules of the C1r catalytic domain were made possible by the production of recombinant segments of varying sizes.[71,72] In contrast to

CCP_1–CCP_2–SP, which associated as a homodimer, the shorter CCP_2–SP and SP segments were monomeric, underlining the essential role of CCP_1 in the assembly of the dimer. Each C1r construct was expressed with the wild-type sequence and with a point mutation, either at the Arg–Ile activation site (Arg446Gln) or at the catalytic Ser residue (Ser637Ala). All wild-type species, whatever their size, were recovered as two-chain, active proteases, which indicates that activation had occurred during biosynthesis. In contrast, both types of mutants were recovered in the zymogen form and proved refractory to activation upon incubation. These data provided the first unambiguous experimental evidence that C1r activation, for several decades a controversial issue, [6] was indeed a self-activation process.[71]

4.4.3 A Low-resolution C1 Model

The elucidation of the domain structure of C1r and C1s, combined with electron microscopy studies,[40,63] led to a model of the C1s–C1r–C1r–C1s tetramer in which the C1r catalytic regions are located in the centre, the corresponding regions of C1s occupying both ends.[52] Given these locations, and on the basis of neutron scattering and electron microscopy analyses of the C1 complex,[31,73,74] there followed the concept that, upon interaction with C1q, C1s–C1r–C1r–C1s adopts a compact, 'figure-of-eight' shape conformation that permits physical contact between the catalytic regions of C1r and C1s, a prerequisite for C1s cleavage by active C1r. This concept provided the basis for most of the low-resolution C1 models proposed at this period.[63,75–77]

4.5 The Era of Structural Biology

4.5.1 A Three-dimensional C1 Model

Structural biology has been used over the past decade to generate more detailed information about the structure of C1 at the atomic level. The dissection strategy used was based on recombinant expression of modular segments from each of the three C1 subunits and resolution of their three-dimensional structure by NMR spectroscopy[78] and X-ray crystallography.[21,79–82] Resolution of the X-ray structure of the C1q globular head revealed a compact, almost spherical heterotrimeric assembly held together mainly by non-polar interactions, with a Ca^{2+} ion bound at the apex.[21] The structure also showed the that the three subunits are arranged in the order A, B, C clockwise when the head is viewed from the top, allowing a three-dimensional model of the collagen-like triple helix of C1q to be derived (see Figure 4.1B). The crystal structure of the proenzyme form of the C1r catalytic domain showed a head-to-tail homodimer held together by interactions between the CCP_1 module of one monomer and the SP domain of its counterpart, with a large central opening.[80] The structure of the C-terminal CCP_2–SP segment of the C1s catalytic domain was determined in the active form, and indicated that CCP_2 is orientated perpendicularly to the surface of the SP domain and closely interacts with it by means of a rigid interface.[79] The X-ray structure of the CUB_1–EGF interaction domain of C1s was also solved, revealing a head-to-tail homodimer that involved interactions between the CUB_1 module of one monomer and the EGF module of its counterpart, strongly stabilized by Ca^{2+} ions.[82] This structure led to a three-dimensional model of the C1r–C1s CUB_1–EGF heterodimer, which in the C1 complex connects C1r to C1s and mediates interaction with C1q. Based on this work and other data (reviewed by Arlaud *et al.*[83]), a structural model of the C1q–C1r–C1s interface was derived, in which one of the C1q collagen-like triple helices fits into a groove along the transversal axis of the C1r–C1s CUB_1–EGF heterodimer.[82] Altogether, the information provided so far by X-ray crystallography accounts for about two-thirds of the structure of the three C1 subunits, and has allowed us to generate a refined, three-dimensional version of the low-resolution C1 model proposed initially,[76] in which the figure-of-eight -shape C1s–C1r–C1rC1s tetramer folds around two opposite pairs of C1q stems[84] (Figure 4.3).

4.5.2 C1r Activation and the Triggering Signal

The X-ray structure of the zymogen C1r catalytic domain provided precise insights into the mechanism that underlies C1 activation. From a functional standpoint, the most intriguing feature of the observed head-to-tail structure was that the catalytic site of one monomer and the activation site of the other lie at opposite ends of the dimer[80] (Figure 4.4A). This configuration is clearly

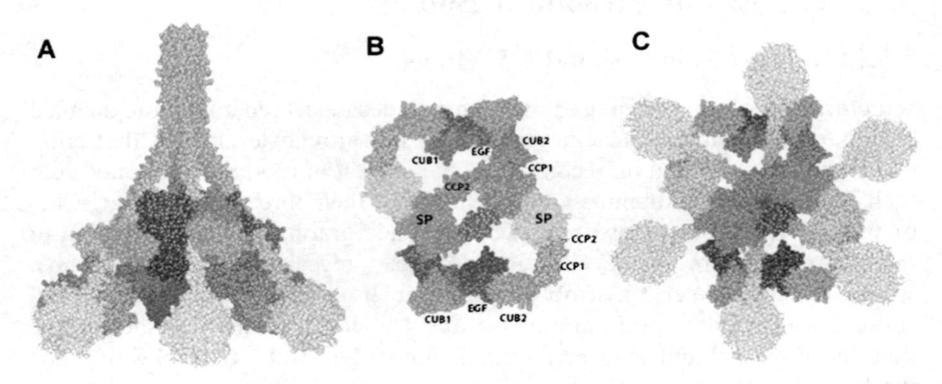

Figure 4.3 Three-dimensional model of the human C1 complex. (A) Side view. (B) Bottom view of the tetramer alone. Labels indicate the location of individual modules within one C1r and one C1s subunit. The colour-coding used differentiates the two C1r subunits. Their catalytic domains associate as a homodimer that forms the core of the tetramer. (C) Bottom view of the C1 complex. From Gaboriaud *et al.*[84]

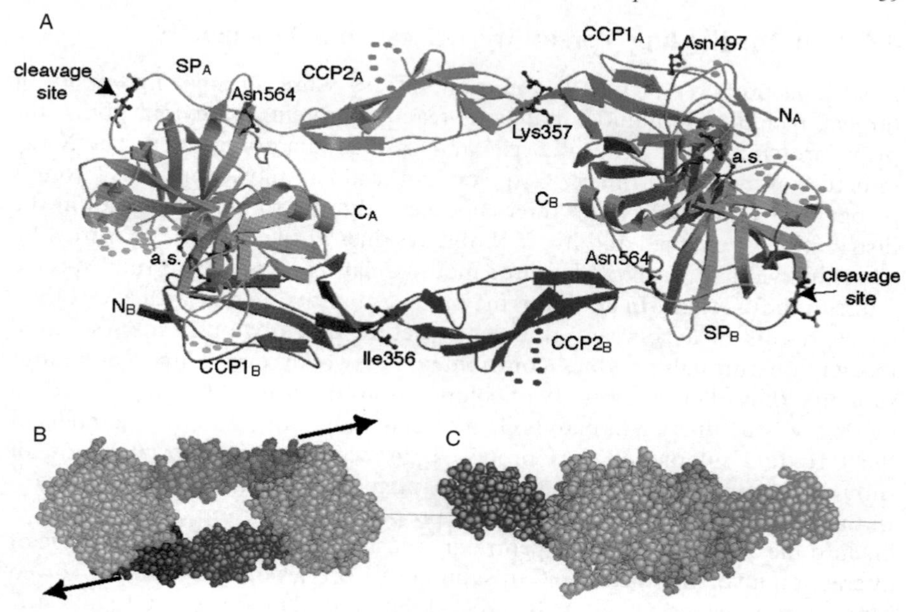

Figure 4.4 X-ray structure of the C1r catalytic domain and its implications in the C1 activation mechanism. (A) Head-to-tail homodimeric structure of the zymogen CCP_1–CCP_2–SP C1r segment. The residues at the cleavage site and at the catalytic site (a.s.) are shown. N_A, N_B and C_A, C_B represent the N- and C-terminal ends of monomers A and B. (B) Resting head-to-tail configuration of the C1r catalytic domain. Arrows illustrate the tension necessary to achieve the transient conformation (C) required to activate one SP domain by its counterpart. Modified from Budayova-Spano *et al.*[80]

not compatible with C1r self-activation, which requires cleavage of the susceptible Arg–Ile bond of each monomer by the catalytic Ser residue of its counterpart,[71] and probably represents a resting state of the molecule designed to prevent spontaneous C1 activation. This led us to propose that the signal that triggers C1r activation in C1 is a mechanical stress transmitted through the C1q–C1r–C1s interface from a C1q stem to the C1r catalytic region when C1 binds to a target, disrupting the head-to-tail structure and allowing the activating contact between the two SP domains[80] (Figures 4.4B and 4.4C). Such a mechanism is consistent with the presence of a semi-flexible hinge in C1q, as discussed above, and is obviously rendered possible by the large opening in the centre of the C1r catalytic region. Further support for this hypothesis arose from the crystal structure of the active C1r CCP_2–SP segment, in which the SP domains are packed in such a way that they form an enzyme-product-like complex, probably similar to the one that occurs upon activation of one C1r molecule by its counterpart (Figure 4.4C). This transient state is fully reminiscent of the intermediary form C1r* hypothesized in 1978 by Dodds and co-workers.[9]

4.5.3 C1q: Binding Versatility Arises from Modularity

A striking property of C1q lies in its ability to sense an amazing variety of targets, including immunoglobulins, C-reactive protein, β-amyloid fibrils, the prion protein, DNA, apoptotic cells and various microorganisms. The X-ray structure of the heterotrimeric C1q globular head has allowed insights into this recognition versatility.[21] The three modules exhibit striking differences in the distribution of charged and hydrophobic residues at their surface (Figure 4.5). Thus, individually each of the three modules can be expected to fulfil specific interaction functions. In addition, the modules tightly interact with each other, which results in a compact globular structure that obviously allows ligand recognition through residues contributed by two or even three contiguous subunits, thus offering a variety of combinations in terms of binding. Based on the X-ray structure of a human IgG1 molecule[85] and other data,[86–89] a model of the C1q–IgG interaction was proposed, in which the equatorial region of subunit B of C1q binds at the fragment antigen-binding (Fab)–Fc interface.[21] In the case of C-reactive protein, it was proposed that the top of the C1q head fits into the central pore of the pentraxin, illustrating another possible mode of interaction involving the three C1q subunits.[21] A series of studies that involved expression of the individual A, B and C globular head modules of C1q and site-directed mutagenesis were performed by Kishore, Kojouharova, Reid and others, and provided further insights into the interaction between C1q and its ligands.[90–93] These investigations support the hypothesis that charged residues belonging to the apex of the C1q head (with participation of all three chains) and to the side of module B are crucial for C1q binding to its major ligands, and that their contribution to each interaction is different.[93] Based on theoretical and experimental approaches, it was suggested that the presence or absence of

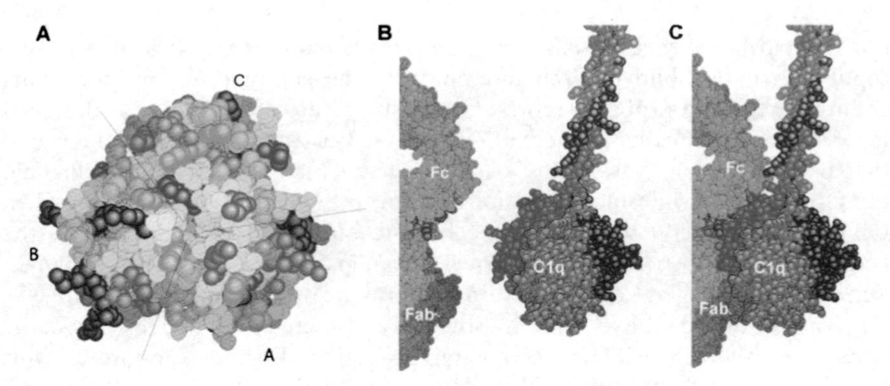

Figure 4.5 The versatile recognition function of C1q. (A) Top view of the C1q globular head, highlighting the different surface patterns of subunits A, B and C. Lines indicate approximate module boundaries. (B, C) Model of the interaction between C1q and human IgG1 b12. The IgG Fc and Fab domains are indicated. The C1q subunits are coloured blue (A), green (B) and red (C). Modified from Gaboriaud *et al.*[21]

the Ca^{2+} ion bound to the apex of the C1q head[21] may influence the target-binding properties of C1q by changing the direction of the electric moment, and thereby have implications in the transmission of the activating signal from C1q to the C1r catalytic domain.[94]

4.6 Conclusions and Perspectives

This overview of nearly 40 years of research on C1 illustrates how this evolved from the pioneering investigations based on the isolation of proteins from plasma and their chemical characterization, through protein engineering and then to X-ray crystallography during the most recent period. Each period has greatly benefited from the previous ones, and many of the early concepts are still valid. Altogether, our current knowledge provides a molecular basis for the mechanisms that underlie the activation and recognition properties of C1. It can be anticipated that the extensive use of protein engineering, combined with more integrative structural biology methods will allow further insights at the atomic level into the finely tuned mechanisms of this sophisticated nanomachine.

References

1. R. R. Porter, *Science*, 1973, **180**, 713.
2. G. E. Connell and R. R. Porter, *Biochem. J.*, 1971, **124**, 53P.
3. M. Colomb and R. R. Porter, *Biochem. J.*, 1975, **145**, 177.
4. I. H. Lepow, G. B. Naff, E. W. Todd, J. Pensky and C. F. Hinz, *J. Exp. Med.*, 1963, **117**, 983.
5. K. B. Reid, D. M. Lowe and R. R. Porter, *Biochem. J.*, 1972, **130**, 749.
6. K. B. Reid and R. R. Porter, *Biochem. J.*, 1976, **155**, 19.
7. G. C. Sellar, D. J. Blake and K. B. Reid, *Biochem. J.*, 1991, **274**, 481.
8. R. B Sim, R. R. Porter, K. B. Reid and I. Gigli, *Biochem. J.*, 1977, **163**, 219.
9. A. W. Dodds, R. B. Sim, R. R. Porter and M. A. Kerr, *Biochem. J.*, 1978, **175**, 383.
10. R. B. Sim, G. J. Arlaud and M. G. Colomb, *Biochem. J.*, 1979, **179**, 449.
11. J. Jackson, R. B. Sim, A. Whelan and C. Feighery, *Nature*, 1986, **323**, 722.
12. G. J. Arlaud and J. Gagnon, *Biochemistry*, 1983, **22**, 1758.
13. A. Erdei and K. B. M. Reid, *Biochem. J.*, 1988, **255**, 493.
14. B. Ghebrehiwet, B. L. Lim, E. L. Peerschke, A. C. Willis and K. B. Reid, *J. Exp. Med.*, 1994, **179**, 1809.
15. U. Kishore, S. K. Gupta, M. V. Perdikoulis, M. S. Kojouharova, B. C. Urban and K. B. Reid, *J. Immunol.*, 2003, **171**, 812.
16. K. B. Reid, *Biochem. J.*, 1974, **141**, 189.
17. K. B. Reid, *Biochem. J.*, 1976, **155**, 5.
18. K. B. Reid, *Biochem. J.*, 1979, **179**, 367.
19. K. B. Reid, J. Gagnon and J. Frampton, *Biochem. J.*, 1982, **203**, 559.
20. K. B. Reid, *Biochem. Soc. Trans.*, 1983, **11**, 1.

21. C. Gaboriaud, J. Juanhuix, A. Gruez, M. Lacroix, C. Darnault, D. Pignol, D. Verger, J. C. Fontecilla-Camps and G. J. Arlaud, *J. Biol. Chem.*, 2003, **278**, 46974.
22. E. Shelton, K. Yonemasu and R. M. Stroud, *Proc. Natl. Acad. Sci. U.S.A.*, 1972, **69**, 65.
23. H. R. Knobel, W. Villiger and H. Isliker, *Eur. J. Immunol.*, 1975, **5**, 78.
24. B. Brodsky-Doyle, K. R. Leonard and K. B. Reid, *Biochem. J.*, 1976, **159**, 279.
25. K. Yonemasu and R. M. Stroud, *Immunochemistry*, 1972, **9**, 545.
26. B. Tissot, F. Gonnet, A. Iborra, C. Berthou, N. Thielens, G. J. Arlaud and R. Daniel, *Biochemistry*, 2005, **44**, 2602.
27. N. C. Hughes-Jones and B. Gardener, *Mol. Immunol.*, 1979, **16**, 697.
28. J. Tschopp, W. Villiger, A. Lustig, J.-C. Jaton and J. Engel, *Eur. J. Immunol.*, 1980, **10**, 529.
29. K. B. Reid, R. B. Sim and A. P. Faiers, *Biochem. J.*, 1977, **161**, 239.
30. R. C. Siegel and V. N. Schumaker, *Mol. Immunol.*, 1983, **20**, 53.
31. C. J. Strang, R. C. Siegel, M. L. Phillips, P. H. Poon and V. N. Schumaker, *Proc. Natl. Acad. Sci. U.S.A.*, 1982, **79**, 586.
32. P. H. Poon, V. N. Schumaker, M. L. Phillips and C. J. Strang, *J. Mol. Biol.*, 1983, **168**, 563.
33. D. C. Hanson, R. C. Siegel and V. N. Schumaker, *J. Biol. Chem.*, 1985, **260**, 3576.
34. G. B. Naff and O. D. Ratnoff, *J. Exp. Med.*, 1968, **128**, 571.
35. I. Gigli, R. R. Porter and R. B. Sim, *Biochem. J.*, 1976, **157**, 541.
36. N. R. Cooper, *Adv. Immunol.*, 1985, **37**, 151.
37. R. B. Sim and R. R. Porter, *Biochem. Soc. Trans.*, 1976, **4**, 127.
38. T. Barkas, G. K. Scott and J. E. Fothergill, *Biochem. Soc. Trans.*, 1973, **1**, 1219.
39. R. J. Ziccardi and N. R. Cooper, *J. Immunol.*, 1977, **118**, 2047.
40. J. Tschopp, W. Villiger, H. Fuchs, E. Kilcherr and J. Engel, *Proc. Natl. Acad. Sci. U.S.A.*, 1980, **77**, 7014.
41. G. J. Arlaud, J. Gagnon and R. R. Porter, *Biochem. J.*, 1982, **201**, 49.
42. G. J. Arlaud and J. Gagnon, *Biosci. Rep.*, 1981, **1**, 779.
43. J. Gagnon and G. J. Arlaud, *Biochem. J.*, 1985, **225**, 135.
44. G. J. Arlaud, A. C. Willis and J. Gagnon, *Biochem. J.*, 1987, **241**, 711.
45. G. J. Arlaud and J. Gagnon, *FEBS Lett.*, 1985, **180**, 234.
46. S. P. Leytus, K. Kurachi, K. S. Sakariassen and E. W. Davie, *Biochemistry*, 1986, **25**, 4855.
47. A. Journet and M. Tosi, *Biochem. J.*, 1986, **240**, 783.
48. P. E. Carter, B. Dunbar and J. E. Fothergill, *Biochem. J.*, 1983, **215**, 565.
49. S. E. Spycher, H. Nick and E. E. Rickli, *Eur. J. Biochem.*, 1986, **156**, 49.
50. C. M. Mackinnon, P. E. Carter, S. J. Smith, B. Dunbar and J. E. Fothergill, *Eur. J. Biochem.*, 1987, **169**, 547.
51. M. Tosi, C. Duponchel, T. Meo and C. Julier, *Biochemistry*, 1987, **26**, 8516.
52. C. L. Villiers, G. J. Arlaud and M. G. Colomb, *Proc. Natl. Acad. Sci. U.S.A.*, 1985, **82**, 4477.

53. T. F. Busby and K. C. Ingham, *Biochemistry*, 1987, **26**, 5564.
54. N. M. Thielens, C. A. Aude, M. B. Lacroix, J. Gagnon and G. J. Arlaud, *J. Biol. Chem.*, 1990, **265**, 14469.
55. T. F. Busby and K. C. Ingham, *Biochemistry*, 1990, **29**, 4613.
56. J.-F. Hernandez, B. Bersch, Y. Pétillot, J. Gagnon and G. J. Arlaud, *J. Peptide Res.*, 1997, **49**, 221.
57. I. D. Campbell and P. Bork, *Curr. Opin. Struct. Biol.*, 1993, **3**, 385.
58. N. M. Thielens, K. Enrié, M. Lacroix, M. Jaquinod, J.-F. Hernandez, A. F. Esser and G. J. Arlaud, *J. Biol. Chem.*, 1999, **274**, 9149.
59. S. Lakatos, *Biochem. Biophys. Res. Commun.*, 1987, **149**, 378.
60. N. M. Thielens, C. Illy, I. M. Bally and G. J. Arlaud, *Biochem. J.*, 1994, **301**, 509.
61. C. Illy, N. M. Thielens and G. J. Arlaud, *J. Prot. Chem.*, 1993, **12**, 771.
62. S. W. Tsai, P. H. Poon and V. N. Schumaker, *Mol. Immunol.*, 1997, **34**, 1273.
63. V. Weiss, C. Fauser and J. Engel, *J. Mol. Biol.*, 1986, **189**, 573.
64. V. Rossi, I. Bally, N. M. Thielens, A. F. Esser and G. J. Arlaud, *J. Biol. Chem.*, 1998, **273**, 1232.
65. M. Matsumoto, K. Nagaki, H. Kitamura, S. Kuramitsu, S. Nagasawa and T. Seya, *J. Immunol.*, 1989, **142**, 2743.
66. L. V. Medved, T. F. Busby and K. C. Ingham, *Biochemistry*, 1989, **28**, 5408.
67. V. Rossi, C. Gaboriaud, M. B. Lacroix, J. Ulrich, J. Fontecilla-Camps, J. Gagnon and G. J. Arlaud, *Biochemistry*, 1995, **34**, 7311.
68. G. J. Arlaud, J. Gagnon, C. L. Villiers and M. G. Colomb, *Biochemistry*, 1986, **25**, 5177.
69. M. Lacroix, C. A. Aude, G. J. Arlaud and M. G. Colomb, *Biochem. J.*, 1989, **257**, 885.
70. M. Lacroix, V. Rossi, C. Gaboriaud, S. Chevallier, M. Jaquinod, N. M. Thielens and G. J. Arlaud, *Biochemistry*, 1997, **36**, 6270.
71. M. Lacroix, C. Ebel, J. Kardos, J. Dobo, P. Gal, P. Zavodszky, G. J. Arlaud and N. M. Thielens, *J. Biol. Chem.*, 2001, **276**, 36233.
72. J. Kardos, P. Gal, L. Szilagyi, N. M. Thielens, K. Szilagyi, Z. Lorincz, P. Kulcsar, G. J. Arlaud and P. Zavodszky, *J. Immunol.*, 2001, **167**, 5202.
73. J. Boyd, D. R. Burton, S. J. Perkins, C. L. Villiers, R. A. Dwek and G. J. Arlaud, *Proc. Natl. Acad. Sci. U.S.A.*, 1983, **80**, 3769.
74. S. J. Perkins, C. L. Villiers, G. J. Arlaud, J. Boyd, D. R. Burton, M. G. Colomb and R. A. Dwek, *J. Mol. Biol.*, 1984, **179**, 547.
75. M. G. Colomb, G. J. Arlaud and C. L. Villiers, *Philos. Trans. R. Soc. London B*, 1984, **306**, 283.
76. G. J. Arlaud, M. G. Colomb and J. Gagnon, *Immunol. Today*, 1987, **8**, 106.
77. V. N. Schumaker, P. Zavodszky and P. H. Poon, *Annu. Rev. Immunol.*, 1987, **5**, 21.
78. B. Bersch, J.-F. Hernandez, D. Marion and G. J. Arlaud, *Biochemistry*, 1998, **37**, 1204.
79. C. Gaboriaud, V. Rossi, I. Bally, G. J. Arlaud and J. C. Fontecilla-Camps, *EMBO J.*, 2000, **19**, 1755.

80. M. Budayova-Spano, M. Lacroix, N. M. Thielens, G. J. Arlaud, J. C. Fontecilla-Camps and C. Gaboriaud, *EMBO J.*, 2002, **21**, 231.
81. M. Budayova-Spano, W. Grabarse, N. M. Thielens, H. Hillen, M. Lacroix, M. Schmidt, J. C. Fontecilla-Camps, G. J. Arlaud and C. Gaboriaud, *Structure*, 2002, **10**, 1509.
82. L. A. Gregory, N. M. Thielens, G. J. Arlaud, J. C. Fontecilla-Camps and C. Gaboriaud, *J. Biol. Chem.*, 2003, **278**, 32157.
83. G. J. Arlaud, C. Gaboriaud, N. M. Thielens, M. Budayova-Spano, V. Rossi and J. C. Fontecilla-Camps, *Mol. Immunol.*, 2002, **39**, 383.
84. C. Gaboriaud, N. M. Thielens, L. A. Gregory, V. Rossi, J. C. Fontecilla-Camps and G. J. Arlaud, *Trends Immunol.*, 2004, **25**, 368.
85. E. O. Saphire, P. W. Parren, R. Pantophlet, M. B. Zwick, G. M. Morris, P. M. Rudd, R. A. Dwek, R. L. Stanfield, D. R. Burton and I. A. Wilson, *Science*, 2001, **293**, 1155.
86. E. E. Idusogie, L. G. Presta, H. Gazzano-Santoro, K. Totpal, P. Y. Wong, M. Ultsch, Y. G. Meng and M. G. Mulkerrin, *J. Immunol.*, 2000, **164**, 4178.
87. E. E. Idusogie, P. Y. Wong, L. G. Presta, H. Gazzano-Santoro, K. Totpal, M. Ultsch, Y. G. Meng and M. G. Mulkerrin, *J. Immunol.*, 2001, **166**, 2571.
88. M. Hezareh, A. J. Hessel, R. C. Jensen, J. G. J. Van de Winkel and P. W. Parren, *J. Virol.*, 2001, **75**, 12161.
89. G. Marqués, L. C. Anton, E. Barrio, A. Sanchez, A. Ruiz, F. Gavilanes and F. Vivanco, *J. Biol. Chem.*, 1993, **268**, 10393.
90. U. Kishore, S. K. Gupta, M. V. Perdikoulis, M. S. Kojouharova, B. C. Urban and K. B. Reid, *J. Immunol.*, 2003, **171**, 812.
91. M. S. Kojouharova, M. G. Gadjeva, I. G. Tsacheva, A. Zlatarova, L. T. Roumenina, M. I. Tchorbadjieva, B. P. Atanasov, P. Waters, B. C. Urban, R. B. Sim, K. B. Reid and U. Kishore, *J. Immunol.*, 2004, **172**, 4351.
92. A. S. Zlatarova, M. Rouseva, L. T. Roumenina, M. Gadjeva, M. Kolev, I. Dobrev, N. Olova, R. Ghai, J. C. Jensenius, K. B. Reid, U. Kishore and M. S. Kojouharova, *Biochemistry*, 2006, **45**, 9979.
93. L. T. Roumenina, M. M. Ruseva, A. Zlatarova, R. Ghai, M. Kolev, N. Olova, M. Gadjeva, A. Agrawal, B. Bottazzi, A. Mantovani, K. B. Reid, U. Kishore and M. S. Kojouharova, *Biochemistry*, 2006, **45**, 4093.
94. L. T. Roumenina, A. A. Kantardjiev, B. P. Atanasov, P. Waters, M. Gadjeva, K. B. Reid, A. Mantovani, U. Kishore and M. S. Kojouharova, *Biochemistry*, 2005, **44**, 14097.

CHAPTER 5

Complement Components C3 and C4

S. K. ALEX LAW

School of Biological Sciences, Nanyang Technological University,
60 Nanyang Drive, Singapore, 637551, Singapore

5.1 The Road to Oxford

I stayed in the Immunochemistry Unit from 1981 to 2002. It all started with an incident that happened in Hong Kong in 1980. I was working in the Department of Genetics, Washington University School of Medicine, with Paul Levine, who was my PhD thesis supervisor at Harvard. I went to Hong Kong for my brother's wedding and, rather unexpectedly, the US Consulate refused to issue me a visa to return to the US. I was stranded in Hong Kong for 6 months before I was granted a 6-month temporary visa so that I could pack up and leave the US. I returned to the US in January 1981, and one of my priorities was to secure a position in a laboratory outside the US to continue work on complement. It was natural that I wrote to Professor Rodney Porter at Oxford (the Prof.). He sought a one-year Wellcome Travelling Fellowship for me. I arrived at Oxford on 1st September 1981.

When I was a graduate student in Paul Levine's laboratory at Harvard, I found that the complement protein C3 bound covalently to cell surfaces. The work was first submitted to *The Journal of Biological Chemistry*, but was rejected. Eventually, we succeeded in convincing Alvin Pappenheimer to communicate the manuscript to *The Proceedings of the National Academy of Sciences, USA*. The paper was published in 1977,[1] and I received my PhD in 1978.

Molecular Aspects of Innate and Adaptive Immunity
Edited By Kenneth BM Reid and Robert B Sim

Published by the Royal Society of Chemistry, www.rsc.org

In the same year, Paul received an offer to move to the Washington University School of Medicine in St Louis and I went with him. Our work had been met with mixed enthusiasm; it was against the conventional wisdom at the time. Whereas some congratulated us for a most significant piece of work in complement biochemistry, others were convinced that it was some kind of an artefact. Furthermore, we were often challenged that even if C3b bound covalently to the cell surface, how did we know that this form was biologically active? In retrospect, the covalent binding of C3b to cell surfaces by reacting with simple hydroxyl groups, and perhaps amino groups, was the only way that C3b could bind indiscriminately to all cell surfaces. The discovery of the internal thioester in C3 provided the precondition for the covalent binding reaction. It must be a record with respect to the number of publications, in 1980 and 1981, on the thioester in C3, C4 and α2-macroglobulin and related topics,[2–22] including three (on C4) from the Porter group in the MRC Immunochemistry Unit[2,14,15] three (on C3) from Bob Sim's group in the Unit[16,21,22] and three from myself and Paul Levine.[9,10,20] Tack, Janatova and Prahl showed there was a thioester in native C3, and also that the thioester was between the sulfhydryl group of a cysteine residue and the γ-carbonyl group of a glutamyl residue with two intervening residues of glycine and a glutamic acid in between. It was later shown that the glutamyl residue was coded as a glutamine when the cDNA of C3 was cloned and sequenced by George Fey.[23] Thus, by the time I went to Oxford, the covalent binding story was officially accepted as many textbooks started to publish a variation of the mechanism shown in Figure 5.1.

Was there more to establish with the chemistry of the binding reaction? In 1966, Müller-Eberhard *et al.*[24] had described the labile binding site of C3. A key feature was that upon activation on a cell surface, the activated C3b would have

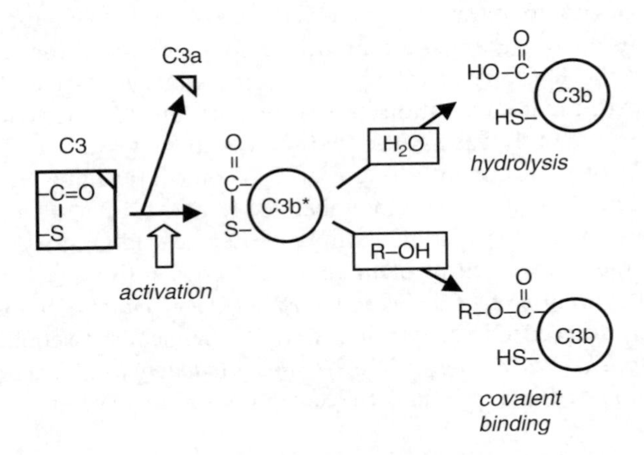

Figure 5.1 The covalent binding reaction of C3. Most immunology textbooks have had a model similar to this one since the early 1980s, and it has not been updated since the chemical mechanism was worked out in the 1990s. This version is sufficient to understand the importance of the covalent binding of C3 to target cells in immunology without the details of the chemistry.

the capacity to bind to the same cell, but not to bystander cells since the exposure of the binding site was transient, *i.e.* if binding did not take place within a very short time, the binding capacity of C3b would be irreversibly lost. The activated C3b had a very short half-life ($t_{1/2}$) which could not be measured accurately, although it was estimated later to be as short as 60 μs.[21] This description fitted well with the textbook version of the binding reaction in 1980. The labile binding site was the internal thioester. Activation would result in the exposure of the thioester, which either reacted with nucleophiles on the cell surface *via* an ester bond, or was hydrolyzed. However, an isolated thioester is rather stable with a $t_{1/2}$ in the order of several tens of minutes in an aqueous environment at physiological temperature and pH. It is $\sim 10^7$ times longer than that for the exposed thioester in C3b. It took me another 15 years to resolve this discrepancy.

5.2 The Autolytic Cleavage Reaction

In the late 1970s, an unusual behaviour was described for α2-macroglobulin: when heated to 90 °C, the single chain would split into two.[25] A similar cleavage was characterized in C3 and this, together with limited sequence data around the thioester site, was the first indication that C3, C4 and α2-macroglobulin are homologues.[4,22] An intact thioester was required because autolytic cleavage cannot be demonstrated in C3b, nor when C3 is inactivated by small nucleophiles.[22] Ammonia-induced inactivation of complement activity has been known for a long time,[26] but was not appreciated until 1980, when it was understood as the reaction with the internal thioesters of C4 and C3. The chemistry of the heat-induced chain-splitting reaction was elucidated[12,16] in that the backbone amino group of the glutamine residue would attack the thioester to form an internal pyroglutamic acid, followed by the hydrolysis of the peptide bond preceding the residue (Figure 5.2). Khan and Erickson[27,28] synthesized the 15-member thiolactone ring of the thioester proteins and showed that the hydrolysis of the peptide

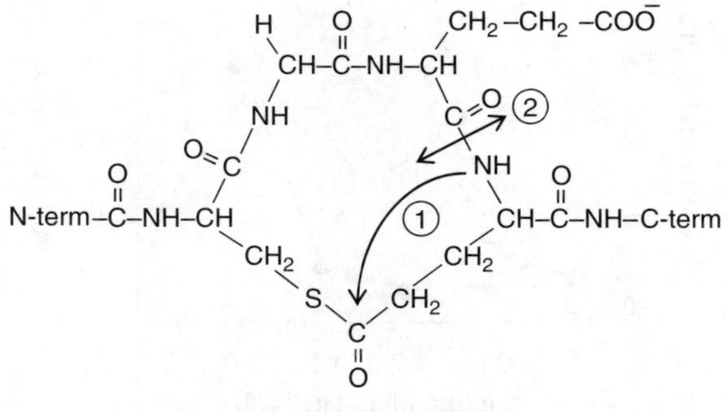

Figure 5.2 The internal thioester from the four residues of –Cys–Gly–Glu–Gln–. The two steps of the autolytic cleavage reaction are the formation of the internal pyroglutamic acid (1) and the hydrolysis of the preceding peptide bond (2).

bond was the preferred reaction over the hydrolysis of the thioester. Indeed, it became rather confusing as to how the three reactions, namely, the autolytic cleavage reaction (or denaturation split), the hydrolysis of the thioester and the covalent binding of the thioester to molecules other than water were related.

In 1981, David Isenman, in Müller-Eberhard's group at Scripps, published a paper on the conformational change of C3 upon activation.[29] The change was measurable by the change in circular dichroism, or by the binding of a chemical reagent, 8-anilino-1-napthalenesulfonate, to the exposed hydrophobic region of the protein. I was interested to find if the proteolytic cleavage of C3 to C3a to C3b was an absolute requirement for the covalent binding of C3b to substrates, or was only a means to change the conformation of C3 to allow such a reaction to take place. The latter was shown to be the case.[30] When C3 was treated with potassium bromide (KBr) or potassium thiocyanate (KSCN), it could bind glycerol without having to be cleaved with a proteolytic enzyme into C3a and C3b. KBr and KSCN had been known, and used as reagents, to 'inactivate' C3, but they would not induce the autolytic cleavage reaction. I also included in my experiments guanidine hydrochloride, sodium dodecyl sulfate (SDS) and urea. It was the guanidine hydrochloride experiment that provided a relationship between the covalent binding reaction and the autolytic cleavage reaction.

C3 was mixed with radioactively labelled glycerol and guanidine hydrochloride was added to different final concentrations. At low concentrations, nothing happened. At increasing concentrations, covalent binding of glycerol to C3 was observed. Yet at even high concentrations, the autolytic cleavage reaction gradually took over glycerol binding (Figure 5.3). I concluded that autolytic cleavage was a property of the naked thioester as shown by Khan and

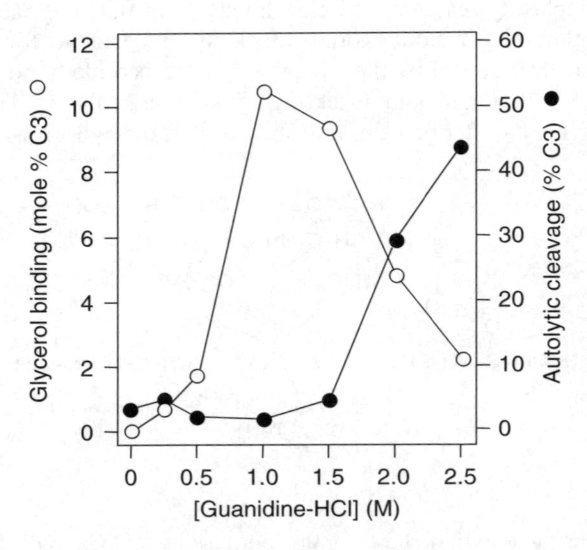

Figure 5.3 The autolytic cleavage reaction and covalent binding reaction of C3. At about 2 M of guanidine–HCl both reactions can proceed (from Law,[30] with permission of the *Biochemical Journal*).

Erickson,[28] which was therefore observed if C3 was denatured very quickly. If C3 was activated without being denatured, covalent binding took place.

5.3 Red Wine and the Isotypes of C4

My Wellcome Travelling Fellowship was for one year. I needed some financial support if I was stay in Oxford in 1982. The Lister Institute had decided to close down its research facility but to fund, generously, research fellowships to promising young investigators. I applied with the support of the Prof. I was among the inaugural batch of the Lister Institute Research Fellows, which included Alec Jefferys, now Sir Alec and famous for his pioneer DNA fingerprinting work, and Judy Armitage, who became a colleague in the Department of Microbiology at Oxford. My proposed project was to search for other thioester proteins in leukocytes. Today, only CD109 has been described as a leukocyte antigen with a thioester.[31] With a rather restricted expression profile, it was not surprising that I failed to identify it or, indeed, any other thioester proteins.

In the 1970s, it was established that the incidence of some autoimmune diseases could be linked with certain haplotypes of the major histocompatibility complex (MHC). MHC is about 3.5 Mb, located in human chromosome 6, with the class II genes at the centromeric end and the class I genes at the telomeric end. The two C4 genes, C4A and C4B, are arranged at tandem in the MHC class III region, a 1 Mb segment between the class I and class II regions.[32] Both C4A and C4B are highly polymorphic, with over 30 different allotypes described in the early 1980s. There are very few crossovers within the MHC, and thus the genes within are often inherited as haplotypes. In particular, C4 null alleles for both C4A and C4B are present at the level of about 10%.[33] The association of the haplotypes with C4A or C4B null alleles to certain autoimmune diseases, such as systemic lupus erythematosus,[34] demanded the study of the structure and functions of the various allotypes of C4.[35]

A coordinated effort was launched in the Unit headed by the Prof. Mike Carroll was the first to clone part of the C4 cDNA[36] and, later, joined forces with Duncan Campbell and David Bentley to map the C4, C2 and factor B genes in the MHC class III region.[32] Tertia Belt (Softley), and later Yung Yu, continued with the DNA sequencing of the C4 allotypes.[37–39] C4 typing was carried out by John Edwards' group in the Department of Genetics. About the same time, Edith Sim had improved the C4 typing protocol to make the identification of C4 allotypes from sera easier.[40] It had been known that the two isotypes of C4, C4A and C4B, were functionally distinct in that C4B has a higher specific haemolytic activity, by about three-fold, than C4A.[41] Yet Tertia Belt had shown that the two isotypes had very similar sequences.[37] Alister Dodds was trying to determine the cause of this functional difference. I was on the sideline, searching for the elusive thioester proteins.

Tertia talked to me quite frequently about her work. One day, she showed me the differences found in the C4d region of C4A and C4B. I could not help but notice the difference in two residues in C4A and C4B, which was leucine–aspartic acid in C4A and isoleucine–histidine in C4B. The reason I picked up on these two residues was that when the C3 sequence was aligned against the C4 sequences, it had isoleucine–histidine as in C4B. Previously, I had studied the binding reaction of C3 and C4, and showed that they were different.[42] (Since the C4 used was purified from untyped serum, it had a >95% probability that it was a mixture of C4A and C4B.) C3 hardly reacted with the amino group of glycine at neutral pH, but showed substantial reaction with the hydroxyl group of glycerol. On the other hand, C4 showed both activities. I remember quite clearly that I remarked that wouldn't it be nice if (the binding properties of) C4B behaved like C3 and C4A behaved completely differently. No one was in a hurry to do the experiment.

The Prof. also talked to me regularly and we pondered on the difference in specific haemolytic activity between C4A and C4B. I suggested to him that perhaps we should look at the covalent binding reactions of C4A and C4B. He was not convinced. One day I asked him if it was OK for me to have a go at the experiment. His response was that it would be great if I was right, but he was sceptical and wagered me a bottle of wine on the result. The scene was set. First I had to find two individuals who were deficient either in C4A and C4B. Jean Gagnon was typed to be C4A3BQO and he agreed to donate his blood. The only C4A-deficient person we had access to was Pall Hersteinsson, the husband of Asta Palsdottir, an Icelandic student in John Edwards' laboratory. Later, Asta would join the Immunochemistry Unit and completed her D. Phil studies there. Pall's C4 type was B1B2. It was not ideal but that was the best we could do at the time. One afternoon I arranged for Jean and Pall to have their blood drawn at the Oxford Blood Transfusion Centre. I compensated them each with two bottles of red wine, to replenish their bodily fluid with a liquid of matching colour, if not content.

The experiment was straightforward enough and, indeed, C4B was very much C3-like, though not identical, but C4A was completely different. C4A did not bind glycerol at all, but did bind glycine efficiently. We, the Prof. and I, were quite elated as we watched the results coming through from the printer in the corridor of the Unit, which was located on the fourth floor of the old Biochemistry building. However, we learned soon after that David Isenman in the University of Toronto had similar results. David published his work in *The Journal of Immunology*,[43] while we published ours in *The EMBO Journal*.[44] When the dust settled, the Prof. gave me a bottle of wine – the content was long ago consumed, but I have kept the bottle as a personal prize I won from the Prof.

5.4 LOO-3, the Anti-C4 Monoclonal Antibody that Only Worked for Us

There were about 10 amino acid differences between C4A3 and C4B1/B2. At the time, it was premature to construct C4 variants by site-directed mutagenesis. Most individuals with C4 null alleles have C4A3 or C4B1 or C4B2. To obtain the rarer forms of C4, we would have to find a way to separate C4A and C4B from the plasma of single individuals. Alister Dodds found that when he purified C4 by affinity chromatography using the LOO-3 antibody, purchased from the Commonwealth Serum Laboratory in Australia, he could elute C4A before C4B using a pH gradient.[45] Thus, we were able to characterize C4A1, A3, A4 and A6, and C4B1, B2, B3 and B5.[46] Combined with the sequencing results by Yung Yu,[39] we pinpointed the relevant residues of about 100 amino acids C-terminal to the thioester. In a hexapeptide at positions 1101 to 1106 all the C4A allotypes had the PCPVLD sequence, whereas all the C4B allotypes had the LSPVIH sequence. The rest of the differences were found in the mixed allotypes and were therefore eliminated for their contribution to the very distinct covalent binding properties of the C4A and C4B isotypes (Figure 5.4).

We were not secretive about the magic of the LOO-3 antibody and had passed on the information to those who were interested. However, when David Isenman tried it, the antibody gave him poor C4-binding activity and no differential binding to C4A and C4B. The same happened to Edith Sim, who had moved to the Department of Pharmacology and wanted to study the C4 isotypes for their possible roles in systemic lupus erythematosus. At the same time, our antibody column had gradually lost its differential specificity for C4A and C4B – but only lost it completely after we had completed our series of experiments. We tracked down the origin of the LOO-3 antibody and found it was created by John Wetherall at the Curtin University of Technology in Perth, Australia. We contacted him and he told us he could not find the original hybridoma. When we tried the LOO-3 from a new company, it did not have the capacity to separate C4A and C4B. Life moved on.

C4 allotypes	k_2/k_0(M^{-1}) glycine	glycerol	Polymorphic residues 328	707	1054	1101	1102	1105	1106	1157	1182	1188	1191	1267
A1	18,800	2.5			G	P	C	L	D	S	T	A	R	A
A3	17,000	1.5	YS	L	D	P	C	L	D	N	T	V	L	S
A4	13,300	1.5	S	P	D	P	C	L	D	N	S	V	L	S
A6	11,300	0.5	Y	P	D	P	C	L	D	N	T	V	L	A
B1	110	16.0	S		G	L	S	I	H	NS	T	A	R	S
B2	90	15.0	S	P	D	L	S	I	H	S	T	A	R	A
B3	110	13.0			G	L	S	I	H	S	T	A	R	A
B5	60	12.5			D	L	S	I	H	S	T	V	L	A

Figure 5.4 The glycine and glycerol binding properties, and the polymorphic residues of the human C4 allotypes. The results were compiled from references Yu *et al.*[39] and Dodds *et al.*[46]

5.5 How to Catch Up from Three Years Behind

Mike Carroll left the Unit in 1984 to set up his own laboratory at Harvard. David Isenman, from the University of Toronto, went and did a sabbatical in Mike's laboratory from 1986 to 1987. They constructed C4 with various combinations of residues in the hexapeptide at 1101 to 1106. The C4 variants were expressed in a transient expression system and their covalent binding capacities to form an amide bond or an ester bond were determined. They published the work in 1990 in *The Proceedings of the National Academy of Sciences, U.S.A.*[47] Essentially, their findings had narrowed down to the residue at position 1106, which was an aspartic acid in C4A and a histidine in C4B. The most remarkable result was that C4 with an alanine at position 1106 behaved like C4A, *i.e.* binding to cell-surface macromolecules *via* a hydroxylamine-resistant amide bond.

At that time, Alister and I were also ready to construct C4 variants by site-directed mutagenesis, but the shortage of manpower had hampered our progress. Alister had never done any molecular biology work, and I was too busy in pressing on with the work on the integrins. It was not until the arrival of Armin Sepp, a student from Estonia on a Soros Scholarship, that we had some 'spare hands' to do the experiments. In addition to the 'standard' variants, we included the hexapeptide from C3 and α2-macroglobulin, with a histidine and an asparagine at position 1106, respectively. We also included the variant with an alanine in that position. We chose to develop stable cell lines so that we could obtain highly purified C4 to measure the glycerol *vs.* glycine binding capacity in a system that we had established. The results were in line with the Carroll–Isenman paper, perhaps with the addition that an asparagine at position 1106 also resulted in a C4 that was C4A-like. I did not feel like publishing something which would basically confirm what Mike and David had published three years previously. I re-examined the problem with the closest scrutiny.

In our *EMBO Journal* paper in 1984,[44] we suggested that the "aspartic acid in C4A could lead to deprotonisation of amino groups at neutral pH and hence explain (its preferred) reactivity with glycine" and "the histidine in C4B could increase the nucleophilicity of alcohols and be responsible for (its preferred) reactivity with hydroxyl groups". What bothered me was not what we suggested for the roles of either the aspartic acid or the histidine in C4A and C4B, respectively, but that we suggested that by changing a single residue we had two different reactions. This was not how proteins work. The new results pointed to the elimination of the aspartic acid since its replacement with alanine or asparagine would not change the binding properties. The interpretation had changed so that if there was a histidine at position 1106, C4 would bind preferentially to glycerol; if not, C4 would bind glycine. The question had also changed: What was the role of the histidine in C4B?

I knew that if we were to understand the chemistry of the covalent binding reaction, we had to move from the system of studying C3b or C4b binding to erythrocytes. The erythrocyte surface was a jungle in which concentrations of substrate groups, either hydroxyl or amino in nature, were impossible to define. I had devised a fluid-phase system with C3 using trypsin as the proteolytic activator, which could convert C3 into C3a and C3b in minutes but would take a long time, hours, to convert C3b into C3c and C3d.[20] (Thus, by limiting the reaction time to a few minutes, we would have a stable product of C3b for analyses. It was even easier for C4, since C1s would only cleave C4 into C4a and C4b and no further.) The system also allowed us to relate the binding efficiency, at a defined concentration of the hydroxyl group binding substrate, which was glycerol, or the amino group binding substrate, which was glycine, to the two rate parameters, k_2 and k_0 (Figure 5.5). k_2 was the second-order reaction rate of the activated C3b species with glycerol or glycine, and k_0 the first-order rate of hydrolysis of the thioester. When we compared the reactivity of the different C4 allotypes with glycerol and glycine,[46] we found the k_2/k_0 ratio most revealing (see Figure 5.5).

We had always assumed that the primary difference would be in k_2 rather than k_0. Indeed, we had been publishing some misleading statements, such as "The average reaction rate of C4A with glycine is about 150 times that of C4B, whereas that of C4B with glycerol is about 10 times that of C4A".[46] In fact, we were not determining rates at all but a ratio of k_2/k_0. k_2 and k_0 were two different parameters and had to be examined separately. It dawned on me that if the histidine was responsible for catalyzing the binding to hydroxyl groups, it should also be responsible for the hydrolysis. Thus, that C4B did not bind efficiently to amino groups was because the histidine was catalyzing both the hydrolysis and binding to hydroxyl groups, and its binding to amino groups was therefore apparently less efficient. In the absence of the histidine in C4A, hydrolysis was not catalyzed and the more nucleophilic amino group would therefore appear to react more effectively with the exposed thioester. How was this hypothesis to be tested experimentally?

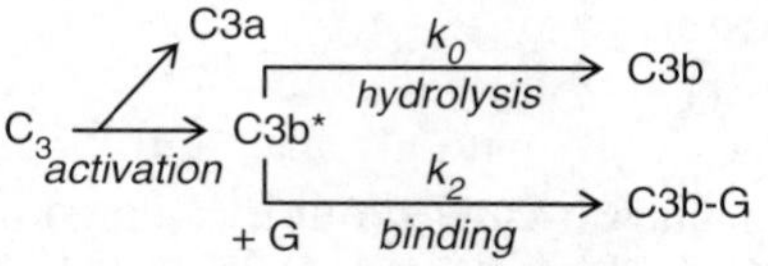

C3a
$C_3 \xrightarrow{\text{activation}}$ C3b* $\xrightarrow[\text{hydrolysis}]{k_0}$ C3b
\+ G $\xrightarrow[\text{binding}]{k_2}$ C3b-G

$$BE = \frac{k_2[G]}{k_0 + k_2[G]}$$

$$\frac{k_2}{k_0} = \frac{1}{[G]} \times \frac{BE}{(1 - BE)}$$

Figure 5.5 The k_2/k_0 ratio as a parameter to compare the covalent binding properties of C3 to different substrates. G can either be glycine or glycerol in our experiments. BE is the binding efficiency, which can be defined as the fraction of C3b labelled with the substrate molecule in a particular reaction, and can be determined experimentally. Since [G] is known, the k_2/k_0 value (M^{-1}), can be calculated.

We had always regarded the active species of C3 or C4 as having a very short $t_{1/2}$. In light of the new hypothesis, it might be true for C4B but not for C4A. I mentally went through the experimental procedure of the binding reaction, which included the mixture of C4 with the radioactively labelled glycine or glycerol. C1s was added and the reaction was allowed to proceed to completion. The proteins were precipitated with trichloroacetic acid (TCA) and, after solubilization, separated by SDS polyacrylamide gel electrophoresis (PAGE). After staining, the α' band of C4b was excised and the radioactivity associated with the band determined. For a time-course experiment, we also had to measure the rate of activation, *i.e.* the cleavage rate of C4 by C1s. Fortunately, C1s only cleaved a single bond in C4 (to give C4a and C4b), and the addition of TCA would stop the reaction. The amount of C4 activated, *i.e.* the conversion to C4a and C4b, could be estimated from the relative staining of the α and α' bands of C4. The experiment was possible in principle. What was the $t_{1/2}$ of C4A at which a meaningful measurement could be made? I derived the equations that governed the reaction, and simulated the results using simple computer programming. (I was a physics graduate from Caltech and had earned some spending money by writing computer programmes in my undergraduate days. I was rusty both with differential equations and programming. But I was not taking an undergraduate exam and so did not have to score an A within 3 hours. I was doing research, which meant that I could take my time as long as I got it right.) What came out was that if the $t_{1/2}$ of C4A was around 10 seconds we would have a chance. I did not worry about that of C4B since 'too fast to be measured' was an acceptable conclusion. Alister and I set out to do the experiments, and the $t_{1/2}$ of the activated C4A was, indeed, approximately 10 seconds.[48]

In my opinion, this was the conceptual breakthrough in our understanding of the binding reaction with a catalyzed hydrolysis of the thioester that is the centre stage of the reaction.

5.6 C4K and C4Y

The next set of experiments was carried out by Xiangdong Ren, a student from China on an Oxford University Run Run Shaw Scholarship – the award of which was based mostly on his command of the English language. Ren had already determined the composition of C4 genes in sheep and cattle, which are the only non-primate mammals that have genes for both C4A and C4B[49] – most other mammals have only a single C4 gene coding for a C4B-like protein.[50] To consolidate his thesis, he was to continue with the covalent binding studies by further site-directed mutagenesis. He constructed five C4 variants with an arginine, cysteine, lysine, serine and tyrosine at position 1106. It was the lysine (C4K) and tyrosine (C4Y) variants that provided the final set of data for the formulation of the chemical mechanism of the covalent binding reaction.

How did the histidine catalyze the hydrolysis of the thioester? There were two possibilities: it either acted as a base to catalyze the reaction, or as a nucleophile which attacked the thioester directly to form an acyl–imidazole intermediate. One hypothesis was that if the histidine acted as a nucleophile, its substitution with a lysine would have the ε-amino group of the lysine attacking the thioester, resulting in an amide bond between the acyl group of the thioester and the ε-amino group of the lysine. If we digested the resultant C4b with trypsin, we should find two peptides cross-linked by the amide bond. The cross-linked peptide could be traced since the released thiol could be labelled with radioactive iodoacetic acid. We already had one piece of information that was supportive of this hypothesis. The covalent binding of C4K was very low to both glycine and glycerol, as the hypothesis would have predicted.

Tony Willis was responsible for the purification of the peptide and its analysis. In addition to C4K, he also analyzed C4B as a control. He told me that he found one radioactive peak in each sample, but they eluted at different positions from a high performance liquid chromatography (HPLC) column. When he sequenced the peptides in the peaks, he found the results were as expected (Figure 5.6). I was afraid that there would be some residual radioactive signal of the C4B peptide in the C4K HPLC fractions. On close examination, we found none. It was a big relief since the hypothesis of the covalent binding reaction, or hydrolysis, that started with an internal nucleophilic attack to the thioester suggested a 100% reaction. Some residual hydrolyzed thioester would not be consistent with this hypothesis.

We did not know what to expect from C4Y. The standard covalent binding results suggested something interesting. It was the only C4 variant, other than C4B, that showed any significant reaction with glycerol. Thus, the tyrosine could, to a certain extent, play the role of the histidine in the reaction. Ren had started writing up his thesis and I was left to myself to do the experiments. I persisted with the determination of the $t_{1/2}$ of the activated C4Y, which I found to be around two minutes. To this day I do not know why I was so determined to do this particular experiment, except that I did not like to leave questions unanswered, especially when I had the experimental set-up to obtain the answer,

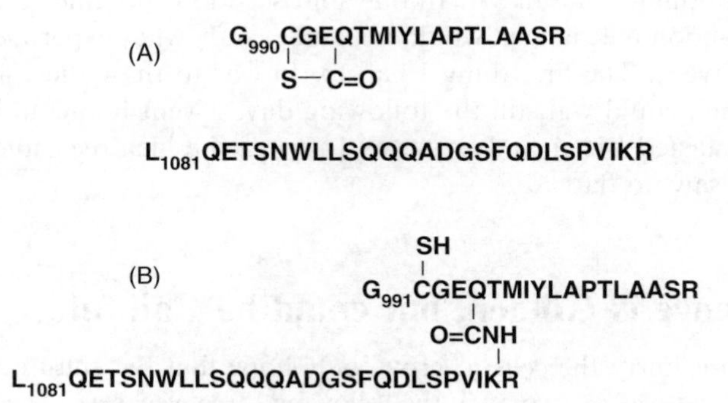

Figure 5.6 The peptides of C4K (*i.e.* C4B with the histidine at position 1106 substituted by a lysine). (A) The two peptides in native C3K. (B) The cross-linked peptides after C4K has been activated by C1s.

but I did not quite know the significance of this result either. It was in the autumn of 1993, and soon after I left for Kyoto for the Complement Workshop.

Upon my return to Oxford, Ren told me that Alister had this idea that if the $t_{1/2}$ of C4Y was about 20 minutes we may be able to detect the kinetics of the appearance of the free thiol and tell whether it followed the kinetics of activation or of binding. If it followed activation, then the thioester was broken first before the binding reaction took place, which would suggests an intermediate. If it followed binding, we had to think of something else. It was a great idea. The appearance of the free thiol did follow the kinetics of activation. There was an intermediate.

These were definitive results, and we were faced with two new questions:

(1) The histidine was therefore not a base for this reaction. Were we back to square one and had to search for another base?
(2) The stabilities of the thioester and the acyl–imidazole in water were comparable. Why should the reaction start with the apparently redundant step in transferring the acyl group from the thiol to the imidazole?

I was not in a hurry to publish these definitive but inconclusive results.

5.7 The Best Beer that I had Ever Tasted

It was in the late autumn of 1994. My integrin work was not progressing at a pace I liked. The complement work was at a standstill – we had some excellent results, but I did not know what to do with them. It was a Wednesday afternoon and, as I was getting quite fed up, went home – something I had never done. I took a nap and woke up at around 6 PM. I checked the fridge, which was well stocked with food, but I had ran out of beer. I drove to the supermarket at Kidlington and bought some beer, among other things. I let my mind wander on the way home. Then the idea struck me.

There is nothing quite like it. In one split second it became so clear. I knew what the reaction mechanism would be, and exactly what experiments I needed to do to prove it. The first thing I had to do was to thaw the C4Y clone and grow it, which could wait till the following day. I went home and enjoyed the beer, which tasted better and better as I turned the idea over and over in my mind and I saw no flaw.

5.8 Silence is Golden, but could be Painful

One could not hurry the cells to grow, only hope that the culture was not contaminated. That was my worry in the following three weeks as I tried to grow the C4Y clone to obtain sufficient C4Y for the experiment. In the meantime, I also purified some C4A and C4B from typed sera (remember that our LOO-3 column no longer worked). I was selfish – I did not involve Alister. I wanted not only to

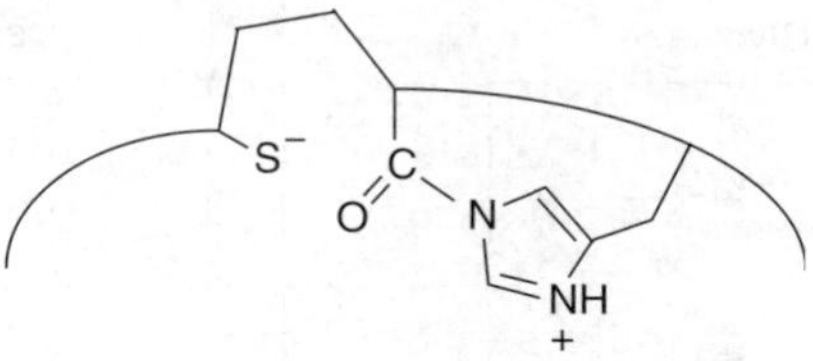

Figure 5.7 The acyl–imidazole intermediate of C4B and C3. The attack of the thioester by the histidine releases the thionate anion, which then acts as a base for the hydrolysis of the acyl–imidazole bond.

have the credit, but above all I wanted to have the pleasure of doing the experiment myself. There was a drawback – I was not to talk about this with anyone for a month. Of course, one reason for not talking was that I could be wrong, just in case, but the story was too good to be wrong – certainly in retrospect.

I reasoned that when the histidine attacked the thioester, it freed the thiol. In its thionate anion form it is in the right place to be the base to activate water to attack the acyl–imidazole intermediate (Figure 5.7). If the thiol was blocked, the binding would not continue. Reagents, *e.g.* iodoacetamide, took minutes at a reasonable concentration, such as 10 mM, to react with the thiol. The $t_{1/2}$ of C4B was in the order of 1 ms (still no reliable estimate) and it would require unrealistically high concentrations of iodoacetamide to have any effect. C4Y was different, with a $t_{1/2}$ of two minutes, and thus a reasonable concentration of iodoacetamide would competitively inhibit the reaction. It had been known from the work of Jarmila Janatova and Brian Tack that the quantitative titration of the released thiol of C3 would require the presence of the detecting reagent before the thiol was exposed, otherwise it would be oxidized and became undetectable.[8] Thus, the thioester was to serve the dual functions of keeping the acyl group active and also the thiol group from oxidation. Upon activation, the histidine attacked the thioester, keeping the acyl group active, and also freeing the thionate anion to act as a base. The thionate anion is required to be available for a fraction of a second, and only required to work once.

The results were as expected (Figure 5.8). The work was published in 1996.[51]

5.9 A Bet that I both Won, and Lost

David Isenman had the same idea. He created C3K (the critical histidine substituted with a lysine). He claimed that he could activate C3K to bind to erythrocytes, and that the binding was hydroxylamine sensitive. In his set-up, C3K was metabolically radiolabelled in a transient expression system. After binding to erythrocytes, the membrane proteins were analyzed by SDS-PAGE followed by autoradiography. Covalent binding was demonstrated by the presence of radioactive bands with molecular weights greater than the native C3 band when run under non-reducing conditions. If the bond was an ester, treatment of the samples with alkaline hydroxylamine would reduce the

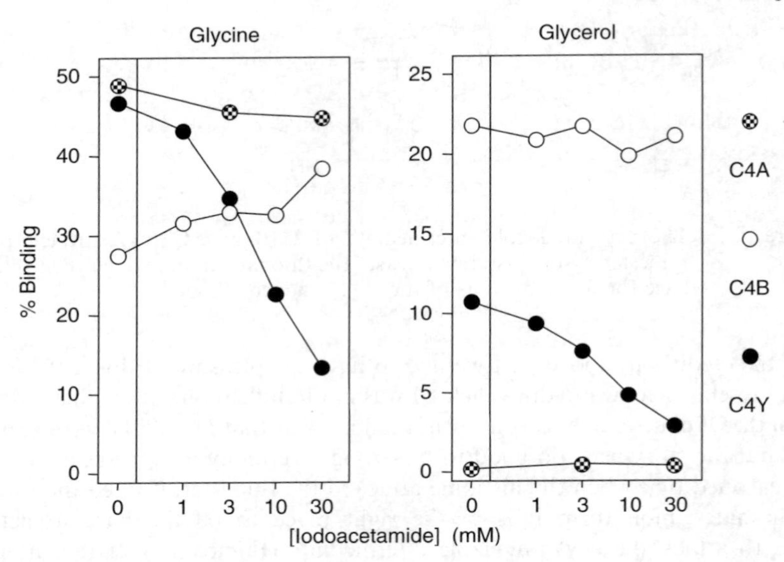

Figure 5.8 The inhibition of the binding of glycine and glycerol to C4Y (*i.e.* C4B with the histidine at position 1106 substituted by a tyrosine) with iodoacetamide. The reaction of iodoacetamide with the thionate anion (see Figure 5.7) is too slow to have any effect on the binding reaction of C4B. Since the binding reaction of C4Y has a half-life of approximately two minutes, the inhibitory effect of iodoacetamide can be demonstrated. The results are from the original experiment obtained in December 1994. Another set of experiments to include higher concentrations of iodoacetamide was done on the recommendation of the reviewers, and was subsequently published.[51]

radioactive bands to the single native C3 band. If David was correct, this would mean that replacing the histidine with lysine in C3 would not affect its binding to hydroxyl groups. This result clearly contradicted our conclusion on C4B. The alternative, that C4B and C3 have different chemistry in the covalent binding, was unthinkable – the significance of the proposed reaction mechanism would be severely compromised.

I asked to see David's result, which he sent to me without hesitation. I did see the high molecular weight bands, but I saw no bands upon hydroxylamine treatment. I pointed this out to David, who said he was also worried about this, and that was why he had not written up the results for publication. We met up in the Complement Workshop in Boston in 1996, and decided to join forces to tackle the C3 problem. To add some spice to our collaboration, I suggested that whoever was right would be the senior author in the resultant publication. We shook hands on it.

Mihaela Gadjeva, a Bulgarian student on a Soros/Foreign Office and Commonwealth (FCO) scholarship, was given the project. The progress was not as smooth as that for C4 because C4 was more negatively charged and eluted from the ion exchange column at a position rather free of other proteins. C3 was eluted in the midst of many other proteins, so to obtain sufficient amount of pure C3 took much more effort. At the end, we obtained sufficient C3K to demonstrate the presence of the same cross-linked peptides upon activation. I was right and we published the work in *The Journal of Immunology* in 1998, and I was the senior author of the paper.[52]

In retrospect, it was the wrong bet. Of course it went without saying that whoever was right had to be the senior author. I could not see myself as the senior author if I was wrong, nor would David think of being the senior author if I was right. The bet should have been on the publication charges, which amounted to a few hundred pounds. As senior author, I had to pay from the Unit's kitty. We live and learn.

5.10 Final Comments

I stopped working on the thioester proteins after the publication in 1998, with the exception of using iC3b as a ligand for CR3 and CR4, the $\alpha M\beta 2$ and $\alpha X\beta 2$ integrins, or the CD11b/CD18 and CD11c/CD18 antigens, respectively. Before the C3 work[52] we were so confident that we wrote a review article and published it in *Protein Science*.[53] To this day, I think it summarized all the science in this particular investigation since the description of the inactivation of C4 by ammonia.[26] This article, however, provides more of the background of my personal involvement in the work, in particular with the work done in the MRC Immunochemistry Unit. I am proud to have brought this work to a satisfactory conclusion.

We had tried to crystallize C3 only half-heartedly and therefore unsuccessfully. In 2005, the structures of C3 and C3c were solved.[54] The thioester was there, in the native C3, and was protected by hydrophobic groups, of a different domain, such that the histidine was far removed from the thioester.[55] Upon activation, the thioester domain was freed from the rest of C3, and the histidine was no longer prohibited from attacking the thioester. The structures supported our chemical mechanism.

Soon after I joined Oxford, the Prof. confided in me one day that no matter how much sequencing is done (on the thioester proteins), it would not demonstrate the thioester or the covalent binding. I took it that it was his way to express his approval of my work. Now I can take this line of thought further to suggest that no matter how much structural work is done, it cannot reveal the reaction mechanism.

It has often been said that successful scientific research is about asking the right questions. During this work, I had often pondered what the right questions were. Before the last series of experiments, I had asked the two questions:

(1) If the histidine was not the base, where was it?
(2) Why should there be an apparently redundant step in the forming of the acyl–imidazole intermediate?

The final revelation came in a split second, but that split second could have come at any time. Did I ask the right questions? Now I know that I did, but I could only be certain after I had obtained the right answers.

Acknowledgements

My 20 years in the Unit was a memorable period. I would be lying if I were to say that there had been no depressing times, but I can confidently say that they are but an insignificant dent to all the positive moments that I had with the Unit. The Prof. was inspirational and very down to earth. One can be a famous top-of-the-world scientist and be a human at the same time. I remember the day when we were invited to help him plant some trees on his farm. He worked together with us and, of course, Julia had prepared a very substantial lunch to replenish our energy. After finding that I was trying to grow some vegetables in my garden, she went on to dig from a big pile of horse manure a small bag for me to take home. (I was not much of a gardener and I am afraid that her effort was wasted.)

Ken Reid and his family made me feel very much at home in Oxford. His style as a Director was much different to that of the Prof., but it was nonetheless very successful. I remember one piece of advice and encouragement he gave me when I was considering my present job in Singapore and doubting my quality as an administrator. He said something to the effect that I can always act sensibly and therefore do a job better than most, with or without administrator experience. I am glad to say that I have performed in my Nanyang Technological University (NTU) post up to his expectation.

Alister Dodds was my closest colleague on the covalent binding story in the Unit. He was excellent in instrumentation and had contributed to making the most out of whatever experiments we were doing. My three students involved in this project were completely different and they contributed differently. Armin Sepp was incredibly sharp and understood the problems of the covalent binding reaction, as they were in the early 1990s, after discussion with me for an afternoon. By that time, I had been working on the problem for 15 years. Among all the students that I had known from China, including over 100 in the School of Biological Sciences at NTU, X. D. Ren still has the best command of the English language. Ren is also the only person I know who expressed a wish to have a body shape like mine when he becomes my age – and that includes myself. Mihaela Gadjeva was also different: she was a walking textbook in my laboratory and dazzled others with all the information in her head. She also amazed us by the amount of chocolate she could consume.

Two other 'students' were also important to me. Asta Palsdottir became the Prof.'s student when she transferred from Genetics. After the Prof.'s tragic death in 1985 she became, officially, the student of Ken Reid. Ken did not know much of C4 and so I was the one to help her. During that period, I learned much about C4 genetics from her. I only hope that she learned as much C4 biochemistry from me. The other one is, of course, Tertia Belt (Softley). She shared with me her scientific findings in the early 1980s, which may have played an important part in my first experiment to segregate the covalent binding properties of C4A and C4B. I often thought that without that experiment[44] it would have been very difficult to locate the catalytic histidine from the more than 1700 amino acids in C4. We might have picked that up from the work of David Isenman,[43] but we would have been two years behind, and it would not have been quite the same.

References

1. S. K. Law and R. P. Levine, *Proc. Natl. Acad. Sci. USA*, 1977, **74**, 2701–2705.
2. R. D. Campbell, A. W. Dodds and R. R. Porter, *Biochem. J.*, 1980, **189**, 67–80.
3. J. B. Howard, M. Vermeulen and R. P. Swenson, *J. Biol. Chem.*, 1980, **255**, 3820–3823.
4. J. B. Howard, *J. Biol. Chem.*, 1980, **255**, 7082–7084.
5. J. B. Howard, *Proc. Natl. Acad. Sci. USA*, 1981, **78**, 2235–2239.
6. J. P. Gorski and J. B. Howard, *J. Biol. Chem.*, 1980, **255**, 10717–10720.
7. J. Janatova, B. F. Tack and J. W. Prahl, *Biochemistry*, 1980, **19**, 4479–4485.
8. J. Janatova, P. E. Lorenz, A. N. Schechter, J. W. Prahl and B. F. Tack, *Biochemistry*, 1980, **19**, 4471–4478.
9. S. K. Law, N. A. Lichtenberg, F. H. Holcombe and R. P. Levine, *J. Immunol.*, 1980, **125**, 634–639.
10. S. K. Law, N. A. Lichtenberg and R. P. Levine, *Proc. Natl. Acad. Sci. USA*, 1980, **77**, 7194–7198.
11. M. K. Pangburn and H. J. Müller-Eberhard, *J. Exp. Med.*, 1980, **152**, 1102–1114.
12. R. P. Swenson and J. B. Howard, *J. Biol. Chem.*, 1980, **255**, 8087–8091.
13. B. F. Tack, R. A. Harrison, J. Janatova, M. L. Thomas and J. W. Prahl, *Proc. Natl. Acad. Sci. USA*, 1980, **77**, 5764–5768.
14. R. D. Campbell, J. Gagnon and R. R. Porter, *Biosci. Rep.*, 1981, **1**, 423–429.
15. R. D. Campbell, J. Gagnon and R. R. Porter, *Biochem. J.*, 1981, **199**, 359–370.
16. S. G. Davies and R. B. Sim, *Biosci. Rep.*, 1981, **1**, 461–468.
17. K. J. Gadd and K. B. M. Reid, *Biochem. J.*, 1981, **195**, 471–480.
18. R. A. Harrison, M. L. Thomas and B. F. Tack, *Proc. Natl. Acad. Sci. USA*, 1981, **78**, 7388–7392.
19. J. Janatova and B. F. Tack, *Biochemistry*, 1981, **20**, 2394–2402.
20. S. K. Law, T. M. Minich and R. P Levine, *Biochemistry*, 1981, **20**, 7457–7463.
21. R. B. Sim, T. M. Twose, D. S. Paterson and E. Sim, *Biochem. J.*, 1981, **193**, 115–127.
22. R. B. Sim and E. Sim, *Biochem. J.*, 1981, **193**, 129–141.

23. M. H. de Bruijn and G. H. Fey, *Proc. Natl. Acad. Sci. USA*, 1982, **82**, 708–712.
24. H. J. Müller-Eberhard, A. P. Dalmasso and M. A. Calcott, *J. Exp. Med.*, 1966, **124**, 33–54.
25. P. C. Harpel, M. B. Hayes and T. E. Hugli, *J. Biol. Chem.*, 1979, **254**, 8669–8678.
26. J. Gordon, H. R. Whitehead and A. Wormall, *Biochem. J.*, 1926, **20**, 1028–1035.
27. S. A. Khan and B. W. Erickson, *J. Am. Chem. Soc.*, 1981, **103**, 7374–7376.
28. S. A. Khan and B. W. Erickson, *J. Biol. Chem.*, 1982, **257**, 11864–11867.
29. D. E. Isenman, D. I. C. Kells, N. R. Cooper, H. J. Müller-Eberhard and M. K. Pangburn, *Biochemistry*, 1981, **20**, 4458–4467.
30. S. K. A. Law, *Biochem. J.*, 1983, **211**, 381–389.
31. M. Lin, D. R. Sutherland, W. Horsfall, N. Totty, E. Yeo, R. Nayar, X. F. Wu and A. C. Schuh, *Blood*, 2002, **99**, 1683–1691.
32. M. C. Carroll, R. D. Campbell, D. R. Bentley and R. R. Porter, *Nature*, 1984, **307**, 237–241.
33. G. Hauptmann, J. Goetz, B. Uring-Lambert and E. Grosshans, *Prog. Allergy.*, 1986, **39**, 232–249.
34. A. H. L. Fielder, M. J. Walport, J. R. Batchelor, R. I. Rynes, C. M. Black, I. A. Dodi and G. R. V. Hughes, *Brit. Med. J.*, 1983, **286**, 425–428.
35. R. R. Porter, *Mol. Biol. Med.*, 1983, **1**, 161–168.
36. M. C. Carroll and R. R. Porter, *Proc. Natl. Acad. Sci. USA*, 1983, **80**, 264–267.
37. K. T. Belt, M. C. Carroll and R. R. Porter, *Cell*, 1984, **36**, 907–914.
38. K. T. Belt, C. Y. Yu, M. C. Carroll and R. R. Porter, *Immunogenetics*, 1985, **21**, 173–180.
39. C. Y. Yu, K. T. Belt, C. M. Giles, R. D. Campbell and R. R. Porter, *EMBO. J.*, 1986, **5**, 2873–2881.
40. E. Sim and S. J. Cross, *Biochem. J.*, 1986, **239**, 763–767.
41. Z. L. Awdeh and C. A. Alper, *Proc. Natl. Acad. Sci. USA*, 1980, **77**, 3576–3580.
42. S. K. A. Law, T. M. Minich and R. P. Levine, *Biochemistry*, 1984, **23**, 3267–3272.
43. D. E. Isenman and J. R. Young, *J. Immunol.*, 1984, **25**, 577–584.
44. S. K. A. Law, A. W. Dodds and R. R. Porter, *EMBO. J.*, 1984, **3**, 1819–1823.
45. A. W. Dodds, S. K. A. Law and R. R. Porter, *EMBO. J.*, 1985, **4**, 2239–2244.
46. A. W. Dodds, S. K. A. Law and R. R. Porter, *Immunogenetics*, 1986, **24**, 279–285.
47. M. C. Carroll, D. M. Fathallah, L. Bergamaschini, E. M. Alicot and D. E. Isenman, *Proc. Natl. Acad. Sci. USA*, 1990, **87**, 6868–6872.
48. A. Sepp, A. W. Dodds, M. J. Anderson, R. D. Campbell, A. C. Willis and S. K. A. Law, *Protein Sci.*, 1993, **2**, 706–716.

49. X. D. Ren, A. W. Dodds and S. K. A. Law, *Immunogenetics*, 1993, **37**, 120–128.
50. A. W. Dodds and S. K. A. Law, *Biochem. J.*, 1990, **265**, 495–502.
51. A. W. Dodds, X. D. Ren, A. C. Willis and S. K. A. Law, *Nature*, 1996, **379**, 177–179.
52. M. Gadjeva, A. W. Dodds, A. Taniguchi-Sidle, A. C. Willis, D. E. Isenman and S. K. A. Law, *J. Immunol.*, 1998, **161**, 985–990.
53. S. K. A. Law and A. W. Dodds, *Protein Sci.*, 1997, **6**, 263–274.
54. B. J. C. Janssen, E. G. Huizinga, H. C. A. Raajkmakers, A. Roos, M. R. Daha, K. Nilsson-Ekdahl, B. Nilsson and P. Gros, *Nature*, 2005, **437**, 505–511.
55. B. Nagar, R. G. Jones, R. J. Diefenbach, D. E. Isenman and J. M. Rini, *Science*, 1998, **280**, 1277–1281.

CHAPTER 6

Complement Control Proteins and Receptors: From FH to CR4

ROBERT B. SIM,[a] BERYL E. MOFFATT,[a] JACQUELINE M. SHAW[a, b] AND JANEZ FERLUGA[a]

[a] MRC Immunochemistry Unit, Department of Biochemistry, South Parks Rd, Oxford OX1 3QU, UK; [b] MRC Human Immunology Unit, Weatherall Institute of Molecular Medicine, John Radcliffe Hospital, Oxford OX3 9DX, UK

6.1 Introduction

The MRC Immunochemistry Unit has conducted world-leading structure–function research on many of the complement control proteins and receptors. The characterization of factor H (FH) and factor H-like protein 1 (FHL-1), and of factor I (FI), C4bp and properdin, and the identification of complement receptor 4 (CR4) and calreticulin, the C1q/collectin receptor–adapter, were among the more extensive of these studies. The references cited in this review are deliberately chosen mainly from work published from the MRC Unit, or from collaborations with the Unit. Many other outstanding research groups also, of course, contributed to our current knowledge of the structure and functions of the proteins discussed below.

Molecular Aspects of Innate and Adaptive Immunity
Edited By Kenneth BM Reid and Robert B Sim

Published by the Royal Society of Chemistry, www.rsc.org

6.2 Factor H

Factor H (Figure 6.1) was first described in about 1965 as β1H, named because of its electrophoretic mobility in a system in which human serum proteins were assigned α, β or γ mobility.[1] Its complement regulatory role, as a cofactor for FI, and a decay-accelerator for the alternative pathway convertases, was described in 1976 by Whaley, Weiler and colleagues (Figures 6.2 and 6.3).[2–4] We did not work on it until about 1980 to 1981, and first became interested in it as a contaminant in C3 preparations, at a time when we were preparing C3 fragments[5,6] in large amounts to use as ligands for investigating C3 receptors. Richard DiScipio conducted C3b–FH binding studies[7] and, with Bob Sim, published a protein chemical characterization of FH in 1982.[8] This established the N-terminal sequence and showed that FH is a very elongated monomeric glycoprotein of 155 kDa. At this time, the research emphasis in the Unit was moving towards molecular biology, using methods introduced by Mike Carroll and Duncan Campbell, with impetus from Rod Porter and training from George Brownlee's laboratory. So we moved to begin the cDNA sequencing of FH, which was undertaken by Jean Ripoche and Tony Day, with very extensive protein-sequence backup by Tony (A. C.) Willis. This took a long time with the methods available then, and with the complication of the alternative splicing to form FH and the protein now named FHL-1 (Figure 6.1). Partial sequences were published,[9–12] and it was recognized that FH was made up of CCP (complement control protein) domains, which had first been identified as a

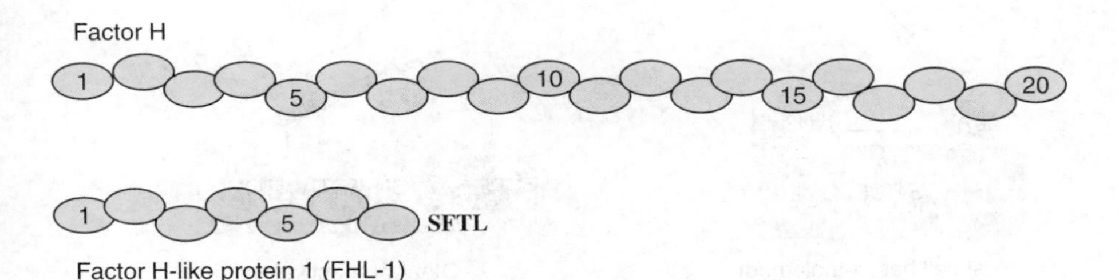

Figure 6.1 FH and FHL-1. FH is made up of 20 CCP domains, each of about 60 amino acids, and about 3.6 nm long. The protein contains up to three binding sites for C3 or its breakdown products. The major binding site is in CCPs1–4, with subsidiary sites at CCP19–20 and possibly in the region CCP7–10. FH also binds charged ligands, like heparin. The major heparin-binding sites are in CCP 19–20, CCP7 and there may be another site between CCP7 and CCP19. FH also binds to specific proteins on the surface of many microorganisms. CCP6, for example, has a binding site for a protein of *Neisseria meningitidis*. FHL-1 is an alternative splicing product from the FH gene. It has CCPs1–7, which are identical to those in FH, followed by a four amino acid sequence, SFTL. Additional genes, linked to the FH gene, produce a series of proteins called FH-related proteins (FHRs) 1 to 5. These contain four to seven CCPs, which are not identical to those in FH, but are highly homologous. The functions of FHRs are not known.

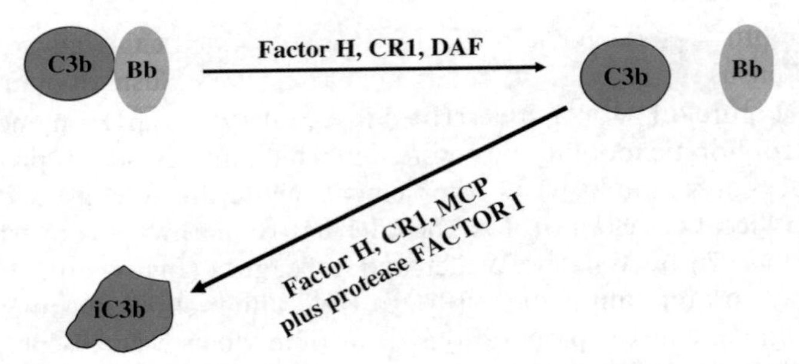

Figure 6.2 FH regulates the alternative pathway C3 convertase, C3bBb. FH has two activities involved in binding to C3b. The first is *decay–acceleration activity*, in which FH (or its homologous cell-surface proteins DAF or CR1) interact with C3bBb and accelerate its decay (*i.e.* promote dissociation of Bb from C3b). The second activity is *factor I-cofactor activity*, in which C3b forms a complex with FH (or its homologous cell-surface proteins MCP or CR1) and FI cleaves C3b in the complex, to form iC3b. For the classical pathway C3 convertase, C4b2a, FH has no regulatory activity, but C4bp fills the same role as FH, accelerating the decay of C4b2a and acting as FI-cofactor for the breakdown of C4b into C4c plus C4d.

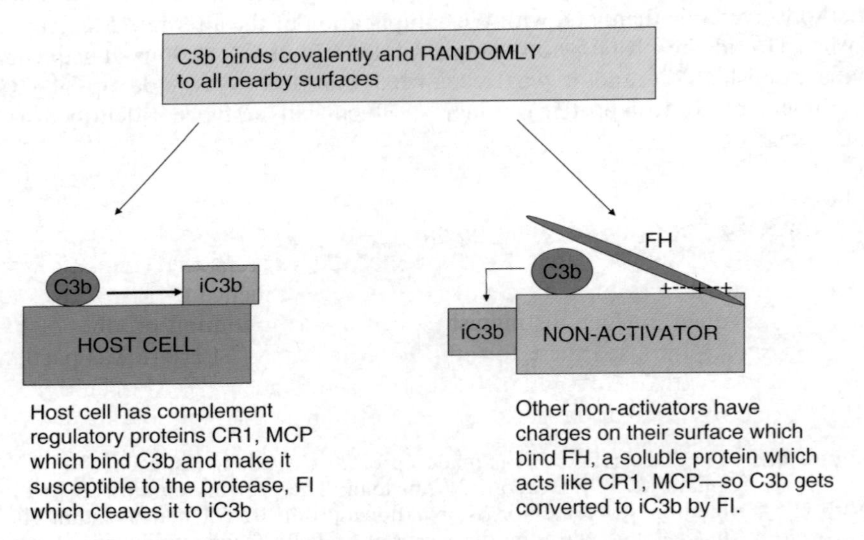

Figure 6.3 FH controls the activation and amplification of the complement alternative pathway on surfaces. C3b is generated constantly at a low rate in blood plasma. The C3b binds covalently and randomly to all surfaces exposed to the blood. On host cells, the cell-surface regulatory proteins CR1 and MCP bind the C3b and make it susceptible to degradation by FI. On surfaces which do not have CR1 or DAF, FH may bind to the surface *via* recognition of charge clusters; this increases the apparent avidity for the surface-bound C3b, and FI degrades the C3b in complex with FH into iC3b. On surfaces which do not have suitable charge clusters, FH does not bind, so instead FB binds C3b and is converted into C3bBb by the protease Factor D. C3bBb activates more C3, activating or amplifying the alternative pathway.

repeating motif in the sequence of factor B (FB).[13] Each CCP [also called SCR (short consensus repeat or short complement repeat) or Sushi domain] is about 60 amino acids long, with two internal disulfide bridges. The existence of more than one size of mRNA (4.3, 1.8 and 1.2–1.5 kb) for FH or FH-related (FHR) sequences indicated a greater degree of complexity than had been expected.[14] These were later resolved as mRNAs for FH, FHL-1 and FHRs. The complete cDNA sequences of both FH and FHL-1 were established in 1988.[15] The cDNA sequence was supported by amino acid sequencing of about 40% of FH (necessary to sort out FH from FHL-1), and the *N*-glycan attachment sites were characterized. A common polymorphism in FH, a Tyr/His interchange at position 402 (or 384, in the processed secreted protein), was described by us, also in 1988,[16] and has since become the subject of intensive research because it was shown to be a risk factor in age-related macular degeneration (AMD).[17]

During this period, substantial progress was made in making the assay and purification of FH and FI easier, because of the development of monoclonal antibodies against them by Edith Sim and Li Min Hsiung,[18,19] with the help of Mike Puklavec and Neil Barclay. The anti-FH antibodies were used in an early localization of the main C3b-binding site.[20] Beryl Moffatt optimized the use of these antibodies in affinity purification[21] and enzyme-linked immunosorbent assay (ELISA).

The gene localization for human FH (and C4bp) to chromosome 1q was done in collaboration with Ellen Solomon's laboratory,[22] and later the same was achieved for bovine FH.[23] The human genes were localized to a region already known as the regulation of complement activation (RCA) cluster in which are encoded CR1, CR2 and DAF, all of which are homologues of FH and C4bp.

Some further work was done on the characterization of FHL-1 and the FHRs. Val Dee showed that FHL-1 really did exist in the circulation, by making antibodies against its unique C-terminal sequence (SFTL – the only part of its sequence which distinguishes it from a degradation product of FH; see Figure 6.1) and used these for affinity purification.[24] FHR-1 was partially characterized, with Marc Fontaine's laboratory in Rouen.[25,26] Marcia McAleer then started to work on the gene sequence of human FH, and obtained about half of the sequence of this very large gene, as well as characterizing a useful polymorphism which could be used to trace inheritance[27] in a Sicilian family. Unfortunately we did not publish this fully at the time, as we had entered a loose collaboration with other groups, and so held back on publication, hoping that the rest of the sequence would be completed by our collaborators. Probably mainly through lack of effective communication, this did not quite work out, although summaries of the data were published.[28,29]

Investigation of the high-resolution three-dimensional (3D) structure of FH was started by Tony Day and Martin Baron who expressed, in a yeast system, CCP16 of FH (chosen because it had a typical sequence, not because it had any known functional significance). They persevered with this difficult work, and formed a collaboration with Paul Barlow and Dave Norman in Iain Campbell's group (Biochemistry Department, Oxford University) to solve the structure by proton NMR.[30,31] This was the first CCP structure solved, and revealed a compact ovoid domain, about 3.8 nm long, that contained five β strands. The collaboration continued, with the main biochemistry and molecular biology input later provided by Alexander Steinkasserer, with CCP5[32] (close to the major C3b-binding site) and with a double module (CCP15 + 16),[33] done to investigate the intermodule interface. We tried to express triple modules, but the expression system was not suitable for this. We also tried CCP13, which was of interest because of its very basic sequence, which suggested it might be a polyanion-binding site; however, the product was unstable to proteolysis during expression. Although often suggested in reviews as a possible polyanion-binding site, no-one has yet obtained any evidence for a distinct function for CCP13. We also expressed CCP3 + 4 for functional studies, as part of the major C3b-binding site, but it was not sufficiently homogeneous for structural studies.

Other structural studies on FH were undertaken by Steve Perkins, including the prediction of secondary structures of CCP modules from sequence alignment and averaging,[34] and low angle X-ray and neutron scattering to determine the low-resolution structure.[35] In this study, it was unexpectedly found that the batches of FH being examined were dimeric. It is still not clear why this was, as at physiological concentration and conditions, FH is monomeric.[8] However, attempting to resolve this problem led to some interesting findings. Janez Ferluga (unpublished results) found that FH is a very good substrate for plasma transglutaminase (FXIIIa) and FH is cross-linked readily by this enzyme, but to form large oligomers, not dimers. Other possible explanations for the dimerization include the findings that the His384 polymorphic variant may have a greater propensity to dimerization than the Tyr384 variant,[36] or that the dimerization may have resulted from Tyr-SO4 recognition. Variability in post-synthetic modification of FH has been studied for some time, *e.g.* in connection with properties of interacting with cells,[37] and FH can undergo tyrosine sulfation[38] (J. M. Shaw, unpublished). Bovine FH is reported to bind to Tyr-SO4,[39] and we are currently trying to show this for human FH. Perhaps FH molecules may form dimers by binding to Tyr-sulfated FH molecules.

After the early 1990s, we turned to work mainly on collectins, a topic which was funded by the British Lung Foundation and the Arthritis and Rheumatism Council. However, work on the ligand-binding properties of FH continued, in conjunction with studies on a homologue of FH, apolipoproten H (APOH; also called β2 glycoprotein 1), which consists of five CCP domains. APOH was cloned and sequenced[40] by Alexander Steinkasserer, and its properties of binding to anionic phospholipids (PLs) and as the antigen for autoantibodies in antiphospholipid syndrome were examined in a productive collaboration with John Jackson and Con Feighery (St James' Hospital, Dublin).[41] The fifth CCP domain of APOH was of particular interest as it mediates PL and autoantibody binding.[42,43] The function of APOH is uncertain, but it is thought to inhibit the initiation of coagulation by binding to exposed anionic PL. FH, however, has a very similar spectrum of binding to that of anionic PL.[43] Although APOH is known to lack the complement alternative pathway control function of FH, FH

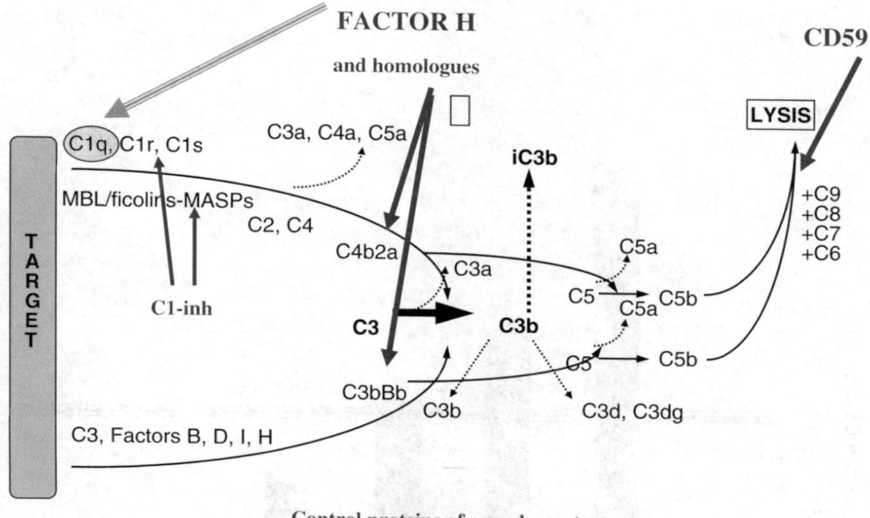

Figure 6.4 The complement system and its regulation. The figure shows the complement alternative, classical and lectin pathways. FH and its homologues C4bp, CR1, DAF and MCP regulate the C3 convertases. Here a new activity for FH is indicated – competition with C1q to control activation of the classical pathway. C1-inhibitor (C1-inh) is a serpin which inhibits C1r, C1s, MASP1 and MASP2. CD59 is a cell-surface protein which inhibits cell lysis by the C5b-9 (membrane attack) complex.

does have quantitatively identical coagulation inhibition properties to those of APOH[44] (and J. Ferluga, unpublished). The work on anionic PL led to some further interesting studies, by Bing bin Yu, Lee Aun Tan and Yu hoi (Jenny) Kang, on the functions of FH. C1q also binds to anionic PL (and activates the classical pathway), and FH competes for this binding.[45] Thus, with anionic PL, FH regulates the classical pathway activation, a role completely distinct from its accepted role as inhibitor of the alternative pathway (Figures 6.4 and 6.5). Anionic PLs, however, are not very interesting to complementologists as classical pathway activators, as there has been relatively little experimental work on this topic. *In vivo*, anionic PLs may be important as complement activators in tissue damage and apoptosis,[46] but this also has not been extensively explored. We looked at some well-established complement classical pathway activators, to see whether the same competition between C1q and FH could be observed. Competition was seen with binding to lipid A, whole *Escherichia coli*[47] and sheep erythrocytes sensitized with IgG or IgM (Y. H. Kang, unpublished). This suggests that FH has a general role in competing with C1q for binding to many (but not all) types of classical pathway activators, and is, in fact, an important regulator of the complement classical pathway as well as of the alternative pathway (Figure 6.5). Most recently, we showed that clotting of fibrinogen in plasma activates the classical pathway[48] (Figure 6.6) and this is

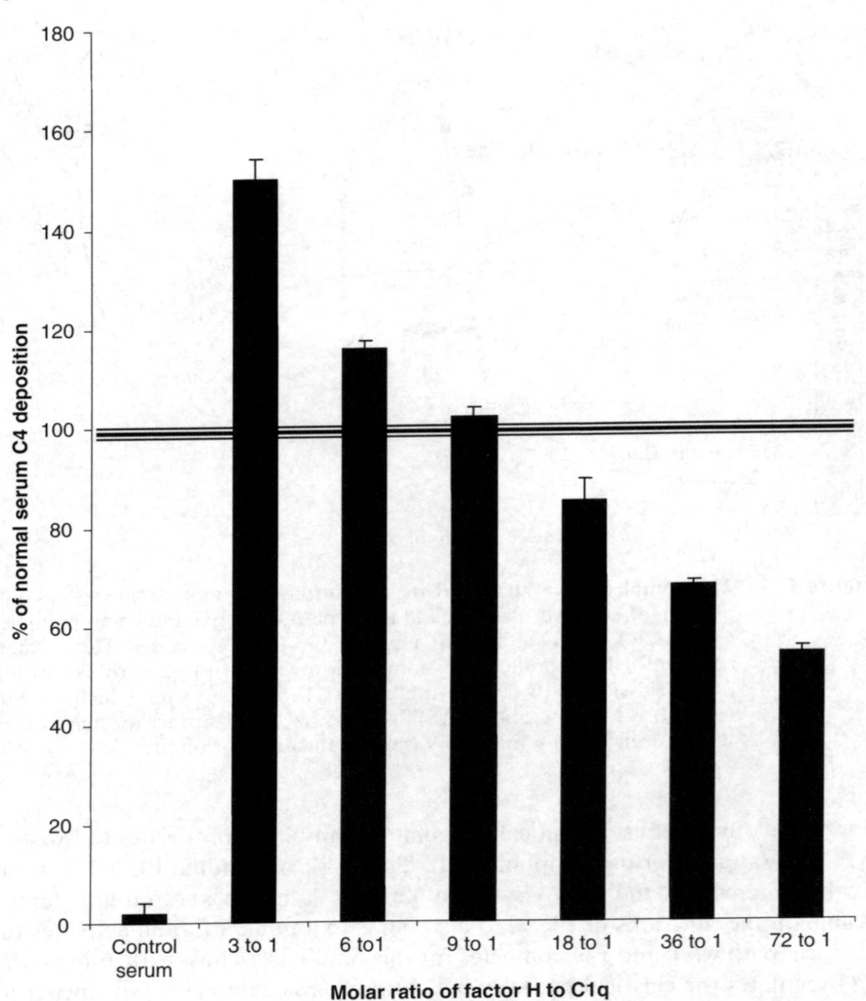

Figure 6.5 An example of C1q–FH balance determining the extent of complement activation. Here the complement activator, coated on microtitre plate wells, is cardiolipin (CL), the anionic PL. Serum was depleted of C1q and FH, then these proteins were added back at different molar ratios. Serum was incubated in the CL-coated wells, then C4b deposition in the wells was measured. C4b deposition is a measure of complement activation *via* C1q. As the quantity of FH increases relative to C1q, so the extent of activation diminishes. The physiological FH:C1q ratio is in range 6:1 to 25:1 (FH concentration varies quite widely in serum).

again influenced by C1q–FH competition. To return to a theme mentioned above, FXIIIa cross-links both FH and C1q into clots during coagulation (Figure 6.7).[48] These associations of FH with clotting require extensive further investigation, particularly in view of the occurrence of FH in platelets.

These new findings on FH stimulated us to take up further studies on the FH structure, to begin to explain why C1q and FH might bind to similar structures on common complement-activating targets. A decade after the earlier studies on FH structure had come to an end, Tony Day and Simon Clark made a recombinant CCP6,7,8 segment of FH. CCP7 was the main interest, as D. L. Gordon and colleagues had shown that this was a heparin-binding site.[49] Its binding properties were explored and its structure first modelled,[50] then determined by low angle scattering[51] and finally by crystallography in complex with a sulfated ligand.[52,53] We were fortunate that while this work was being done the linkage of the common 384 Tyr/His polymorphism (in CCP7) to AMD was published, so it was possible to synthesize the CCP6–8 construct in both the Tyr and His forms, and to compare their structures and ligand-binding properties. This led to the first demonstration that the Tyr/His substitution did have profound effects on binding to a range of ligands.[50] Further structural studies on FH–ligand complexes accompanied by more detailed functional studies may finally allow us to understand more satisfactorily the very diverse binding reactions of FH, for comparison with C1q binding (see G. J. Arlaud, Chapter 4).

FH in species other than human was also studied. Candi Soames obtained substantial cDNA sequence data on bovine FH.[54] At that time we were keen to use sequence comparisons to help us find out what regions of FH were important for interactions with C3b or FI, but sequence data were available only for human and mouse FH (and a little later, for rat). Many mammalian FHs were known to interact with human C3b and human FI, so sequence comparison was a valid means of refining candidate binding sites. Sequence alignments of human FH, mouse FH (which also interacts with human C3b) and bovine FH and comparison with CCP modules whose structures have been determined experimentally, were used to predict residues in the variable sequence loops of CCPs 2–5 and to identify residues of potential importance in human C3b binding and FI cofactor activity. Leu-17 and Gly-20 of CCP2, Ser-17, Ala-19, Glu-21, Asp-23 and Glu-25 of CCP3 and Lys-18 of CCP4 are all conserved between the three species.[54]

Ana Ferreira, Alvaro Diaz and Florencia Irigoin performed an outstanding series of analyses on the resistance to complement-mediated killing of *Echinococcus granulosus* (the parasite which forms hydatid cysts).[55,56] The cyst material used was extracted mainly from bovine tissue, and it was shown that the hydatid cyst sequesters host (bovine in this case) FH.[57] It is now known that many invading bacteria bind host FH, possibly as a way to resist complement attack (*e.g.* Neisseria[58]), but at the time of the study on *E. granulosus* only Streptococcal species were known to bind FH. The hydatid cyst study still seems to be the only demonstration of FH binding to a large multicellular parasite.

Purification of FH from other vertebrate species has been made a lot easier by observations, mainly from Beryl Moffatt, Bing-bin Yu and James Arnold, that cardiolipin and TNP- or DNP-derivatized macromolecules bind FH quite selectively from sera.[59,60] Binding to these ligands is another property in common with C1q.

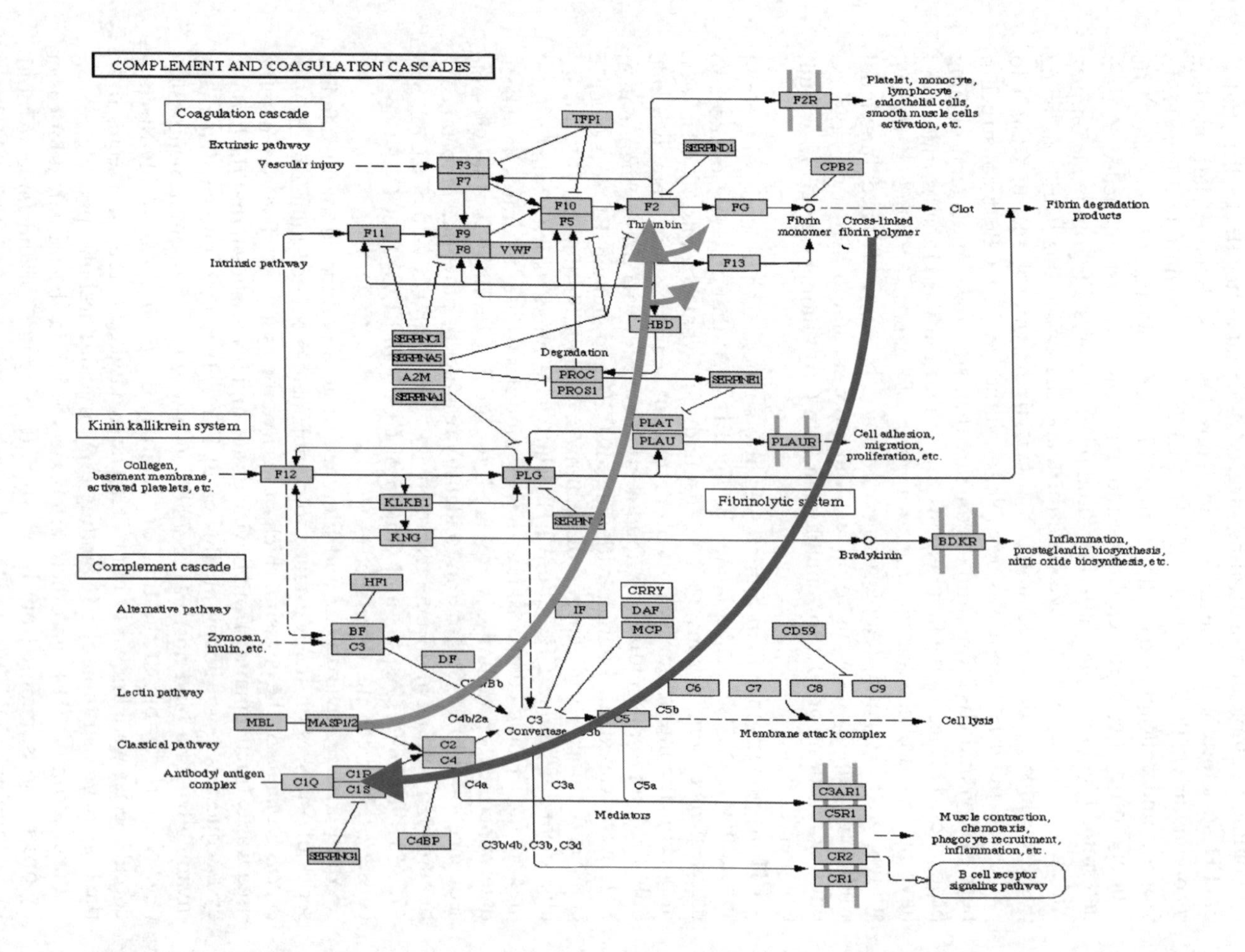
COMPLEMENT AND COAGULATION CASCADES
Coagulation cascade
Extrinsic pathway
Vascular injury
Intrinsic pathway
TFPI
F3
F7
F10
F5
F2
Thrombin
FG
F11
F9
F8
VWF
SERPIND1
F2R
Platelet, monocyte, lymphocyte, endothelial cells, smooth muscle cells activation, etc.
CPB2
Fibrin monomer
Cross-linked fibrin polymer
Clot
Fibrin degradation products
F13
THBD
SERPINC1
SERPINA5
A2M
SERPINA1
Degradation
PROC
PROS1
SERPINE1
PLAT
PLAU
PLAUR
Cell adhesion, migration, proliferation, etc.
Kinin kallikrein system
Collagen, basement membrane, activated platelets, etc.
F12
PLG
KLKB1
KNG
Fibrinolytic system
Bradykinin
BDKR
Inflammation, prostaglandin biosynthesis, nitric oxide biosynthesis, etc.
Complement cascade
HF1
Alternative pathway
Zymosan, inulin, etc.
BF
C3
DF
IF
CRRY
DAF
MCP
CD59
C6
C7
C8
C9
Lectin pathway
MBL
MASP1/2
C4b/2a
C3 Convertase
C5
C5b
Cell lysis
Membrane attack complex
Classical pathway
C2
C4
Antibody/ antigen complex
C1Q
C1R
C1S
C4a
C3a
C5a
Mediators
C3AR1
C5R1
CR2
CR1
Muscle contraction, chemotaxis, phagocyte recruitment, inflammation, etc.
B cell receptor signaling pathway
SERPING1
C4BP
C3b/4b, C3b, C3d

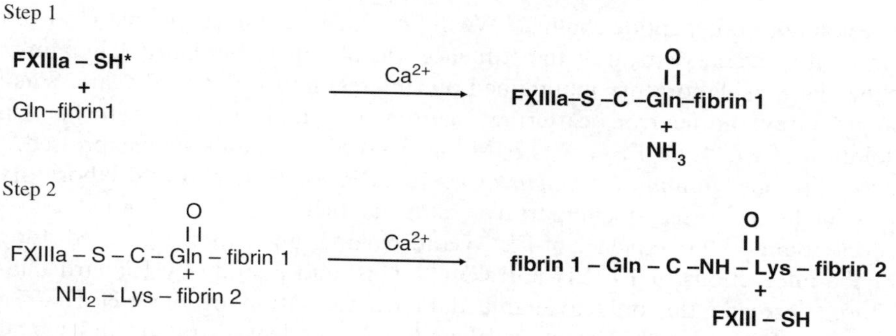

Step 1

$$\text{FXIIIa–SH*} + \text{Gln–fibrin1} \xrightarrow{Ca^{2+}} \text{FXIIIa–S–C(=O)–Gln–fibrin 1} + NH_3$$

Step 2

$$\text{FXIIIa–S–C(=O)–Gln–fibrin 1} + NH_2\text{–Lys–fibrin 2} \xrightarrow{Ca^{2+}} \text{fibrin 1–Gln–C(=O)–NH–Lys–fibrin 2} + \text{FXIII–SH}$$

Figure 6.7 FXIIIa cross-linking of fibrin monomers (denoted as fibrin 1, fibrin 2). In the first step, the active form of FXIII (FXIIIa-SH*) binds a glutamine residue of fibrin to form a thioester bond intermediate and releasing ammonia. Then, the enzyme-substrate complex interacts with a lysine residue on another fibrin molecule, producing an isopetide bond. Ca^{2+} is required for both steps of catalysis. Proteins other than fibrin can participate in this cross-linking – plasminogen, fibronectin, FH and C1q are examples.

6.3 Factor I

Factor I was first identified and studied under other names [Konglutinogen-activating factor (KAF) or C3b-inactivator or C3b/C4b inactivator], and by 1980 it was well-established as the protease which cleaved C3b to form iC3b, and C4b to form C4c plus C4d. The first isolation and characterization in the Immunochemistry Unit was done by Gale Crossley and Rod Porter.[61] Jean Gagnon, Limin Hsiung and J. M. Yuan[62] obtained substantial amino acid sequences for the two polypeptide chains, and the sequences around the active site. In the mid 1980s, we obtained the complete cDNA sequence, in a collaboration in which most of the work was done by Cath Catterall and Tim J. R. Harris in Celltech.[63] A monoclonal antibody against FI, MRCOX21, facilitated purification,[19] as FI is tedious to purify by conventional chromatography. FI was shown to have a multi-domain structure, as expected, with domain types called factor I module (FIM), scavenger receptor cysteine-rich (SRCR), low density lipoprotein receptor type A (LDLRA) in the heavy chain (50 kDa), which is disulfide linked to the light chain (35 kDa), which is a serine protease (SP) domain.[63,64] FI is heavily glycosylated, with three N-linkage glycosylation

---

Figure 6.6 A Kyoto Encyclopedia of Genes and Genomes (KEGG) diagram (http://www.genome.ad.jp/kegg/pathway/hsa/hsa04610.html) showing the complement and coagulation systems, modified to show two new sets of interactions. In one (upward arrow) the MASP1 and MASP2 proteases activate the coagulation system.[132] In the second (downward arrow) the formation of fibrin clots is a powerful stimulus for classical pathway activation.[48] As discussed in the text, both C1q and FH are covalently linked into clots by FXIIIa.

sites on both polypeptide chains.[65] We did not have resources (people!) to work on FI for several years after the sequence was obtained, but Chris Ullmann in Steve Perkins' laboratory maintained our interest in it for several years. Small angle X-ray and neutron scattering was done to establish the overall shape in solution of FI,[66] and FI and its FIM and LDLRA domains were expressed.[67] Serendipitous cloning of *Xenopus laevis* FI in R. A. B. Ezekowitz' laboratory provided much-needed comparative data, as there was very limited cross-species data on the sequence of FI.[64] Candi Soames undertook a detailed study of the interactions of FI, FH and C3b.[68] This, and a study by Richard DiScipio.[69] provide the only available data on formation of the FH–FI–C3b complex. Direct binding between FI and FH was demonstrated, and ligand blotting indicated that FH interacts with the heavy chain of FI. Similarly, direct C3b–FI and C3b–FH binding was characterized. Both FH and FI interact with both chains of C3b in ligand blotting. Binding reactions between all three pairs of components are highly dependent on ionic strength, and showed similar pH optima. Binding assays with all three components present led to the following conclusions:

(a) Binding sites for C3b and FI on FH do not overlap, and binding of FI and C3b to soluble FH promotes the weak FI–C3b interaction.
(b) Similarly, binding sites on FI for C3b and for FH do not overlap, and binding of FH and C3b to FI promotes direct FH–C3b interactions.[70]

FI is an unusual SP in that it circulates in blood in a cleaved (apparently active) form, is not inhibited by any of the plasma protein protease inhibitors and cleaves its soluble protein substrates, C3b and C4b, only when they are bound to a 'FI-cofactor' protein, such as FH or C4bp (see Figure 6.2). We are still uncertain whether FI is able to cleave these soluble substrates at a very low rate in the absence of a cofactor. Since FI has a very low catalytic activity, the tiniest traces of contaminant protease could confuse this issue. However, surface-bound C3b (*i.e.* C3b covalently bound to a complement activator like zymosan) can be cleaved by FI alone,[71] although a cofactor does accelerate cleavage. We were never able to observe reliably cleavage by FI of iC3b into C3dg and C3c, as is reported (at low ionic strength, perhaps using only CR1 as cofactor) in the literature, and we like to think that our FI is the purest available!

Stefanos Tsiftsoglou explored the enzymic properties of FI, and was able to show in the first fully documented study that FI does cleave synthetic substrates (tripeptide aminomethylcoumarins) without the need for a cofactor protein.[72] This was a very useful advance, as it allowed us to assay FI activity rapidly, without using the very slow assay of quantitation of C3b to iC3b conversion. It also confirmed that the enzyme is active without the need for conformational change induced by protein substrate or cofactor. Further studies with a proteolytic fragment of FI, consisting of its SP domain disulfide-linked to a short segment of the FI heavy chain showed that the SP domain ('alone') is enzymically active, and will cleave C3b very slowly, without need for a cofactor; however, it cleaves C3b less specifically (*i.e.* at more sites) than does intact FI in the presence of FH. Therefore, the cofactor restricts the specificity of the FI such that only two sites are cleaved in C3b to form one specific product, iC3b.[73]

FH does influence the enzymic activity of FI in another subtle way, which we have been able to observe only peripherally and non-physiologically, by considering pH optima. FI alone, in the absence of FH, has a pH optimum for the cleavage of tripeptide aminomethylcoumarin substrates of around pH 8.2, as is expected for serine proteases.[72] No cleavage is found at pH 5.5. However, the cleavage of C3b to form iC3b, by FI in the presence of FH, has a very unusual low pH optimum around pH 5.[74] As shown by Elisabetta Zulatto (E. Zulatto and R. B. Sim, unpublished) the addition of FH to a mixture of FI and tripeptide aminomethylcoumarin at low pH does allow the FI to cleave the substrate. Therefore, the interaction of FI with FH does influence the charge relay system of FI (presumably the protonation of the histidine), and therefore does cause conformational change in FI.

6.4 C4b-Binding Protein and Properdin

C4b-binding protein (C4bp), another protein with a similar domain structure and function to that of FH, was first isolated by Nagasawa and Stroud[75] as a cofactor for the breakdown of C3b by FI. In physiological conditions, however, it acts only as a cofactor for C4b breakdown by FI.[76] It consists of a covalent heptamer of 70 kDa polypeptide chains, each of which has eight CCP domains. Some molecules of C4bp have an additional 40 kDa chain (four CCPs) attached. Betty Press and Jean Gagnon[77] in the MRC Unit characterized the pattern of breakdown of C4b by C4bp, determining the N-terminal sequences and size of fragments formed as C4b is cleaved into iC4b then C4c plus C4d. L. P. Chung and Ken Reid worked on the amino acid sequencing of C4bp and 55% of the 70 kDa chain sequence was available in 1985,[78] followed rapidly by the cDNA sequence.[79,80] Jarmila Janatova established the disulfide linking pattern within the CCP domains of C4bp.[81] The laboratory subsequently made contributions to the cDNA sequencing of mouse C4bp.[82]

Properdin, a control protein which appears to act at least partly by stabilizing the alternative pathway convertase C3bBb, was first studied in the MRC Unit by Richard DiScipio, who isolated the protein and carried out hydrodynamic studies, showing that a trimeric as well as a tetrameric form of the protein occurs.[83] Binding interactions with C3b and FB were also examined quantitatively.[7] Ken Reid and Jean Gagnon published initial amino acid sequencing data on human properdin in 1981,[84] but it was not until 1988 that there were sufficient resources to resume the characterization of properdin. Goundis and Reid showed that properdin was made up of homologous domains, now named thrombospondin-like repeats (TSRs).[85] These are each 60 amino acids long, and there are six TSRs in the 50 kDa properdin polypeptide chain. The chains oligomerize to form mainly trimers and tetramers with some dimers, the low-resolution structures of which have been determined by Steve Perkins' laboratory.[86] In collaboration with Yvonne Boyd, the gene for human properdin was localized to Xp11.23-Xp21.1, and for mouse properdin to band A3 of the X chromosome.[87,88] Wider mapping around the human properdin gene was also undertaken.[89] Kath Nolan completed the cDNA sequence and subsequently with Jon Higgins and Dimitrios Goundis, the gene sequence for human properdin.[90–92] Using domain-deletion mutants, and the expression of single TSR domains, Jon Higgins, and later Michael Perdikoulis and Uday Kishore, determined the functions of parts of the sequence – TSR5, for example is the major C3b-binding site.[93,94] Structural aspects of the TSR domains were established in collaboration with Steve Perkins' laboratory.[95,96] Aspects of properdin deficiency were studied by Konrad Kolble,[97] and a method for carrier detection was developed.[98] In collaboration with Wilhelm Schwaeble, properdin was shown to be synthesized in neutrophils[99] and in T lymphocytes[100] and monocytes.[101]

6.5 Complement Receptors

In the early 1980s, we had optimized the production of C3 fragments (C3b, iC3b, C3dg, C3d and C3c) and these were used in affinity chromatography and cell-binding studies, by Vivek Malhotra, Kingsley Micklem and Edith Sim, to clarify the identity and specificity of C3 fragment receptors on leukocytes. This work led to the identification of the integrin p150,95 (CD11c, CD18) as an iC3b receptor.[102,103] It is a homologue of the integrin MAC1 (CD11b, CD18), which was already known as an iC3b receptor. These are sometimes called CR3 (MAC1) and CR4 (p150,95), although both of these integrins recognize many additional ligands which are not associated with complement. We also carried out structural and functional studies on CR1, 2, 3 and 4, either in house or in collaboration with Simon Gordon's laboratory (Sir William Dunn School of Pathology, Oxford).[71,104–108] Interestingly, we found that CR1 (and FH) do not distinguish between native, active C3 and its cleavage product, C3b; CR1 and FH bind equally well to both.[71,109] We did not report this in much detail, as it was against the dogma of the time, which was that CR1 and FH must recognize something which is present in C3b but not in C3, *i.e.* the signal for FH or CR1 binding is the activation of C3. However, this is not so. FH makes transient complexes with C3 and C3b; however, FI only acts on C3b. Thus FH carries out 'surveillance' of C3 in plasma, by sampling all the C3 present; if the C3b is activated, FI cleaves it to iC3b. If C3 is not activated, FH simply dissociates. For CR1 binding the signal is not activation of C3, but multiple presentation of C3, as occurs when several copies of C3b become fixed to a complement activator. At this time, several other groups, notably those of D. T. Fearon and G. D. Ross were very actively studying C3 receptors, and as we did not have the resources to follow either the sequence and structure of these receptors or their cell biology, we did not go further with C3 receptors.

Alain Sobel had studied C1q binding to cells in the 1970s, and reported that C1q bound to all human cell types tested except erythrocytes. The first studies of C1q receptor activity in the Immunochemistry Unit were done by Anna Erdei with Ken Reid.[110,111] Rajneesh Malhotra followed these up and isolated, in sufficient quantity for characterization, a protein from tonsil lymphocytes and from U937 cells which bound C1q[112,113] Anna Erdei's studies had shown (by ^{35}S amino acid incorporation) that this protein was made by the cells and was present on the cell surface (shown by surface radioiodination). Limited protein-sequence data showed the protein was closely related to a cDNA sequence identified by others, in error, as RoSSA antigen,[114] but was later shown to be calreticulin.[115] Calreticulin was not accepted by others as a potential C1q receptor at the time, as calreticulin had been characterized as a soluble (not membrane bound) intracellular protein possibly involved in calcium ion storage, or as a chaperone. Indeed, the reaction to our findings was surprisingly hostile, so we did not pursue them as we should have. The attention of researchers interested in C1q receptor activity switched to the more attractively named C1qRP (or CD93),[116] but it was never convincingly demonstrated that this receptor had any direct interaction with C1q, and it was eventually shown to be an adhesion receptor.[117] Returning to calreticulin, there were, even in 1993, many reports that it is found on the surface of different cells. Arosa *et al.*,[118] for example, showed it was bound to the T lymphocyte surface *via* major histocompatibility complex (MHC) class 1 heavy chain. Confirmation of the role of calreticulin as a receptor (or, rather, as an adaptor or co-receptor, binding C1q and linking it to a cell-surface protein), was provided by P. M. Henson's laboratory, who showed that calreticulin bound to the transmembrane protein CD91 (the α2macroglobulin receptor) acted as a C1q receptor, which mediated the uptake of C1q-coated particles into phagocytes.[119,120] The role of calreticulin-CD91 as a phagocytic uptake receptor seems now to be widely accepted (see, *e.g.* Takemura *et al.*,[121] who report adiponectin binding *via* calreticulin-CD91). However, it may be that calreticulin interacts with cell surfaces *via* several different molecules. CD91 has a limited cell distribution, while (as noted above) Sobel found that C1q bound to a very wide range of cell types. Arosa *et al.* reported cell-surface calreticulin associated with HLA class I heavy chain,[118] which could certainly occur on most cells, and calreticulin binding to cell surfaces *via* CD59, or the scavenger receptors SREC and SR-A, has also been reported.[122]

A major point of interest in our studies on C1q receptor was the demonstration that it (calreticulin) bound not only C1q, but to other structurally similar, collagen-containing molecules, namely the collectins, SP-A, MBL and conglutinin.[123–125] We were surprised by this finding, as we had originally tested the collectins with the idea that they might be good negative controls for C1q-binding specificity. We did not detect SP-D binding at the time, but this has since been shown by others, and adiponectin, another protein with a quaternary structure like that of C1q, also binds.[121] The region of calreticulin and of the collagen segments of the ligands responsible for the binding was defined, in collaboration with Wilhelm Schwaeble and Steffen Thiel.[126–128] Recent papers on the roles of collectins and C1q in the processing of apoptotic cells are generally consistent with the existence of a single 'collectin receptor', which also binds C1q and adiponectin. Other receptor activity for these proteins may well exist. Soren Moestrup, for example, showed that megalin, a protein in the same family as CD91, acts as a C1q receptor (but not a receptor for the collectins) and might mediate uptake and lysosomal degradation of C1q-bearing particles.[115] Megalin, however, has a limited tissue distribution, and is obviously not a major mediator of C1q-dependent uptake into cells. Other recent reports suggest a role for an integrin as a collectin receptor.[129]

6.6 Complement Research into its Third Century

Complement activity was first described in the 1890s, and great progress was made up to the mid 1930s in defining components and mechanisms (*e.g.* the inactivation of C4 by ammonia was known by 1924). In the 1950s and 1960s, immune adherence was explored, the alternative pathway defined and the components C1–C9 of the classical pathway were separated. Ever since the MRC Unit started work on complement, it has been predicted that discoveries in complement were 'over' and that there was nothing more useful to be learned. At first, this was largely because the concept of redundancy was not understood (*i.e.* that several molecules or molecular systems could operate physiologically to achieve similar results). Thus, for example, it was often said that since many human complement deficiencies were symptomless, complement could not be critical for survival. By the mid 1990s, however, when most of the soluble proteins and receptors known at the time had been cloned, sequenced and expressed, and funding for complement research in the USA had diminished, there was perhaps a more justified feeling that a twilight zone was approaching. However, the new findings on mannan-binding lectins (MBLs), MBL-associated serine proteinases (MASPs), and ficolins, potential roles for complement in the clearance of apoptotic cells and autoimmunity, and the recent findings of complement polymorphisms in AMD have emphasized how much remains to be explored in terms of the detailed mechanisms and the integration of our knowledge of molecules and structures into the higher order complexity of cells, tissues and whole organisms. Recent findings on complement roles in the central nervous system (to which we have already had some input[130,131]) will provide a major boost for complement research over the next decade.

Acknowledgements

R. B. Sim thanks Vivek Malhotra, Tony Day, Marcia McAleer, Rajneesh Malhotra, Konrad Kolble, Val Dee, Samantha Williams, Candi Soames, Alvaro Diaz, Guy Stuart, Tim Hickling, Bing bin Yu, Mayumi Kojima, Lee Aun Tan, Julia Presanis, Stefanos Tsiftsoglou, James Arnold, Simon Clark, Yu hoi (Jenny) Kang, Carolina Salvador-Morales, Anders Krarup and Maria Carroll, all former or current postgraduate students, and also all the visitors and postdoctoral researchers in the laboratory for their energy, innovation and hard work, and for making it a great place to work in.

References

1. U. R. Nilsson and H. J. Mueller-Eberhard, *J. Exp. Med.*, 1965, **122**, 277–298.
2. K. Whaley and S. Ruddy, *J. Exp. Med.*, 1976, **144**, 1147–1163.
3. K. Whaley and S. Ruddy, *Science*, 1976, **193**, 1011–1013.
4. J. M. Weiler, M. R. Daha, K. F. Austen and D. T. Fearon, *Proc. Natl. Acad. Sci. U.S.A.*, 1976, **73**, 3268–3272.
5. E. Sim, A. B. Wood, L. M. Hsiung and R. B. Sim, *FEBS Letters*, 1981, **132**, 55–60.
6. E. Sim and R. B. Sim, *Biochem. J.*, 1981, **198**, 509–518.
7. R. G. DiScipio, *Biochem. J.*, 1981, **199**, 485–496.
8. R. B. Sim and R. G. DiScipio, *Biochem. J.*, 1982, **205**, 285–293.
9. R. B. Sim, V. Malhotra, J. Ripoche, A. J. Day, K. J. Micklem and E. Sim, *Biochem. Soc. Symp.*, 1986, **51**, 83–96.
10. J. Ripoche, A. J. Day, A. C. Willis, K. T. Belt, R. D. Campbell and R. B. Sim, *Biosci. Rep.*, 1986, **6**, 65–72.
11. K. B. M. Reid, D. R. Bentley, R. D. Campbell, L. P. Chung, R. B. Sim, T. Kristensen and B. F. Tack, *Immunology Today*, 1986, **7**, 230–234.
12. A. J. Day, J. Ripoche, A. Lyons, B. McIntosh, T. J. Harris and R. B. Sim, *Biosci. Rep.*, 1987, **7**, 201–207.
13. B. J. Morley and R. D. Campbell, *EMBO J.*, 1984, **3**, 153–157.
14. J. Ripoche, A. J. Day, B. Moffatt and R. B. Sim, *Biochem. Soc. Trans.*, 1987, **15**, 651–652.
15. J. Ripoche, A. J. Day, T. J. R. Harris and R. B. Sim, *Biochem. J.*, 1988, **249**, 593–602.
16. A. J. Day, A. C. Willis, J. Ripoche and R. B. Sim, *Immunogenetics*, 1988, **27**, 211–214.
17. A. O. Edwards, R. Ritter, 3rd, K. J. Abel, A. Manning, C. Panhuysen and L. A. Farrer, *Science*, 2005, **308**, 421–424.
18. E. Sim, M. S. Palmer, M. Puklavec and R. B. Sim, *Biosci. Rep.*, 1983, **3**, 1119–1131.
19. L. Hsiung, A. N. Barclay, M. R. Brandon, E. Sim and R. R. Porter, *Biochem. J.*, 1982, **203**, 293–298.
20. J. Alsenz, T. F. Schulz, J. D. Lambris, R. B. Sim and M. P. Dierich, *Biochem. J.*, 1985, **232**, 841–850.
21. R. B. Sim, A. J. Day, B. E. Moffatt and M. Fontaine, *Methods Enzymol.*, 1993, **223**, 13–35.
22. S. Hing, A. J. Day, S. J. Linton, J. Ripoche, R. B. Sim, K. B. M. Reid and E. Solomon, *Ann. Human Genet.*, 1988, **52**, 117–122.
23. J. L. Williams, D. H. Lester, V. M. Teres, W. Barendse, R. B. Sim and C. J. Soames, *Mamm. Genome*, 1997, **8**, 77–78.

24. V. M. Dee, A. Tse, A. C. Willis and R. B. Sim, *Complement and Inflammation*, 1990, **7**, 158–159.
25. M. A. McAleer, J. Ripoche, A. J. Day, B. E. Moffatt, M. Fontaine and R. B. Sim, *Complement and Inflammation*, 1989, **6**, 366–367.
26. M. Fontaine, M. J. Demares, V. Koistinen, A. J. Day, C. Davrinche, R. B. Sim and J. Ripoche, *Biochem. J.*, 1989, **258**, 927–930.
27. M. A. McAleer, G. Hauptmann, M. Brai, G. Misiano and R. B. Sim, *Complement*, 1987, **4**, 191–191.
28. E. H. Weiss, M. C. Jung, U. Spengler, G. Pape, A. Steinkasserer and R. B. Sim, *Mol. Immunol.*, 1993, **30**, 62–62.
29. W. Schwaeble, C. M. Stover, S. Hanson, E. H. Weiss and R. B. Sim, *Immunopharmacology*, 2000, **49**, 13–13.
30. P. N. Barlow, M. Baron, D. G. Norman, A. J. Day, A. C. Willis, R. B. Sim and I. D. Campbell, *Biochemistry*, 1991, **30**, 997–1004.
31. D. G. Norman, P. N. Barlow, M. Baron, A. J. Day, R. B. Sim and I. D. Campbell, *J. Mol. Biol.*, 1991, **219**, 717–725.
32. P. N. Barlow, D. G. Norman, A. Steinkasserer, T. J. Horne, J. Pearce, P. C. Driscoll, R. B. Sim and I. D. Campbell, *Biochemistry*, 1992, **31**, 3626–3634.
33. P. N. Barlow, A. Steinkasserer, D. G. Norman, B. Kieffer, A. P. Wiles, R. B. Sim and I. D. Campbell, *J. Mol. Biol.*, 1993, **232**, 268–284.
34. S. J. Perkins, P. I. Haris, R. B. Sim and D. Chapman, *Biochemistry*, 1988, **27**, 4004–4012.
35. S. J. Perkins, A. S. Nealis and R. B. Sim, *Biochemistry*, 1991, **30**, 2847–2857.
36. A. N. Fernando, P. B. Furtado, S. J. Clark, H. E. Gilbert, A. J. Day, R. B. Sim and S. J. Perkins, *J. Mol. Biol.*, 2007, **368**, 564–581.
37. J. Ripoche, A. Erdei, D. Gilbert, A. Al Salihi, R. B. Sim and M. Fontaine, *Biochem. J.*, 1988, **253**, 475–480.
38. R. Malhotra, M. Ward, R. B. Sim and M. I. Bird, *Biochem. J.*, 1999, **341**, 61–69.
39. Y. Sakakibara, M. Suiko, P. H. Fernando, T. Ohashi and M. C. Liu, *Cytotechnology*, 1994, **14**, 97–107.
40. A. Steinkasserer, C. Estaller, E. H. Weiss, R. B. Sim and A. J. Day, *Biochem. J.*, 1991, **277**, 387–391.
41. J. Guerin, C. Feighery, R. B. Sim and J. Jackson, *Clin. Exp. Immunol.*, 1997, **109**, 304–309.
42. A. Steinkasserer, P. N. Barlow, A. C. Willis, Z. Kertesz, I. D. Campbell, R. B. Sim and D. G. Norman, *FEBS Lett.*, 1992, **313**, 193–197.
43. Z. Kertesz, B. B. Yu, A. Steinkasserer, H. Haupt, A. Benham and R. B. Sim, *Biochem. J.*, 1995, **310**, 315–321.
44. J. Ferluga, B. Yu, J. Guerin, J. Jackson and R. B. Sim, *Mol. Immunol.*, 1998, **35**, 375–375.
45. L. A. Tan, U. Kishore, B. B. Yu and R. B. Sim, *submitted*, 2008.
46. H. Paidassi, P. Tacnet-Delorme, V. Garlatti, C. Darnault, B. Ghebrehiwet, C. Gaboriaud, G. J. Arlaud and P. Frachet, *J. Immunol.*, 2008, **180**, 2329–2338.

47. L. A. Tan, U. Kishore, A. Yang and R. B. Sim, *submitted*, 2008.
48. Y. H. Kang, C. O'Kane and R. B. Sim, *submitted*, 2008.
49. R. J. Ormsby, T. S. Jokiranta, T. G. Duthy, K. M. Griggs, T. A. Sadlon, E. Giannakis and D. L. Gordon, *Mol. Immunol.*, 2006, **43**, 1624–1632.
50. S. J. Clark, V. A. Higman, B. Mulloy, S. J. Perkins, S. M. Lea, R. B. Sim and A. J. Day, *J. Biol. Chem.*, 2006, **281**, 24713–24720.
51. A. Fernando, S. Clark, R. B. Sim and S. J. Perkins, *Mol. Immunol.*, 2006, **43**, 143–143.
52. B. E. Prosser, S. Johnson, P. Roversi, S. J. Clark, E. Tarelli, R. B. Sim, A. J. Day and S. M. Lea, *Acta. Crystallogr Sect. F. Struct. Biol. Cryst. Commun.*, 2007, **63**, 480–483.
53. B. E. Prosser, S. Johnson, P. Roversi, A. P. Herbert, B. S. Blaum, J. Tyrrell, T. A. Jowitt, S. J. Clark, E. Tarelli, D. Uhrin, P. N. Barlow, R. B. Sim, A. J. Day and S. M. Lea, *J. Exp. Med.*, 2007, **204**, 2277–2283.
54. C. J. Soames, A. J. Day and R. B. Sim, *Biochem. J.*, 1996, **315**, 523–531.
55. A. Diaz, F. Irigoin, F. Ferreira and R. B. Sim, *Immunopharmacology*, 1999, **42**, 91–98.
56. A. M. Ferreira, F. Irigoin, M. Breijo, R. B. Sim and A. Diaz, *Parasitol Today*, 2000, **16**, 168–172.
57. A. Diaz, A. Ferreira and R. B. Sim, *J. Immunol.*, 1997, **158**, 3779–3786.
58. M. C. Schneider, R. M. Exley, H. Chan, I. Feavers, Y. H. Kang, R. B. Sim and C. M. Tang, *J. Immunol.*, 2006, **176**, 7566–7575.
59. B. B. Yu, B. E. Moffatt, A. C. Willis and R. B. Sim, *Exp. Clin. Immunogenetics*, 1997, **14**, 39.
60. J. N. Arnold, M. R. Wormald, D. M. Suter, C. M. Radcliffe, D. J. Harvey, R. A. Dwek, P. M. Rudd and R. B. Sim, *J. Biol. Chem.*, 2005, **280**, 29080–29087.
61. L. G. Crossley and R. R. Porter, *Biochem. J.*, 1980, **191**, 173–182.
62. J. M. Yuan, L. M. Hsiung and J. Gagnon, *Biochem. J.*, 1986, **233**, 339–345.
63. C. F. Catterall, A. Lyons, R. B. Sim, A. J. Day and T. J. Harris, *Biochem. J.*, 1987, **242**, 849–856.
64. L. M. Kunnath-Muglia, G. H. Chang, R. B. Sim, A. J. Day and R. A. Ezekowitz, *Mol. Immunol.*, 1993, **30**, 1249–1256.
65. S. A. Tsiftsoglou, J. N. Arnold, P. Roversi, M. D. Crispin, C. Radcliffe, S. M. Lea, R. A. Dwek, P. M. Rudd and R. B. Sim, *Biochim. Biophys. Acta.*, 2006, **1764**, 1757–1766.
66. S. J. Perkins, K. F. Smith and R. B. Sim, *Biochem. J.*, 1993, **295**, 101–108.
67. C. G. Ullman, D. Chamberlain, A. Ansari, V. C. Emery, P. I. Haris, R. B. Sim and S. J. Perkins, *Mol. Immunol.*, 1998, **35**, 503–512.
68. C. J. Soames and R. B. Sim, *Biochem. J.*, 1997, **326**, 553–561.
69. R. G. DiScipio, *J. Immunol.*, 1992, **149**, 2592–2599.
70. C. J. Soames and R. B. Sim, *Biochem. Soc. Trans.*, 1995, **23**, 53S.
71. V. Malhotra and R. B. Sim, *Biochem. Soc. Trans.*, 1984, **12**, 781–782.
72. S. A. Tsiftsoglou and R. B. Sim, *J. Immunol.*, 2004, **173**, 367–375.

73. S. A. Tsiftsoglou, A. C. Willis, P. Li, X. Chen, D. A. Mitchell, Z. Rao and R. B. Sim, *Biochemistry*, 2005, **44**, 6239–6249.
74. E. Sim and R. B. Sim, *Biochem. J.*, 1983, **210**, 567–576.
75. S. Nagasawa and R. M. Stroud, *Immunochemistry*, 1977, **14**, 749–756.
76. E. Sim, A. B. Wood, L. M. Hsiung and R. B. Sim, *FEBS Lett.*, 1981, **132**, 55–60.
77. E. M. Press and J. Gagnon, *Biochem. J.*, 1981, **199**, 351–357.
78. L. P. Chung, J. Gagnon and K. B. Reid, *Mol. Immunol.*, 1985, **22**, 427–435.
79. L. P. Chung, D. R. Bentley and K. B. Reid, *Biochem. J.*, 1985, **230**, 133–141.
80. S. J. Lintin, A. R. Lewin and K. B. Reid, *FEBS Lett.*, 1988, **232**, 328–332.
81. J. Janatova, K. B. Reid and A. C. Willis, *Biochemistry*, 1989, **28**, 4754–4761.
82. T. Kristensen, R. T. Ogata, L. P. Chung, K. B. Reid and B. F. Tack, *Biochemistry*, 1987, **26**, 4668–4674.
83. R. G. Discipio, *Mol. Immunol.*, 1982, **19**, 631–635.
84. K. B. Reid and J. Gagnon, *Mol. Immunol.*, 1981, **18**, 949–959.
85. D. Goundis and K. B. Reid, *Nature*, 1988, **335**, 82–85.
86. Z. Sun, K. B. Reid and S. J. Perkins, *J. Mol. Biol.*, 2004, **343**, 1327–1343.
87. D. Goundis, S. M. Holt, Y. Boyd and K. B. Reid, *Genomics*, 1989, **5**, 56–60.
88. E. P. Evans, M. D. Burtenshaw, S. H. Laval, D. Goundis, K. B. Reid and Y. Boyd, *Genet. Res.*, 1990, **56**, 153–155.
89. M. P. Coleman, J. C. Murray, H. F. Willard, K. F. Nolan, K. B. Reid, D. J. Blake, S. Lindsay, S. S. Bhattacharya, A. Wright and K. E. Davies, *Genomics*, 1991, **11**, 991–996.
90. K. F. Nolan and K. B. Reid, *Biochem. Soc. Trans.*, 1990, **18**, 1161–1162.
91. K. F. Nolan, W. Schwaeble, S. Kaluz, M. P. Dierich and K. B. Reid, *Eur. J. Immunol.*, 1991, **21**, 771–776.
92. K. F. Nolan, S. Kaluz, J. M. Higgins, D. Goundis and K. B. Reid, *Biochem. J.*, 1992, **287**, 291–297.
93. J. M. Higgins, H. Wiedemann, R. Timpl and K. B. Reid, *J. Immunol.*, 1995, **155**, 5777–5785.
94. M. V. Perdikoulis, U. Kishore and K. B. Reid, *Biochim. Biophys. Acta.*, 2001, **1548**, 265–277.
95. S. J. Perkins, A. S. Nealis, P. I. Haris, D. Chapman, D. Goundis and K. B. Reid, *Biochemistry*, 1989, **28**, 7176–7182.
96. K. F. Smith, K. F. Nolan, K. B. Reid and S. J. Perkins, *Biochemistry*, 1991, **30**, 8000–8008.
97. K. Kolble and K. B. Reid, *Int. Rev. Immunol.*, 1993, **10**, 17–36.
98. K. Kolble, A. J. Cant, A. C. Fay, K. Whaley, M. Schlesinger and K. B. Reid, *J. Clin. Invest.*, 1993, **91**, 99–102.
99. U. Wirthmueller, B. Dewald, M. Thelen, M. K. Schafer, C. Stover, K. Whaley, J. North, P. Eggleton, K. B. Reid and W. J. Schwaeble, *J. Immunol.*, 1997, **158**, 4444–4451.

100. W. Schwaeble, W. G. Dippold, M. K. Schafer, H. Pohla, D. Jonas, B. Luttig, E. Weihe, H. P. Huemer, M. P. Dierich and K. B. Reid, *J. Immunol.*, 1993, **151**, 2521–2528.
101. W. Schwaeble, H. P. Huemer, J. Most, M. P. Dierich, M. Strobel, C. Claus, K. B. Reid and H. W. Ziegler-Heitbrock, *Eur. J. Biochem.*, 1994, **219**, 759–764.
102. K. J. Micklem and R. B. Sim, *Biochem. J.*, 1985, **231**, 233–236.
103. V. Malhotra, N. Hogg and R. B. Sim, *Eur. J. Immunol.*, 1986, **16**, 1117–1123.
104. K. J. Micklem, R. B. Sim and E. Sim, *Biochem. J.*, 1984, **224**, 75–86.
105. K. Micklem, E. Sim and R. B. Sim, *FEBS Lett.*, 1985, **189**, 195–201.
106. R. A. Ezekowitz, R. B. Sim, G. G. MacPherson and S. Gordon, *J. Clin. Invest.*, 1985, **76**, 2368–2376.
107. J. M. Blackwell, R. A. Ezekowitz, M. B. Roberts, J. Y. Channon, R. B. Sim and S. Gordon, *J. Exp. Med.*, 1985, **162**, 324–331.
108. R. B. Sim, *Biochem. J.*, 1985, **232**, 883–889.
109. E. Sim and R. B. Sim, *Biochem. J.*, 1981, **198**, 509–518.
110. A. Erdei and K. B. Reid, *Biochem. J.*, 1988, **255**, 493–499.
111. A. Erdei and K. B. Reid, *Mol. Immunol.*, 1988, **25**, 1067–1073.
112. R. Malhotra, K. B. M. Reid and R. B. Sim, *Biochem. Soc. Trans.*, 1988, **16**, 735–736.
113. R. Malhotra and R. B. Sim, *Biochem. J.*, 1989, **262**, 625–631.
114. R. Malhotra, A. C. Willis, J. C. Jensenius, J. Jackson and R. B. Sim, *Immunology*, 1993, **78**, 341–348.
115. R. B. Sim, S. K. Moestrup, G. R. Stuart, N. J. Lynch, J. H. Lu, W. J. Schwaeble and R. Malhotra, *Immunobiology*, 1998, **199**, 208–224.
116. P. Eggleton, K. B. Reid and A. J. Tenner, *Trends Cell Biol.*, 1998, **8**, 428–431.
117. E. P. McGreal, N. Ikewaki, H. Akatsu, B. P. Morgan and P. Gasque, *J. Immunol.*, 2002, **168**, 5222–5232.
118. F. A. Arosa, O. de Jesus, G. Porto, A. M. Carmo and M. de Sousa, *J. Biol. Chem.*, 1999, **274**, 16917–16922.
119. C. A. Ogden, A. deCathelineau, P. R. Hoffmann, D. Bratton, B. Ghebrehiwet, V. A. Fadok and P. M. Henson, *J. Exp. Med.*, 2001, **194**, 781–795.
120. R. W. Vandivier, C. A. Ogden, V. A. Fadok, P. R. Hoffmann, K. K. Brown, M. Botto, M. J. Walport, J. H. Fisher, P. M. Henson and K. E. Greene, *J. Immunol.*, 2002, **169**, 3978–3986.
121. Y. Takemura, N. Ouchi, R. Shibata, T. Aprahamian, M. T. Kirber, R. S. Summer, S. Kihara and K. Walsh, *J. Clin. Invest.*, 2007, **117**, 375–386.
122. J. J. Walters and B. Berwin, *Traffic*, 2005, **6**, 1173–1182.
123. R. Malhotra, S. Thiel, K. B. Reid and R. B. Sim, *J. Exp. Med.*, 1990, **172**, 955–959.
124. R. Malhotra, R. B. Sim and K. B. Reid, *Biochem. Soc. Trans.*, 1990, **18**, 1145–1148.
125. R. Malhotra, J. Lu, U. Holmskov and R. B. Sim, *Clin. Exp. Immunol.*, 1994, **97**(Suppl 2), 4–9.

126. G. R. Stuart, N. J. Lynch, J. Lu, A. Geick, B. E. Moffatt, R. B. Sim and W. J. Schwaeble, *FEBS Lett.*, 1996, **397**, 245–249.
127. G. R. Stuart, N. J. Lynch, A. J. Day, W. J. Schwaeble and R. B. Sim, *Immunopharmacology*, 1997, **38**, 73–80.
128. R. Malhotra, S. B. Laursen, A. C. Willis and R. B. Sim, *Biochem. J.*, 1993, **293**, 15–19.
129. M. M. Zutter and B. T. Edelson, *Immunobiology*, 2007, **212**, 343–353.
130. M. K. H. Schafer, W. J. Schwaeble, C. Post, P. Salvati, M. Calabresi, R. B. Sim, F. Petry, M. Loos and E. Weihe, *J. Immunol.*, 2000, **164**, 5446–5452.
131. K. R. Mayilyan, D. R. Weinberger and R. B. Sim, *Drug News Perspect.*, 2008, **21**, in press.
132. A. Krarup, R. Wallis, J. S. Presanis, P. Gal and R. B. Sim, *PLoS ONE*, 2007, **2**, e623.

CHAPTER 7

Biology and Genetics of Complement C4

MICHAEL C. CARROLL

Immune Disease Institute, Harvard Medical School, Boston, MA 02115, USA

7.1 Introduction

This review is organized in two parts. In the first part, I attempt to highlight my initial efforts to clone complement component C4 while a post-doctoral fellow with Professor Rodney R. Porter in the MRC Immunochemistry Unit and the Biochemistry Department at Oxford University from 1980 to 1985. The cloning of human C4 not only provided its protein sequence, but revealed the structural basis for its complicated genetics. Although unanticipated at the time, cloning of C4 also paved the way for our later functional studies in mice in Boston using gene targeting in embryonic stem cell technology. The second part of the review focusses on the biology of C4 and our continuing efforts to understand why humans deficient in C4 almost always develop systemic lupus erythematosus (SLE).

7.2 Early Days in Oxford

I first met Professor Porter ('Prof.') in the home of my PhD mentor, Don Capra, on the Prof.'s visit to UT Southwestern Medical School in Dallas, TX. At the time, I was struggling with my thesis project to purify mouse complement C4. As I explained the complication of degradation, he quickly pointed out that I needed to use a strong irreversible protease inhibitor such as diisopropylfluorophosphate (DFP). This solved the problem and resulted in my first manuscript that he

Molecular Aspects of Innate and Adaptive Immunity
Edited By Kenneth BM Reid and Robert B Sim

Published by the Royal Society of Chemistry, www.rsc.org

graciously communicated to the *Proceedings of the National Academy of Sciences.*[1] This was in the mid-70s and given my desire to live abroad, I asked him for a position in his laboratory as a post-doctoral fellow when I completed my thesis. Fortunately, he approved and, with a fellowship from the American Arthritis Foundation, I travelled to Oxford to begin my training in the summer of 1980.

On my arrival, Prof. suggested that I work on complement C3 receptors as this was a new and exciting topic in the field of complement research. The project was most reasonable as Prof. Porter and his MRC Immunochemistry Unit at Oxford were internationally recognized for their expertise in the biochemistry of complement proteins (and immunoglobulin). However, I had set my mind on another project, namely to clone complement C4. At the time cloning of mammalian genes was relatively new and the tools of molecular biology were very limited. However, I had made a few attempts to isolate RNA from mouse liver while still in Dallas and was intrigued by the prospects of cloning complement genes.

I am forever grateful for Prof. Porter's encouragement and flexibility in allowing me to proceed with this project. Given the biochemistry focus of the Unit at the time, he could have easily required that I work on a project more in line with other members of the unit. Instead, he assigned me a small laboratory in the South Parks Road Biochemistry building (room 26) just around the corner from his office. This laboratory was to be my home for all of my working hours for the next 12 months. Here, I learned (with kind advice and encouragement from Sue and Alan Kingsman) to isolate RNA from mouse liver and translate it into C4 protein. This was a trying but exciting period, and I succeeded only with support from members of the Unit and the Prof. who would often stop by the laboratory at the end of the day to check on my progress.

Near the end of my first year, Prof. suggested that I visit George Brownlee, the newly appointed Professor of Chemical Pathology at the Dunn School of Pathology. After explaining my efforts to synthesize and clone cDNA encoding complement C4, Professor Brownlee offered me a small corner in his laboratory. His group focussed on a number of exciting projects, including development of cloning and screening procedures for mammalian cDNA. With the approval of Prof. Porter, I moved into the Chemical Pathology Department where I would work for the next two years. This was an exciting period as new techniques and reagents for molecular biology were being rapidly developed, many by the Brownlee group. One of the more important technologies set up in the lab was the chemical synthesis of oligonucleotides to screen cDNA clones. As protein sequence data on human C4 were available from the Unit, this approach provided a specific probe to identify clones of C4 cDNA. With advice and a lot of help from members of the Brownlee group, I synthesized and screened a modest cDNA library from human liver RNA. Several clones hybridized positive with the C4-specific oligonucleotide and were confirmed as encoding human C4 by DNA sequence analysis. These results represented the first cloned human complement gene.[2]

To accelerate our progress, Prof. assigned one of his talented biochemistry students, Tertia Belt, to characterize several full-length C4 cDNA clones and this work led to the complete primary sequence of human C4.[3] This sequence characterization also led to the identification of both isotypic and allelic differences that had puzzled geneticists for years.

With the successful cloning of C4, I had the tools to begin dissection of the genetics of C4. About this time members of the MRC Immunochemistry Unit were involved in cloning other complement genes and the procedures were becoming more standardized. As the nature of the problems changed more to genetics, Ken Reid kindly offered bench space in his laboratory in the Unit on the fourth floor of the biochemistry building. Given the location of the human C4 genes within the human leukocyte antigen (HLA) region and evidence of genetic association with certain diseases, especially SLE, there was much interest in understanding the genetics of the apparent isotype and allotype differences. At this point I sought help from Professor John Edwards, Head of Genetics. His group was interested in human genetics in general and, in particular, understanding genes associated with disease. He offered to collaborate and provided much needed advice on human genetics. From these discussions it became clear that we needed a molecular map of the C4 and surrounding HLA region to better understand the genetics. To explore the C4 genetics, one of his talented graduate students, Asta Palsdottir, offered to work jointly between our group in the MRC Unit and his group in genetics. Interestingly, one novel observation we made was to identify a restriction length polymorphism linked to the inactive C4A6 allele.[4]

Our early results from restriction endonuclease analysis with C4 DNA probes yielded confusing results, in part because of the large size of the genes, approximately 20 kbp each, and the high level of sequence identity among the two isotypes and alleles. At this point, I visited the laboratory of Dr Frank Grosveld at Mill Hill who kindly offered to help me prepare cosmid libraries of genomic DNA extracted from the liver DNA of HLA-typed individuals. With the technical advice of Victor Kouissis I was successful in preparing a cosmid library that contained inserts of approximately 40 kbp. This was a breakthrough as it allowed us to identify clones that included the complete sequence of C4. The clones also included intact promoters and the necessary sequence for expression in cell lines, as we later did in Boston. Importantly, the cosmid clones allowed us to prepare highly detailed restriction maps of the C4 locus that could be used as a basis for typing blood cells of individual donors of known HLA class I and II type.

One of the enigmas at the time was the number of genes in the C4 locus. Genetic studies suggested that C4 was duplicated and that there were two isotypes, C4A and C4B. However, there was variability among individuals. By isolating overlapping clones from the genomic library of known HLA and C4 type, we determined that most individuals had two closely linked C4 genes separated by approximately 10 kbp. We later found that this intervening region included other genes, such as 21-hydroxylase.[5,6] Interestingly, the donor from which the library was prepared had three C4 genes. Restriction mapping of

DNA from donors suggested that the C4 locus underwent relatively recent recombination and that about 10% of the population had only one C4 locus, while a small percentage had three C4 genes. Critical to distinguishing C4A from C4B isotypic differences was the published work of Ulf Hellman and colleagues of Uppsala, who sequenced tryptic fragments of the C4d region of C4 prepared from a donor with known blood group Chido (C4B) and Rodgers (C4A) antigens.[7,8] Comparison of their protein sequence with our derived gene sequence allowed us to distinguish the C4A and C4B genes.

As work with the C4 genetics progressed, two other members of the MRC Immunochemistry Unit were characterizing the genes encoding factor B (Duncan Campbell) and C2 (David Bentley). Using the same cosmid libraries from which the C4 genes were isolated, they identified and characterized the genes encoding human Bf and C2, respectively.[9,10] Significantly, we found cosmid clones that overlapped and included all three genes. This led to a molecular map of over 100 kbp that included two C4 genes, Bf and C2. This cloned region represented the largest stretch of human DNA mapped at that time.[11,12] As noted above, further work demonstrated that this region of the HLA includes additional genes, which proved to be a fertile ground for the elegant genetic mapping by Yung Yu, who was a graduate student of Rodney Porter.

7.3 Linkage of Innate and Adaptive Immunity

A question that I carried with me on moving to Boston in 1985 and starting my laboratory at the Children's Hospital, Harvard Medical School, was "How was C4 involved in host protection to lupus"? I felt that the answer should lie in our new knowledge of the genetics of C4.[13] The dominant idea at the time was that early classical pathway complement was critical for the clearance of immune complexes (ICs).[14,15] Thus, total deficiency in C1q or C4 would lead to the accumulation of IC and the pathological events associated with SLE. Moreover, genetic studies indicated that in certain populations partial deficiencies, *i.e.* C4A null, were also predisposing to SLE, which suggested a more specific role for complement. Functional studies from Alex Law and Alister Dodds,[16] in the Oxford MRC Immunochemistry Unit, and David Isenman,[17] at Toronto, had demonstrated that the two isoforms of complement were chemically different; C4A preferentially formed amide linkages while C4B rapidly formed a covalent bond with hydroxyl groups. This suggested that the nature of the target antigen could determine whether C4A or C4B was important in clearance.

Clearance of ICs was a reasonable explanation for a contribution of complement to the effector phase of autoimmunity, but did not explain the larger question as to how complement contributed to the genetic predisposition to lupus. While the link between complement and self-reactive antibodies was not readily apparent, a connection between early complement and humoral immunity was beginning to emerge. The early studies by Victor Nussenzweig suggested that complement might be involved in B cell responses based on their observation of C3 receptors on germinal centre B cells.[18] More direct evidence of a role for complement in humoral immunity came from the observation that transient depletion of C3 substantially reduced the antibody response to thymus-dependent antigens in mice.[19] Likewise, humans,[20] guinea pigs[21] and dogs[22] deficient in C4 or C3 also failed to develop strong humoral responses on immunization with protein antigens. Therefore, it was becoming apparent that the classical pathway of complement was involved in adaptive immunity.

A breakthrough in our understanding of complement in adaptive immunity came from the studies of Doug Fearon and colleagues with their identification of a co-receptor complex on the surface of mature B cells.[23,24] Binding of C3d-coated antigen to the co-receptor complex, *i.e.* CD21, CD19 and CD81, effectively lowered the threshold for the activation of B cells. This led to the novel concept that complement receptors CD21/CD35 'linked innate and adaptive immunity'. To demonstrate the importance of the concept *in vivo*, mice were immunized with protein antigen coupled directly with multiple copies of C3d. It was found that substantially less protein was required to induce an optimal memory response when coated with C3d.

Our approach to gain a better understanding of how complement was involved in B cell immunity was to develop lines of mice deficient in C3, C4 and their receptors CD21 and CD35. Characterization of the humoral response in the 'knock-out' mice demonstrated a clear impairment in antibody responses and a failure to develop memory of protein antigens.[25–27] These findings were extended to pathogens such as herpes simplex virus (HSV), demonstrating that classical pathway complement was critical for an efficient B cell memory response to microorganisms.[28–30] Interestingly, mice deficient in C3, C4 or CD21/CD35 all had a similar impairment in their response to infectious HSV, which illustrates that classical pathway complement enhanced antibody responses primarily *via* the CD21/CD35 receptors. Through these and related studies by others, it became clear that CD21 and CD35 functioned not only as a co-receptor on B cells, but also were important in the retention of antigen on follicular dendritic cells (FDCs) within secondary lymphoid tissues.[31,32] Moreover, CD21/CD35 expression on mature B cells provides an important transport function, *i.e.* transport of C3-coated ICs to the FDC both in the spleen and peripheral lymph nodes (PLNs).[33,34]

7.4 C4 and Autoimmunity

With the availability of mice bearing deficiency in early complement pathway proteins, a role for complement in autoimmune disease was revisited. Walport and colleagues found that their line of C1q-deficient mice developed a lupus-like phenotype, *i.e.* accumulation of apoptotic blebs and increased anti-nuclear antibodies (ANA), when crossed onto a mixed B6 × 129 background.[35] Likewise, we found that our C4−/− mice on a similar background also developed a lupus-like phenotype.[36] Notably, as the mice were backcrossed onto C57 BL/6 (B6), a background that is normally non-autoimmune, the lupus phenotype was eliminated, which suggests that background genes also influence susceptibility. Through the more recent studies of Wakeland and colleagues, it is apparent that autoimmunity and, particularly, SLE is a multigenic disease and the background has an important role in disease susceptibility.[37] These observations suggested that the deficient mice could be used as animal models to identify a mechanism of how complement participated in host protection to lupus.

An example of the importance of background comes from the well-established lpr-model. Mice deficient in the Fas receptor (CD95, *lpr* locus) develop a severe lupus phenotype when bred onto the MRL background. By contrast, when crossed onto the mixed 129 × B6 or B6 alone, they have a relatively mild phenotype at least until six months of age. To test a role for complement in the development of lupus, mice deficient in C4, C3 or receptors CD21/CD35 were bred with B6.*lpr* mice. The double-deficient mice were characterized for weight of spleen and LNs, expression of ANA and dsDNA antibody titres and deposits of IgG in glomeruli at various ages (10–17 weeks). As expected, the complement-sufficient *lpr* mice had a relatively mild phenotype at the ages examined.[38] Similarly, mice double-deficient in C3 and *lpr* also failed to develop a lupus-like phenotype, although examination of their kidneys at 17 weeks revealed above background IgG deposits.[38] By contrast the *lpr* mice deficient in C4 or CD21/CD35 developed a robust lupus phenotype.

One interpretation (the clearance hypothesis) of the results with the C1q–/– and C4–/– mice was that failure to efficiently clear apoptotic cell blebs released from dying cells led to activation of self-reactive B cells in the periphery and the production of pathogenic autoantibody.[39] Although the clearance hypothesis explained an important role for complement in the removal of apoptotic debris, it did not explain how deficiency could alter B cell tolerance. Thus, the hypothesis did not account for tolerogenic mechanisms that normally prevent self-reactive B cells from differentiation into mature and potentially pathogenic lymphocytes.

Much of our understanding of B cell tolerance is based on the characterization of mice bearing a transgenic (Tg) BCR. Thus, mice bearing a relatively high frequency of B cells with a known specificity provide a model to track their fate under different conditions. For example, the hen egg lysozyme (hel) Tg B cell model has been widely used to characterize tolerance of self-reactive B cells. In particular, breeding of mice bearing the anti-Hel Ig transgene (MD4 line) with mice expressing a soluble form of the antigen (shel) (ML5 line) provided a model to study the fate of the autoreactive cells in the presence of self-antigen.[40] The double Tg mice were used to study the negative selection of self-reactive B cells in the presence and absence of self-antigen. The results from this and similar models led to our current understanding that B cells pass through multiple checkpoints during early differentiation to reach maturity. Cells that engage self-antigen undergo clonal deletion, anergy or receptor editing depending, in large part, on the strength of signal induced through the BCR.[41–45] In the hel/shel double Tg model, immature B cells become anergic and fail to fully mature and are, in general, unresponsive to cross-linking of their BCR. The anergic cells have a relatively short half-life and turnover more rapidly then normal WT mature B cells.[46]

To test if complement C4 and its receptors CD21/CD35 participate in the tolerance of B cells, mice bearing deficiencies in C4, C3 or CD21/CD35 were bred with the MD4 single and double (anti-hel/shel) Tg lines. The mice were characterized for the development of anergy in the presence of shel and complement. The results indicated that hel+ B cells were anergized in the presence of complement, as expected.[38] Similarly, hel+ self-reactive B cells in the C3−/− double Tg mice also failed to mature normally and were unresponsive to antigen. By contrast, hel+ B cells in CD21/CD35−/− or C4−/− chimeric mice did not develop a full anergic phenotype. Thus, the B cells responded when cultured *in vitro* with antigen based on Ca^{2+} flux and upregulation of CD86. Moreover, the self-reactive B cells had a longer half-life *in vivo* than C+ controls and continued to differentiate similar to that of single Tg controls.[38]

7.5 Anti-RNA Mouse Model

The results with the hel Tg model demonstrated, for the first time, a role for complement in the development of the tolerance of B cells to self-antigen. Importantly, the results pointed to a role for C4 and its receptor CD21/CD35 in the regulation of self-reactive B cells at the immature stage. From these findings we proposed the B cell tolerance model,[27] in which complement participates in B cell tolerance by enhancing the interaction of self-antigen with cognate B cells. Thus, based on our earlier findings with complement in humoral immunity, we proposed that it might have a similar role in directing interaction of immature B cells with self-antigen. Since CD21/CD35 are not expressed until the 'transitional stage', which occurs as immature B cells enter the spleen, we proposed that the receptors were expressed on stromal cells within the bone marrow (BM) and participated in the retention of self-antigen to facilitate encounters with the developing B cells.

One limitation of the Tg model is that hen lysozyme is not a normal self-antigen in mammals and the naïve Tg B cells bind antigen with an unusually high affinity. Moreover, the B cells express multiple copies of the classical Ig transgenes and therefore do not undergo normal receptor editing. To examine a role for C4 in a more physiologically relevant model, we collaborated with T. Imansihi-Kari (Tuft's New England Medical School, Boston) to cross our C4−/− mice with the 564 Ig knock-in line (564 Igi). The latter mice were constructed by insertion of rearranged Ig heavy (HC) and light chain (LC) gene-coding sequences into their respective Ig loci. The source of the Ig HC and LC was a B cell hybridoma (564) which had been isolated from the autoimmune strain SWF1. Female mice of this line develop a severe lupus-like phenotype at six months of age.[47,48] The specificity of the hybridoma appeared to be nucleolar antigens. Most recently we found that 564 binds ssRNA, but not DNA (Tsiftsoglou and Carroll, unpublished).

On a B6 background, the 564 mice express autoreactive antibody mostly of the IgM isotype, although they fail to develop a lupus-like disease.[49] Recent published studies suggest that the breaking of tolerance is mediated *via* toll-like receptor 7 (TLR 7) signalling because when the mice were crossed with TLR 7–/– mice, the anti-nucleolar titre dropped to background. Despite the production of 564-specific antibody, the idiotype positive (Id+) B cells are, for the most part, anergic based on several criteria (Alimzhanov, Agyemang and Carroll, unpublished).[49]

The initial comparison of splenic and LN B cells in C4+ and C4–/– 564 Igi mice indicated a significantly higher frequency of Id+ B cells in the complement-deficient line. This suggested that C4 might participate in counter-selection of the self-reactive B cells. Further analyses identified a major negative selection of Id+ B cells in the BM at checkpoint I. Thus, about 50% of the Id+ or self-reactive B cells are either deleted or edited in both C4+ and C4–/– mice. The remaining self-reactive B cells undergo further negative selection at checkpoint II in the spleen. This stage (transitional) represents a major stage of negative selection, as approximately 90% of B cells emerging from the BM are eliminated. It is held that anergic B cells entering the spleen are prevented from entering the mature compartment and undergo cell death. Characterization of the Id+ B cells at this stage indicated a similar frequency of immature (AA4+ CD21–, CD23–) Id+ cells for both C4+ and C4–/– mice. By contrast, the frequency of mature (AA4–, CD21+, CD23+) Id+ B cells was 4–5 time higher in spleens and LNs of C4–/– than C4+/+ 564 Igi mice (Alimzhanov, Agyemang and Carroll, unpublished). These results suggested that C4 was important in the induction of anergy of self-reactive B cells in the 564 Igi model.

To test more directly whether the self-reactive B cells on C4+ or C4–/– backgrounds are anergic, splenic B cells were cultured *in vitro* in the presence of various agonists and the cells assayed by fluorescent activated cell sorting (FACS) for upregulation of CD86 (a marker of activation). Significantly, the Id+ B cells prepared from C4+ 564 mice failed to respond to direct cross-linking of BCR, either with or without CD40 and interleukin 4 (IL-4) signalling. Moreover, stimulation with agonists such as CpG, lipopolysaccharide (LPS) or TLR 7 failed to induce appreciable activation relative to Id– cells in the same culture. By contrast, Id+ B cells prepared from C4–/– 564 donors, responded to cross-linking of BCR and to TLR agonists similar to that of the Id– controls (Alimzhanov, Agyemang and Carroll, unpublished). It is interesting that despite the apparent normal maturation of self-reactive B cells, the C4–/– 564 mice do not develop lupus disease, although examination of the kidneys of C4–/– mice at three months of age reveals deposits of IgG not found in C4+ 564 litter mates. This suggests that C4 is not only involved in the silencing of self-reactive B cells, but also in the clearance of ICs. Studies are in progress to follow the fate of the mice at later time points.

The studies with the C4–/– 564 mice provide direct support for the B cell tolerance model. The finding that self-reactive B cells for a typical lupus antigen (RNA) fail to undergo normal anergy in the absence of C4 suggests that classical pathway complement participates in B tolerance. However, the stage and anatomical site in which complement is involved is not clear. One approach to address these questions is to identify the source of C4 that influences anergy. We reasoned that if local complement synthesis were required, then this would favour tolerance occurring in the BM, where developing B cells are less exposed to the open circulation. The majority of C4 protein in circulation (200–400 μg/ml) is derived from the liver. In addition, activated myeloid cells express many of the complement proteins, including C4. Locally produced C4 and C3 are sufficient to promote complement-enhanced humoral immunity.[50,51] Thus, a similar approach was used in the 564 Igi model to examine the importance of locally produced C4.

Characterization of BM chimeras in which C4+ or C4−/− recipients were reconstituted with BM derived from 564 mice of an identical C4 genotype yielded the expected results, *i.e.* Id+ B cells were anergic in C4+ but not in C4 −/− chimeras. By contrast, when C4−/− mice were reconstituted with BM from C4+ 564 donors, the Id+ B cells developed an anergic phenotype similar to that of the C4+ 564 line. Thus, despite a negligible level of C4 in blood, local production of C4 was sufficient to induce an anergic phenotype. By contrast, in 564 chimeras in which there was a normal level of C4 in the blood, local production was negligible; the Id+ B cells developed a partial anergic phenotype (Alimzhanov, Agyemang and Carroll, unpublished observations). These findings demonstrate the importance of the local production of C4 in the induction of anergy and favour a model in which complement participates in the tolerance of self-reactive B cells in the BM compartment.

7.6 Models for C4 in B Cell Tolerance

A prediction of the tolerance model was that self-reactive B cells in the C4−/− 564 mice would fail to take up self-antigen efficiently. Characterization of Id+ 564 B cells isolated from circulation of C4+ and C4−/− 564 mice by FACS identified a similar level of 564 staining. This suggests that C4 is not required for the uptake of cognate antigen by the Id+ cells. However, C4 could participate at a different step in the pathway. For example, activated C4 could act *via* a co-receptor, much like the CD21/CD19 receptor expressed on mature B cells. Thus, binding of self-antigen by the IgM BCR would lead to activation of the classical pathway on the B cell surface and covalent attachment of C4b to the antigen. The presence of C4 ligand attached to the RNA antigen could have several effects. One, it could promote an increase in BCR signalling that translates into the induction of anergy. Accordingly, in the absence of C4, one would predict a reduced BCR signal, which could allow an escape from tolerance. In support of this model, we have observed that a low frequency of B cells bear C4 on their surface that is not dependent on CD21/CD35. However, a more definitive test is the comparison of C4b on the surface of Id+ *vs* Id− immature B cells isolated from the BM, and this is in progress.

An alternative explanation is that both clearance and tolerance are important in silencing self-reactive B cells. Accordingly, exposure of self-antigen on the surface of apoptotic blebs could lead to activation of classical or lectin pathways and covalent attachment of C4b to the RNA complex. Deposition of C4b facilitates uptake of the complex[39] and could potentially participate in the inhibition of release of activating cytokines such as interferon alpha (IFNα). In support of this model, IFNα is a known inducer of the TLR pathway and increased levels in circulation are a hallmark of lupus patients.

Recently, a number of publications propose an important role for TLR such as TLR 7 and 9 in lupus susceptibility.[52] In particular, TLR 7 that is activated by ssRNA could be relevant to the 564 model and B cell anergy. The *yaa* locus represents an apparent translocation of a piece of the female X onto the Y chromosome. The piece includes the gene that encodes TLR 7 (along with 16 other genes) so that males of this strain express an additional copy. The second copy of TLR 7 explains the lupus-like phenotype of B6.*yaa* mice, as expression of multiple copies of TLR 7 transgenes by B6 mice is sufficient to induce a lupus-like phenotype.[53,54]

To test whether an extra copy of TLR 7 would affect B cell anergy in the 564 mice, they were crossed with the B6.*yaa* line. As observed in C4−/− background, the frequency of mature Id+ B cells is significantly higher in male B6.*yaa* 564 mice than in the female controls (Agyemang, Tsiftsoglou and Carroll, unpublished) Interestingly, the Id+ B cells in the *yaa*.564 mice appear to escape anergy at checkpoint II as observed in the C4−/− 564 mice. It is possible that the TLR and complement pathways intersect either in direct B cell regulation or indirectly *via* effects on macrophages or DC. Additional experiments are in progress to test this possibility.

7.7 Summary

My training period at the MRC Immunochemistry Unit in Oxford instilled a rigour and insight into exploring problems that provided an important foundation for my scientific career. I am forever grateful to Rodney Porter and members of the Unit for their encouragement and help during that important part of my training. The closing of the Unit after such a highly productive period is clearly a loss to basic science; but the underlying rigour will be retained and passed along to the students and trainees by all of us who trained there.

References

1. M. C. Carroll and J. D. Capra, *Proc. Natl. Acad. Sci. USA*, 1978, **75**, 2424–2428.
2. M. C. Carroll and R. R. Porter, *Proc. Natl. Acad. Sci. USA*, 1983, **80**, 204–207.
3. K. T. Belt, M. C. Carroll and R. R. Porter, *Cell*, 1984, **36**, 907–914.
4. A. Palsdottir, S. J. Cross, J. H. Edwards and M. C. Carroll, *Nature*, 1983, **306**, 615–616.

5. M. C. Carroll, R. D. Campbell and R. R. Porter, *Proc. Natl. Acad. Sci. USA*, 1985, **82**, 521–525.
6. M. C. Carroll, A. Palsdottir, K. T. Belt and R. R. Porter, *EMBO J.*, 1985, **4**, 2547–2552.
7. A. Lundwall, U. Hellman, G. Eggertsen and J. Sjoquist, *Mol. Immunol.*, 1982, **19**, 1655–1665.
8. U. Hellman, G. Eggertsen, A. Lundwall, A. Engstrom and J. Sjoquist, *FEBS Letters*, 1984, **170**, 254–258.
9. D. R. Bentley and R. R. Porter, *Proc. Nat. Acad. Sci. USA*, 1984, **81**, 1212–1215.
10. R. D. Campbell and R. R. Porter, *Proc. Nat. Acad. Sci. USA*, 1983, **80**, 4464–4468.
11. M. C. Carroll, T. Belt, A. Palsdottir and R. R. Porter, *Phil. Trans. Roy. Soc. London*, 1984, **306**, 379–388.
12. M. C. Carroll, R. D. Campbell, R. D. Bentley and R. R. Porter, *Nature*, 1984, **307**, 237–241.
13. R. R. Porter, *Molecular biology & medicine*, 1983, **1**, 161–168.
14. J. A. Schifferli, Y. C. Ng and D. K. Peters, *N. Engl. J. Med.*, 1986, **315**, 488–495.
15. W. Emlen, G. Burdick, V. Carl and P. J. Lachmann, *J. Immunol.*, 1989, **142**, 4366–4371.
16. S. K. Law, A. W. Dodds and R. R. Porter, *EMBO J.*, 1984, **3**, 1819–1823.
17. D. E. Isenman and J. R. Young, *J. Immunol.*, 1984, **132**, 3019–3027.
18. V. Nussenzweig, C. Bianco, P. Dukor and A. Eden, *Receptors for C3 on B lymphocytes: possible role in the immune response.*, Academic Press, New York, 1971.
19. M. B. Pepys, *J. Exp. Med.*, 1974, **140**, 126–145.
20. C. G. Jackson, H. D. Ochs and R. J. Wedgewood, *N. Engl. J. Med.*, 1979, **300**, 1124–1129.
21. H. D. Ochs, R. J. Wedgewood, S. R. Heller and P. G. Beatty, *Clin. Immunol. Immunopathol.*, 1986, **40**, 94–104.
22. K. M. O'Neil, H. D. Ochs, S. R. Heller, L. C. Cork, J. M. Morris and J. A. Winkelstein, *J. Immunol.*, 1988, **140**, 1939–1945.
23. D. T. Fearon and R. H. Carter, *Annu. Rev. Immunol.*, 1995, **13**, 127–149.
24. P. W. Dempsey, M. E. Allison, S. Akkaraju, C. C. Goodnow and D. T. Fearon, *Science*, 1996, **271**, 348–350.
25. M. B. Fischer, M. Ma, S. Goerg, X. Zhou, J. Xia, O. Finco, S. Han, G. Kelsoe, R. G. Howard, T. L. Rothstein, E. Kremmer, F. S. Rosen and M. C. Carroll, *J. Immunol.*, 1996, **157**, 549–556.
26. M. B. Fischer, S. Goerg, L. Shen, A. P. Prodeus, C. C. Goodnow, G. Kelsoe and M. C. Carroll, *Science*, 1998, **280**, 582–585.
27. M. C. Carroll, *Nature reviews*, 2004, **4**, 825–831.
28. A. Verschoor, M. A. Brockman, D. M. Knipe and M. C. Carroll, *J. Immunol.*, 2001, **167**, 2446–2451.

29. X. DaCosta, M. Brockman, E. Alicot, M. Ma, M. Fischer, X. Zhou, D. Knipe and M. Carroll, *Proc. Natl. Acad. Sci. USA*, 1999, **96**, 12708.
30. M. A. Brockman, A. Verschoor, J. Zhu, M. C. Carroll and D. M. Knipe, *J. Virol.*, 2006, **80**, 7111–7117.
31. R. A. Barrington, O. Pozdnyakova, M. R. Zafari, C. D. Benjamin and M. C. Carroll, *J. Exp. Med.*, 2002, **196**, 1189–1199.
32. Y. Fang, C. Xu, Y. Fu, V. M. Holers and H. Molina, *J. Immunol.*, 1998, **160**, 5273–5279.
33. T. G. Phan, I. Grigorova, T. Okada and J. G. Cyster, *Nature Immunology*, 2007, **8**, 992–1000.
34. G. Cinamon, M. A. Zachariah, O. M. Lam, F. W. Foss, Jr. and J. G. Cyster, *Nature Immunology*, 2008, **9**, 54–62.
35. M. Botto, C. Dell'Agnola, A. E. Bygrave, E. M. Thompson, H. T. Cook, F. Petry, M. Loos, P. P. Pandolfi and M. J. Walport, *Nat. Genet.*, 1998, **19**, 56–59.
36. E. Paul, O. O. Pozdnyakova, E. Mitchell and M. C. Carroll, *Eur. J. Immunol.*, 2002, **32**, 2672–2679.
37. E. Wakeland, A. Wandstrat, K. Liu and L. Morel, *Curr. Opin. Immunol.*, 1999, **11**, 701–707.
38. A. P. Prodeus, S. Goerg, L. M. Shen, O. O. Pozdnykova, L. Chu, E. Alicot, C. C. Goodnow and M. C. Carroll, *Immunity*, 1998, **9**, 721–731.
39. P. R. Taylor, A. Carugati, V. A. Fadok, H. T. Cook, M. Andrews, M. C. Carroll, J. S. Savill, P. M. Henson, M. Botto and M. J. Walport, *J. Exp. Med.*, 2000, **192**, 359–366.
40. C. Goodnow, J. Cyster, S. Hartley, S. Bell, M. Cooke, J. Healy, S. Akkaraju, J. Rathmell, S. Pogue and K. Shokat, *Adv. Immunol.*, 1995, **59**, 279–369.
41. C. C. Goodnow, *Proc. Natl. Acad. Sci. USA*, 1996, **93**, 2264–2271.
42. R. Pelanda, S. Schwers, E. Sonoda, R. M. Torres, D. Nemazee and K. Rajewsky, *Immunity*, 1997, **7**, 765–775.
43. C. Chen, Z. Nagy, E. L. Prak and M. Weigert, *Immunity*, 1995, **3**, 747–755.
44. H. Li, Y. Jiang, E. L. Prak, M. Radic and M. Weigert, *Immunity*, 2001, **15**, 947–957.
45. H. Wardemann, S. Yurasov, A. Schaefer, J. W. Young, E. Meffre and M. C. Nussenzweig, *Science*, 2003, **301**, 1374–1377.
46. J. G. Cyster and C. C. Goodnow, *Immunity*, 1995, **3**, 691–701.
47. T. L. O'Keefe, S. Bandyopadhyay, S. K. Datta and T. Imanishi-Kari, *J. Immunol.*, 1990, **144**, 4275–4283.
48. S. Ghatak, T. L. O'Keefe, T. Imanishi-Kari and S. K. Datta, *Int. Immunol.*, 1990, **2**, 1003–1012.
49. R. Berland, L. Fernandez, E. Kari, J. H. Han, I. Lomakin, S. Akira, H. H. Wortis, J. F. Kearney, A. A. Ucci and T. Imanishi-Kari, *Immunity*, 2006, **25**, 429–440.
50. M. Gadjeva, A. Verschoor, M. A. Brockman, H. Jezak, L. M. Shen, D. M. Knipe and M. C. Carroll, *J. Immunol.*, 2002, **169**, 5489–5495.

51. A. Verschoor, M. A. Brockman, M. Gadjeva, D. M. Knipe and M. C. Carroll, *J. Immunol.*, 2003, **171**, 5363–5371.
52. A. M. Krieg, *Immunity*, 2007, **27**, 695–697.
53. J. A. Deane, P. Pisitkun, R. S. Barrett, L. Feigenbaum, T. Town, J. M. Ward, R. A. Flavell and S. Bolland, *Immunity*, 2007, **27**, 801–810.
54. P. Pisitkun, J. A. Deane, M. J. Difilippantonio, T. Tarasenko, A. B. Satterthwaite and S. Bolland, *Science*, 2006, **312**, 1669–1672.

Section 3
Collectins and Ficolins in Innate Immunity

CHAPTER 8

The Structure of Mannan-binding Lectin and its Functional Relevance

JINHUA LU[a] AND STEFFEN THIEL[b]

[a] Department of Microbiology, Yong Loo Lin School of Medicine, Immunology Programme, National University of Singapore, 5 Science Drive 2, Singapore 117597; [b] Department of Medical Microbiology and Immunology, University of Aarhus, Wilhelms Meyers Allé, 8000 Aarhus, Denmark

8.1 Introduction

Mannan-binding lectin (MBL) is a protein of the innate immune system found in the blood circulation in a complex with serine proteases in proenzyme form. MBL recognizes a fitting pattern of carbohydrates on many pathogens and binding to these carbohydrate structures activates the associated proteases and, in turn, the complement system. MBL belongs to a family of proteins called collectins which possess both collagen-like regions and C-type carbohydrate domains (CRDs).

In this chapter we discuss some structural characteristics of MBL, a significant part of which was determined during 1988–1990 and involved our work in the laboratory of Kenneth B. M. Reid (Ken) in the MRC Immunochemistry Unit (the Unit), Oxford, UK.

In the summer of 1988 (after a previous very short stay in 1987), Steffen came to Oxford from the University of Odense, Denmark, to conduct part of his PhD project to purify a human conglutinin-like protein. Conglutinin was one of the

Molecular Aspects of Innate and Adaptive Immunity
Edited By Kenneth BM Reid and Robert B Sim

Published by the Royal Society of Chemistry, www.rsc.org

pioneer molecules associated with an immune function – it was first discovered in 1906 by the Nobel laureate Jules Bordet and his colleague Frederick Gay as a bovine serum factor that agglutinates guinea pig erythrocytes which had been pre-reacted with horse antibodies and complement. Since complement was involved in this reaction and since conglutinin contained collagen-like sequences, and since Ken likes collagen-like sequences in non-collagen molecules (he must do, having been involved in sequencing every single one of the Gly–Xaa–Yaa repeats present in the three types of polypeptide chains found in the complement protein C1q),[1] there was some interest in the project in the Unit. Back in 1964 conglutinin was noted to bind to carbohydrate structure.[2] Several attempts to purify the human counterpart using a mannan–Sepharose affinity column led to no detectable human conglutinin. Several peptide sequences were generated based on bovine conglutinin, but cDNA cloning, using degenerate oligonucleotide primers, led to no meaningful clones. Anti-conglutinin antibodies were found to be effective in picking up only albumin clones in cDNA expression libraries and gave no new information. The protein purification involved long stays in the cold room. However, you could not complain when Tony Gascoigne, a head technician in the Unit, was around. He would simply laugh and tell his stories about working at −20 °C when he worked on enzymes of the Krebs cycle in the laboratory of Sir Hans Krebs. Indeed, Tony also had experiences of true winters during the Korean wars. It is known now that conglutinin is, in fact, not present in species other than ruminants,[3] but there was no reason to believe this at that time.

8.2 MBL was Purified as a 'By-product'

While our attention had been focussed upon the 45 kDa area of the sodium dodecyl sulfate polyacrylamide gel electrophoresis (SDS-PAGE) gels, where conglutinin was expected under reducing conditions, another protein was consistently eluted from the columns which migrated as a band of approximately 32 kDa under reducing conditions on SDS-PAGE gels. Amino acid sequencing suggested that it was the human form of MBL. At this time Jinhua came to Oxford from Beijing for his D. Phil. studies and we collaborated on the further refinement of the MBL purification protocol and began to obtain MBL in good purity and yield for biochemical and structural characterization and antibody production. Soon after Steffen arrived in Oxford, Malcolm W. Turner from the Children's Hospital in London gave a seminar in Oxford on his studies of a childhood opsonic defect, *i.e.*, the sera from some children failed to deposit complement on yeast surfaces.[4] Steffen naturally suggested human conglutinin as being a strong candidate for the putative opsonic factor, but this idea was not followed up. Then, shortly afterwards, a PhD student in Turner's laboratory came across the work by Ikeda and colleagues which demonstrated the complement-dependent haemolytic activity of MBL.[5] As we, in Oxford, had constructed an enzyme-linked immunosorbent assay (ELISA) for MBL, we measured the MBL levels in serum samples obtained from a group of these children. It turned out that they were very low in serum MBL concentrations.[6] In fact, addition of purified MBL to these sera could restore the opsonic capacity of these patient sera. These initial studies have since spurred many studies on the associations of low MBL concentrations with a range of diseases and symptoms.[7,8]

8.3 MBL Structure, Genetics and Heterogeneity in Size

After we had managed to make the first reproducible purifications of human MBL we contacted Rupert Timpl in Martinsreid, Germany, to ask him for help with rotary shadowing electron microscopy (EM) pictures of MBL. Rupert and his colleague Hanna Wiedemann received different preparations of MBLs and started to produce pictures of the proteins in the samples. The first pictures showed molecules organized as very neat round doughnut-like structures. These, we found, were actually pictures of an impurity in the MBL samples, the serum amyloid P component (SAP). SAP has the ability to bind directly to the Sepharose beads used during the affinity chromatography procedures we used to purify MBL and were co-eluted with MBL. It was a well-known protein contaminant in the Unit as it was also an impurity found when the subcomponents of the C1 complex were initially purified. It was once actually named C1t (following the alphabetic order of C1q, C1r and C1s for the subcomponents of C1). More EM pictures were produced with further purified MBL, which revealed structures suggestive of a 'bouquet of tulips' (Figure 8.1a).[9] This description of the MBL structure was apparently influenced by the formerly suggested structure of C1q, the recognition molecule of the C1 complex.[10] It has also been suggested that another molecule, surfactant protein A (SP-A), forms a similar overall structure.[11] The suggested MBL structure, based on the EM pictures, was supported by the knowledge of the MBL polypeptide sequence. From the MBL sequence it was clear that it would facilitate the formation of a collagen-like triple helix, *i.e.*, its amino terminal half is dominated by repeats of the Gly–Xaa–Yaa triplet. The mature human MBL polypeptide starts from the N-terminal end with a short cysteine-rich stretch (residues 1–21), followed by repeats of Gly–Xaa–Yaa (residues 22–81), with one interruption (residues 43–44), and then by a neck region (residues 82–115) and a carbohydrate recognition domain (CRD; residues 116–228).

After further analysis of the EM pictures of MBL a bouquet with a somewhat shorter stalk, more like a hub with stalks projecting, was suggested (Figure 8.1b).[12] The stalks were thought to be around 15 nm in length, comparable with the average height of 8.4 Å for a single Gly–Xaa–Yaa repeat (based on the crystal structure of a collagen-like peptide),[13] and the presence of 19 Gly–Xaa–Yaa repeats. In the laboratory of Toshisuke Kawasaki in Kyoto, more EM pictures of MBL were later produced.[14] These led to the suggestion of a wheel-like structure for MBL with a hub in the middle (Figure 8.1c).

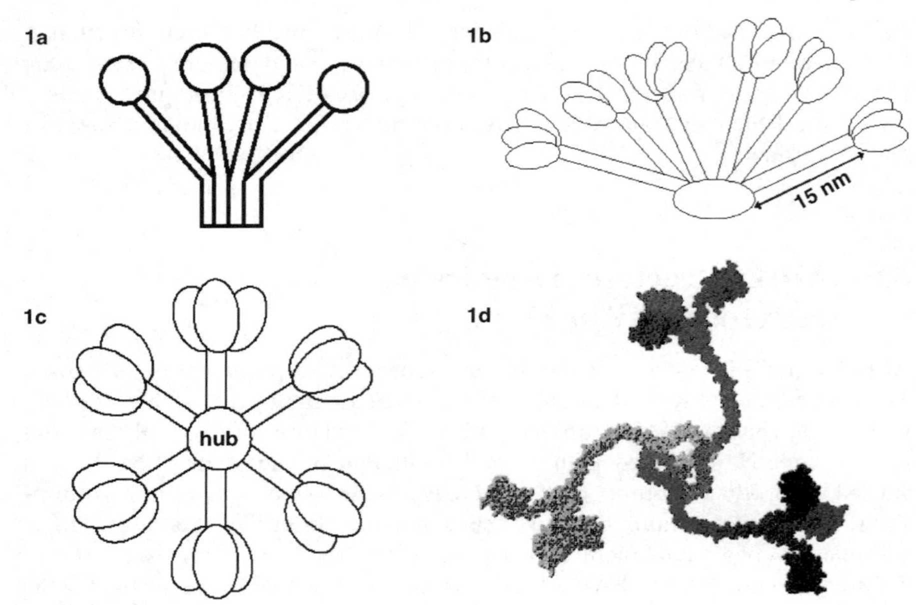

Figure 8.1 Models of the oligomeric structures of MBL. The structural subunits form oligomers and, based on EM studies, we suggested a 'bunch of tulips' like structure with four, as indicated here in (a), five or six oligomers of the structural subunit. This structure has been used by others in a number of papers to depict MBL. We later further elaborated on the EM pictures and suggested the structure shown in (1b). Kawasaki, also based on EM studies, suggested a structure with six structural subunits radiating out from a hub, as in (1c). A similar figure, using primarily four structural subunits, is also used in reviews by, *e.g.*, Thielens, Arlaud and Gaboriaud and colleagues from Grenoble, France. When studying MBL in solution by small angle X-ray scattering analysis, Ming Dong and colleagues interpreted the data to represent structural subunits radiating from a centre (1d).

Recently, MBL was studied by small angle X-ray scattering analysis and the structure that best fitted the data obtained is shown in Figure 8.1d.[15] In contrast to the EM pictures, this structure is based on MBL in solution. Remarkably, it seems that the spider-like structure of MBL seen when adhered to the micra in EM is also seen in solution.

In 1991, it was found that there was a genetic link between a point mutation in the coding region of the MBL gene and low serum levels of MBL.[16] Other mutations in the promoter region and the coding region were later also found to influence the synthesis of MBL.[17] We now also know that the thyroid hormones, T3 and T4, have a strong influence on the transcription and synthesis of MBL.[18]

The naturally occurring mutations found in the collagen-like region of MBL have an impact on the oligomerization of the structural subunits. It was thus found that the low levels of MBL found in individuals who were homozygous for the G54D mutation were of a lower molecular weight (when analyzed by

SDS-PAGE Western blotting) than the MBL found in wild type individuals.[19] This was later confirmed by size-exclusion chromatography of the native molecules, and was also found to be the case for individuals being homozygous for other known mutations, *i.e.*, R52C and G57E.[7,20,21] The smaller number of CRDs present in the MBL produced by individuals homozygous for the mutations mentioned above meant these mutant MBLs are not able to bind mannan (mannan is a preparation of carbohydrate structures from baker's yeast) with high enough strength, *i.e.*, avidity. Therefore, these MBL molecules cannot effectively mediate the initiation of the complement activation on microbial targets.

Knowledge of disulfide bond formation in the N-terminal 21 residues of the MBL polypeptide may lead to a better understanding of the assembly of MBL. Human MBL and rodent MBL-A are the only collectins that contain three cysteine residues in this N-terminal region, whereas the others contain only two cysteines. When recombinant human MBL was studied, a complex and heterogeneous disulfide pattern, as well as free cysteines, was found.[22] The precise disulfide pattern that apparently dictates the assembly of human MBL is not yet known.

The structure of the CRD in MBL was solved in 1994, based on studies on human and rat MBL.[23,24] These structures have formed the basis for studies on similar CRDs in other C-type lectins and C-type lectin-like proteins.[25] The recognition of carbohydrate is based on co-ordination bonds for a calcium ion in the CRD and the 3- and 4-hydroxyl equatorial groups of, *e.g.*, glucose, mannose or derivatives of these sugar molecules. The interaction between the CRD and carbohydrate is very weak, *i.e.*, 10^{-3} M.[26] However, when the clustered CRDs at the end of the collagen helices are allowed to bind carbohydrate simultaneously, a high avidity binding can be achieved. With regard to carbohydrate ligands, this requires densely arrayed carbohydrate patterns, which are common on the surface of microorganisms. Although some theoretical considerations have been attempted, the actual carbohydrate patterns that MBL recognizes, *i.e.*, the number, density and types of monosaccharides, and their three-dimensional organization and so on, have not been described. Nonetheless, a great number of bacteria, viruses and fungi are recognized by MBL.[7]

Three serine protease proenzymes, the MBL-associated serine proteases (MASP-1, -2 and -3), and a non-enzymatic protein, known as MBL-associated protein of 19 kDa (MAp19), have now been found in complex with MBL.[27] Some aspects, of the discovery and properties of these proteases are described in Chapters 9 and 10. The amino acid residues on the MBL molecule that are involved in its interaction with MASPs and/or MAp19 have been identified by mutagenesis studies on MBL. In human MBL, a lysine residue in the collagen-like region (K55), as well as residues in the close vicinity, is clearly important for its interaction with MASPs and/or MAp19; a similar conclusion has been made for rat MBL-A.[28,29] The specific residues of the MASPs that are involved in the interaction have not been published. However, we know that the binding takes place within the first two domains of MASPs, *i.e.*, the so-called CUB1 and EGF domains. When MBL recognizes and binds to a fitting pattern of carbohydrates (often present on microbial cell surfaces), it can lead to the activation of the associated MASPs. MASP-2 has an autoactivating activity and does not need the other MASPs to cleave its proenzyme into an active protease. How ligand recognition by MBL is translated into MASP-2 cleavage and thus its activation is unclear. A conformational change in the MBL collagen-like region has been suggested to lead to MASP-2 activation.

In this chapter we describe our studies on human MBL. In mice, two forms of MBL, MBL-A and MBL-C (a pseudogene *MBL-B* is also present but not expressed) are found. Initially, it was suggested that MBL-C is a liver protein, whereas MBL-A is a plasma protein. Later, when mouse serum was handled at low temperatures with the addition of protease inhibitors, it became clear that both MBL-A and MBL-C were truly plasma proteins.[30] This was confirmed with the use of monoclonal antibodies specific for MBL-A and MBL-C in immune assays. Rats also contain two forms of MBL. We expect that both forms are present in the serum. The rat equivalent to mouse MBL-C is even more fragile than the mouse protein and has not been purified in amounts that allow extensive studies of the native protein. Studies on rat MBLs have thus been conducted primarily on recombinant proteins. Elaborate studies on rat MBLs have been performed by Drickamer, Wallis and co-workers and have been reviewed previously.[29,31] Importantly, they have demonstrated that a rat MBL dimer, which contain six polypeptides or two triple helices, was sufficient for the binding of a MASP-2 dimer.

The recombinant MBL found in culture supernatants of mammalian cell lines, *e.g.*, HEK or CHO cells, has a different oligomeric composition to that of native MBL found in serum. A heterogeneous range of recombinant MBL oligomers is seen, which is in contrast to the high molecular weight MBL forms in plasma.[32] This discrepancy is eliminated after applying the recombinant MBL to affinity columns, as the lower molecular weight forms are not bound to the affinity columns. A product of recombinant MBL is currently used in phase II clinical trials aimed at a therapeutic use of MBL in MBL-deficient cancer patients, who, when undergoing chemotherapy, appear more susceptible to microbial infections than those patients with normal levels of MBL. Steffen has been involved with NatImmune, the biotech company, which has brought this product through the early phases of development.

8.4 Concluding Remarks

Our training in the Unit has had important impacts on our research careers – both of us have been dedicated to the study of innate immunity and have regularly collaborated with the Unit. The expertise in protein chemistry techniques in the Unit, and the application of this expertise to the understanding of the structure–function relationships for a range of proteins in the field of immunology, has profoundly influenced our research directions. This is apparent in the characterization of MBL. Our joint studies in the Unit on MBL, by providing a refined purification protocol, helped determine the classical EM structure for MBL, which was adopted in the popular immunology textbook *Immunobiology* by Janeway and colleagues. That it facilitated the identification of MBL as the serum factor responsible for a childhood opsonic defect with an MBL ELISA, among others, represented one of the best collaborations in our research careers. In this regard, the determination of the C1q structure, which was heavily contributed to by Ken Reid, made the choice of experimental approaches on MBL pleasantly easy.

References

1. K. B. M. Reid, *J. Biochem.*, 1979, **179**, 367.
2. M. A. Leon and R. Yokohari, *Science*, 1964, **143**, 1327.
3. J. Lu, S. B. Laursen, S. Thiel, J. C. Jensenius and K. B. M. Reid, *J. Biochem.*, 1993, **292**, 157.
4. M. W. Turner, J. F. Mowbray and D. R. Roberton, *Clin. Exp. Immunol.*, 1981, **46**, 412.
5. K. Ikeda, T. Sannoh, N. Kawasaki, T. Kawasaki and I. Yamashina, *J. Biol. Chem.*, 1987, **262**, 7451.
6. M. Super, S. Thiel, J. Lu, R. J. Levinsky and M. W. Turner, *Lancet*, 1989, **2**, 1236.
7. R. M. Dommett, N. Klein and M. W. Turner, *Tissue Antigens*, 2006, **68**, 193.
8. S. Thiel, P. D. Frederiksen and J. C. Jensenius, *Mol. Immunol.*, 2006, **43**, 86.
9. J. Lu, S. Thiel, H. Wiedemann, R. Timpl and K. B. M. Reid, *J. Immunol.*, 1990, **144**, 2287.
10. K. B. M. Reid and R. R. Porter, *Biochem. J.*, 1976, **155**, 19.
11. T. Voss, H. Eistetter, K. P. Schäfer and J. Engel, *J. Mol. Biol.*, 1988, **201**, 219.
12. J. Lu, H. Wiedemann, R. Timpl and K. B. M. Reid, *Behring Inst. Mitt.*, 1993, **93**, 6.
13. J. Bella, M. Eaton, B. Brodsky and H. M. Berman, *Science*, 1994, **266**, 75.
14. T. Kawasaki, *Biochim. Biophys. Acta.*, 1999, **1473**, 186.
15. M. Dong, S. Xu, C. L. Oliveira, J. S. Pedersen, S. Thiel, F. Besenbacher and T. Vorup-Jensen, *J. Immunol.*, 2007, **178**, 3016.
16. M. Sumiya, M. Super, P. Tabona, R. J. Levinsky, T. Arai, M. W. Turner and J. A. Summerfield, *Lancet*, 1991, **337**, 1569.
17. H. O. Madsen, P. Garred, S. Thiel, J. A. Kurtzhals, L. U. Lamm, L. P. Ryder and A. Svejgaard, *J. Immunol.*, 1995, **155**, 3013.
18. C. M. Sørensen, T. K. Hansen, R. Steffensen, J. C. Jensenius and S. Thiel, *Clin. Exp. Immunol.*, 2006, **145**, 173.
19. R. J. Lipscombe, M. Sumiya, J. A. Summerfield and M. W. Turner, *Immunology*, 1995, **85**, 660.
20. P. Garred, F. Larsen, J. Seyfarth, R. Fujita and H. O. Madsen, *Genes Immun.*, 2006, **7**, 85.

21. P. D. Frederiksen, S. Thiel, L. Jensen, A. G. Hansen, F. Matthiesen and J. C. Jensenius, *J. Immunol. Methods*, 2006, **315**, 49.
22. P. H. Jensen, D. Weilguny, F. Matthiesen, K. A. McGuire, L. Shi and P. Højrup, *J. Biol. Chem.*, 2005, **280**, 11043.
23. S. Sheriff, C. Y. Chang and R. A. Ezekowitz, *Nat. Struct. Biol.*, 1994, **1**, 789.
24. W. I. Weis and K. Drickamer, *Structure*, 1994, **2**, 1227.
25. S. T. Iobst, M. R. Wormald, W. I. Weis, R. A. Dwek and K. Drickamer, *J. Biol. Chem.*, 1994, **269**, 15505.
26. A genomics resource for animal lectins. http://www.imperial.ac.uk/research/animallectins/
27. S. Thiel, *Mol. Immunol.*, 2007, **44**, 3875.
28. F. Teillet, M. Lacroix, S. Thiel, D. Weilguny, T. Agger, G. J. Arlaud and N. M. Thielens, *J. Immunol.*, 2007, **178**, 5710.
29. R. Wallis, *Immunobiology*, 2007, **212**, 289.
30. H. Liu, L. Jensen, S. Hansen, S. V. Petersen, K. Takahashi, A. B. Ezekowitz, F. D. Hansen, J. C. Jensenius and S. Thiel, *Scand. J. Immunol.*, 2001, **53**, 489.
31. R. Wallis, *Immunobiology*, 2002, **205**, 433.
32. T. Vorup-Jensen, E. S. Sørensen, U. B. Jensen, W. Schwaeble, T. Kawasaki, Y. Ma, K. Uemura, N. Wakamiya, Y. Suzuki, T. G. Jensen, K. Takahashi, R. A. B Ezekowitz, S. Thiel and J. C. Jensenius, *Int. Immunopharmacol.*, 2001, **1**, 677.

CHAPTER 9

Personal Accounts of the Discovery of MASP-2 and its Role in the MBL Pathway of Complement Activation

THOMAS VORUP-JENSEN[a, b, c] AND JENS CHR. JENSENIUS[b]

[a] Biophysical Immunology Laboratory; [b] Institute for Medical Microbiology and Immunology; [c] Interdisciplinary Nanoscience Center, University of Aarhus, DK-8000 Aarhus C, Denmark

9.1 Preamble

In a period spanning more than three decades, the MRC Immunochemistry Unit was to the authors of this chapter a 'friendly giant' in the sciences of the complement system that offered great support and guidance. Our account of how the mannan-binding lectin (MBL) associated serine proteases 2 and 3 (MASP-2 and MASP-3) were discovered and characterized is by no means the complete history of its topic. The field owes its very existence to the seminal work carried out by three Japanese groups, which is only briefly identified below. Our intention is the focus on the involvement of the MRC Immunochemistry Unit in this process. Not putting personal contributions aside, our account shows that the interactions between several people, and certainly those affiliated with the Immunochemistry Unit, created a productive atmosphere that materialized in the discovery of a molecule that nobody had looked for. Such a discovery is not an unusual thing to happen in science and the three

Molecular Aspects of Innate and Adaptive Immunity
Edited By Kenneth BM Reid and Robert B Sim

Published by the Royal Society of Chemistry, www.rsc.org

Princes of Serendip apparently travelled across the territory of many scientific disciplines. But to make sense of the unexpected, which almost by definition carries the stigma of 'not wanted', requires less the good fortune of fairy-tales and more the discipline of knowing the incompleteness of hypothesis-driven research. If the past forms the ground for expectations, unexpected findings are tightly linked with the past. When we began to figure out the path to MASP-2 and put it onto paper, we realized the convoluted nature of the events and found that quite a bit of background information was needed to understand what was going on. We decided to include here some of these aspects of the journey, which we believe some readers may find illuminating and amusing.

9.2 The Oxonian Connection

The fortunate publication on MASP-2 in *Nature* in 1997[1] propelled MASP-2 into its central position for our understanding of the third pathway of complement activation, the MBL or lectin pathway. The process that led to this paper was as convoluted as imaginable. While Steffen Thiel and Jens were the main players in the initial observations, Thomas and several others were instrumental in establishing structural and functional aspects. Central to this were the connections between Denmark and the MRC Immunochemistry Unit in Oxford.

In 1967 Jens travelled to East Grinstead to meet with Morten Simonsen (1921–2002), who conducted his research in the war-time barracks erected as surgical units to patch up Spitfire pilots, which involved transplantations for burn injuries. Morten was famous for his discovery of the graft-*versus*-host reaction (for many years called 'the Simonsen reaction'). He was using the dropped alantoic membrane technique (in fertilized chicken eggs) for his discovery of the unexpected, and still unexplained, stunningly high proportion of T-cells that react against histocompatibility antigens. Jens went back to Copenhagen to set up shop for Morten's return to Denmark. He had the pleasure of going out to buy a hammer and saw with his supervisor, L. T. Mann (a Vietnam veteran on a Fulbright re-establishing program) and finish furbishing the laboratories before Morten's return. He learned the beauty of the fertilized egg as an experimental model while working on Morten's hypothesis that the T cell antigen receptor (TCR) was involved in the protection by serum of the killing of chicken embryos by the intravenous (*i.v.*) injection of histocompatibility antigen preparations. As it turned out this did not involve the TCR, but instead demonstrated the powerful protection by antithrombin III against disseminated intravascular coagulation.[2] As we see below, eggs became an essential tool for finding MASP-2.

Morten insisted that for his PhD Jens should go to his beloved England, and not the USA, as Jens was planning. Morten, a communist saboteur during the war, was banned from entry to the US; he told Jens to contact his friend Rodney R. Porter (1917–1985). Porter had earned his decorations during the campaign up through Italy, and after the war got a rehabilitation PhD stipend with Fred Sanger. Thus it came about that Jens began his career at the famous Fourth Floor of the Biochemistry Building, possibly one of the ugliest in

Oxford, but providing such astonishingly beautiful views over the city. Like Morten, Rod Porter, or simply 'Prof.', was infatuated with the idea of discovering the nature of the TCR, and Jens worked on another great hypothesis, that an excess of this molecule, having no role in serum, might be found excreted in the urine. In practice the project involved the fractionation of enormous volumes of urine from rabbits immunized against the hapten 3-nitro-4-hydoxy-5-iodophenylacetate (NIP), to obtain Prof.'s urinary fragment (PUF). Prof. had just spent his sabbatical year in his own laboratory working on this project. As it turned out, the NIP-binding molecule, with an apparent molecular weight of 35 kDa on gel permeation chromatography, was identical to antigen binding fragment (Fab), attaining a more compact structure when reacting with the hapten.[3] Alan Williams (1945–1992) arrived from Adelaide and with him Jens started on a more direct search for T cell immunoglobulin (Ig), the Ig thought to make up the elusive receptor. Tough times make close friends, and thus the rather frustrating research ensured the continued and fruitful close contact between Jens and the MRC Research Unit, outlasting the death of both the Prof. and Alan, and many of Jens's students subsequently spent years in Oxford, some also doing their DPhils.

Back in Copenhagen with Morten Simonsen, Jens continued the search for the TCR, exploiting the unique feature of the bird immune system. This presents the possibility of completely wiping out B cells by bursectomy through *i.v.* injection of 12-day-old embryos with testosterone. While the results unequivocally disproved an Ig nature for the TCR,[4,5] the experiments were hampered by the chickens hatching with large amounts of circulating immunoglobulin antibody provided by the mother hen through the yolk. Jens realized that immunizing chickens and collecting the eggs would be the ideal source of nearly unlimited amounts of antibodies. Enthusiastic encouragement from Alan Johnstone, another laboratory mate from Oxford, stimulated Jens through experiments towards a reliable procedure for retrieving the antibody from the yolks of which the fat prevented conventional protein purification approaches.[6] As we shall see, the use of chickens to raise antibodies proved essential for the discovery of MASP-2.

9.3 MASP and the Serine Proteases of the Complement System

Before giving the history of how MASP-2 was discovered we review briefly some of the scientific findings made in this area over the past 20 years. MBL (also referred to as mannose or mannan-binding protein or, in some older literature, Ra-reactive factor) is now well-established as an important part of the innate immune system (Jensenius[7] gives an explanation as to why we much prefer 'mannan' for 'mannose'). Following MBL's discovery by Kawasaki *et al.* in 1978[8] it was later found that one biological function of the molecule was to activate the complement system when it bound appropriate carbohydrate ligands.[9] In 1992 Teizo Fujita and Misao Matsushita identified in human plasma a serine protease in complex with MBL and named it MBL-associated serine protease (MASP).[10] Based on murine plasma samples a similar finding had been made earlier by Kawakami, who named it p100 [referring to the apparent molecular weight on sodium dodecyl sulfate polyacrylamide gel

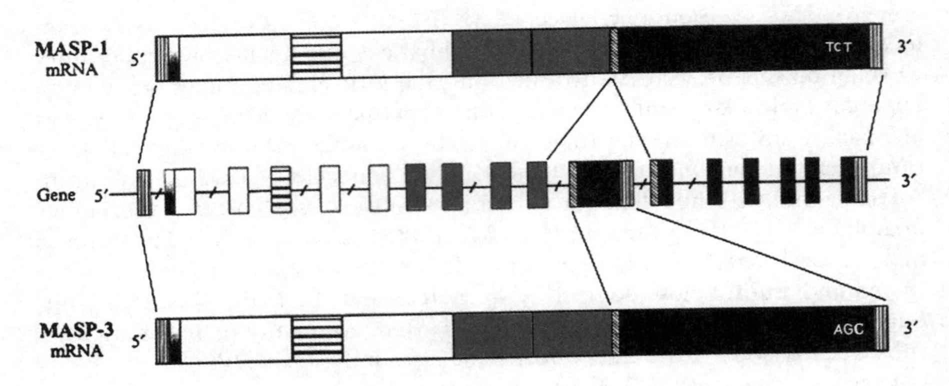

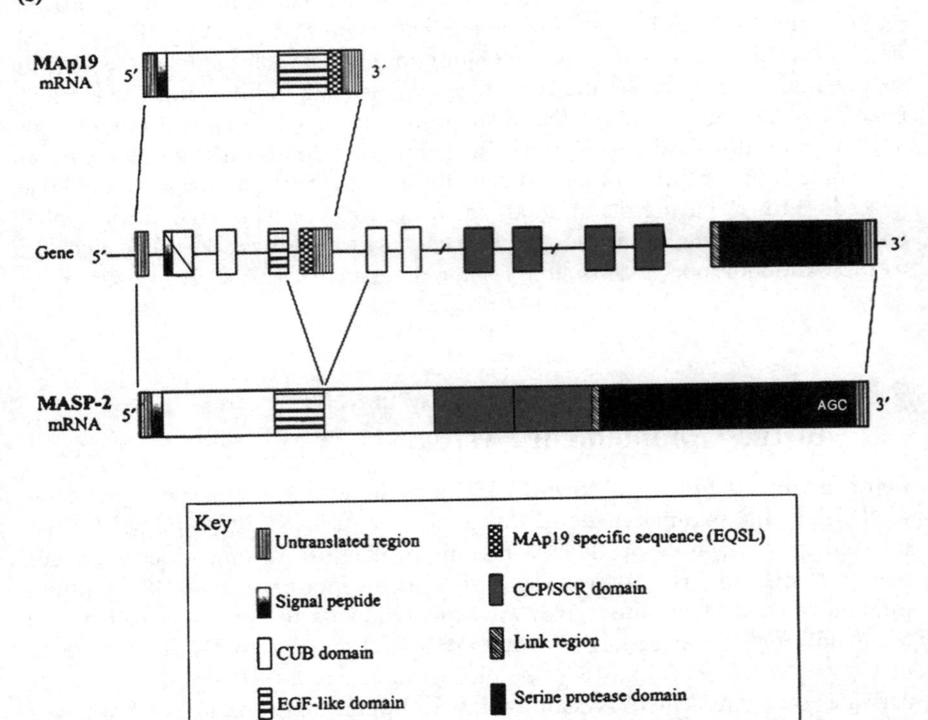

electrophoresis (SDS-PAGE)].[11] We now know that MBL forms complexes with four different molecules, which include three different MASPs and a smaller fragment derived from the *MASP2* gene, named MAp19.[1,12,13] The similarity to C1r and C1s (see below) quickly suggested a role for the MASPs in complement activation through the binding of MBL–MASP complexes to carbohydrate ligands. The early experiments, which indicated a role for C1r and C1s in complement activation through MBL,[14,15] were not supported by subsequent investigations. Indeed, in serum only the MASPs and MAp19, and not C1r or C1s, seems to form complexes with MBL.[16] In contrast to the selective binding of C1r and C1s to C1q only, it has been shown that the MASPs also form complexes with ficolins.[17] Originally thought to be part of the 'lectin pathway' of complement activation it is now clear that the ficolins appear to bind acetylated molecules.[18] The strong binding of ficolins to *N*-acetyl glucosamine is consequently not the result of structures intrinsic to carbohydrates, but that of acetylation.

The MASPs share their domain organization with the serine proteases of the C1 complex, C1r and C1s (Figure 9.1). Several other reviewers have discussed the similarities and differences between these proteases,[19–21] so only a brief outline is given here. The order of domains in the primary structure is, starting from the N-terminal, one complement Uegf–BMP-1 (CUB) domain followed by an epidermal growth factor (EGF)-like domain, a second CUB domain, two

Figure 9.1 Schematic representation of the genes that encoded human MBL-associated proteins and the transcription products. (A) The gene encoding MASP-1/3 is located at chromosome 3q27-q28 and encompasses 17 exons. The first exon encodes the 5′ untranslated region of the mRNA and is followed by nine exons that encode a signal peptide and the five N-terminal domains of the proteins. As indicated, these exons for the A chain are combined identically in the mRNA that encode MASP-1 and -3. These exons are further combined with either a MASP-3-specific exon encoding the link region, the protease domain and the 3′ untranslated region or with the remaining six exons encoding the homologous regions of MASP-1. Thus, as a result of the gene structure and alternative splicing, the first five domains of mature MASP-1 and MASP-3 are identical. (B) The gene that encodes MAp19/MASP-2 encompasses 12 exons and is located on chromosome 1 (1p36). The 5′ untranslated region encoded by the first exon is combined with the following three exons that encode a signal peptide, a CUB domain and an EGF-like domain. As shown, the formation of a MAp19-specific mRNA is obtained by the addition of a fifth exon that encodes the four MAp19-specific amino acids (EQSL) and the 3′ untranslated region. MASP-2 mRNA is obtained by alternative splicing that combines the first four exons with the remaining seven exons, thus removing the MAp19-specific exon. Hence, mature MAp19 and MASP-2 share the first CUB domain and the EGF-like domain. The base-triplet that encodes the active serine of the protease domain is indicated. The regions of the gene and mRNA that code for the individual protein domains are shown and a key given. Introns above 3000 bp are indicated (//). This illustration was part of a review paper on the MASPs written by Steen Vang Petersen *et al.*,[50] who spent part of his PhD studies (1999–2000) with Bob Sim in the MRC Immunochemistry Unit.[51]

complement control protein (CCP) domains, and a C-terminal serine protease domain. At least for MASP-1 and MASP-2 it was demonstrated that a small linker region, which separates the serine protease domain from the remainder of the protein, requires cleavage in the conversion of the proenzyme MASP into the enzymatically active protein.[10,22] Following activation of the enzyme the B chain (serine protease domain) and the A chain (CUB–EGF–CUB–CCP–CCP fragment) are linked by a disulfide bond. This structural feature of activation is shared with C1r and C1s.

With regard to the functionality of the domains it is now well-established that the CUB–EGF–CUB part of the molecule mediates the complex formation with MBL.[23] The CUB domains, moreover, seem to be involved in the dimerization of MASP, as revealed by X-ray crystallography studies on this part of MASP-2.[24,25] The dimerization could play a role in the autocatalytic activity of MASP-2. The MASPs form tight complexes with MBL and ficolins. Both biochemical and structural features distinguish these complexes from the C1 complex. The complex between MBL and the MASPs is broken only by the application of buffers with a high ionic strength and a strong chelator of divalent metal ions, such as ethylenediaminetetraacetic acid (EDTA),[10] while the complexes remain intact in the presence of either high salt concentrations or EDTA. By contrast, the C1 complex dissociates readily in EDTA and in media with high ionic strength. Further, when binding to targets, MBL binding is insensitive to salt strength, while C1q binding is (usually) abolished by high salt content. This difference was used to establish an assay that enables the direct testing of complement activation through MBL–MASP without a confounding influence from the classical pathway.[26] The plasma concentrations of the constituents of the MBL–MASP complexes are subject to considerable variation between individuals, with the MBL concentration ranging from near-complete deficiency to maximum levels of approximately 10 μg/ml. With the recent discovery that mutations in the human *MASP2* gene leads to functional deficiency in complement activation through MBL[27] the assay contributes significantly to identifying clinically significant variations between individuals.

9.4 Gene Characterization in the Pre-genomic Age: Discovering MASP-2

With the sequence of the human genome published in 2001 our methods of gene characterization have changed dramatically. The word 'characterization' is here chosen carefully because the post-genomic age lacks the possibility of discovering new genes in the most primitive sense of determining stretches of unknown nucleic acid sequence. In the mid-1990s this was far from the case and any survey of protein or DNA databases carried little chance of being helpful unless one had rather detailed information about what to search for. Many scientists will still remember that under these conditions the discovery of unknown genes and their protein products was an endeavour of endless biochemical fractionation, and all too many bands on stained SDS-PAGE gels or Western blots. This was followed by arduous protein sequencing and screening of cDNA libraries with radioactive DNA probes, which could still fail to identify your gene of interest for simple technical reasons that needed to be identified. The protein now known as MASP-2 was found in 1995, when biochemistry as a mean of identifying gene products was in its twilight and database-guided approaches on the rise. A few more years and the DNA databases would contain sequence information with many of the necessities to make a sound judgement on the nature of the protein encoded; a decade more and the human genome would be a searchable database.

The finding of MASP-2 grew out of a long-standing interest in lectins on which Jens had based his laboratory. Jens took up his new position as associated professor in Odense (1978), close to the university hospital, and threw himself into clinically oriented research to establish a useful technique for estimating circulating immune complexes (ICs), thought to be involved in a number of immune diseases, not least cancer. Having constructed such an assay[28] the results were to be compared with the results of some of the other 10 to 20 available IC assays. The chosen assay was the conglutinin-binding assay, and Jens became fascinated by the highly selective binding activity of this protein. The binding of conglutinin is restricted to the glycan group on the alpha chain of C3, but it binds only when the glycan is present on C3 in the form of iC3b.[29,30] Rodney Porter had, after a few years of foraging into cellular immunology, decided to leave this field to others and to focus instead on the complement system, a research area taken up at the time when K. B. M. (Ken) Reid arrived at Oxford (1969). Ken stayed with complement after taking over from the Prof., and thus it was only natural to contact Ken when Jens's student, Steffen Thiel, was ready to go abroad for a research year. In his masters degree Steffen had described the human equivalent to conglutinin which, despite Peter Lachmann's considerable resourcefulness, until then had been found only in the bovidae. Conglutinin happens to be the very first mammalian lectin described.[31,32] Of course, we now know that only the vanishing cat's smile had been observed, but the work nevertheless produced a nicely printed document from the US Patent Office, the laboratory's first adventure into the modern world of translational research. Nevertheless, some papers on human conglutinin were published, one in a memorial issue for the Prof.[33] The most interesting paper on conglutinin described the protective effect of bovine conglutinin against *Salmonella typhimurium* infections in mice. A dramatic increase in survival was observed.[34] Bafflingly, the activity decreased on successive purifications. In hindsight, the likely cause was the removal of a different, contaminating lectin, MBL, which was co-purified with conglutinin through the initial carbohydrate-affinity steps.

A central observation leading to the belief that humans have conglutinin was provided by chicken antibovine conglutinin antibodies (retrieved from eggs), which reacted with a likely candidate in human serum. The size and the calcium-dependent binding to oligosaccharides, like zymosan (yeast cell wall) and agarose, are shared with MBL, and contributed to the mistaken identification. In Oxford, Steffen Thiel embarked on studies of MBL, which most elegantly led to the discovery of MBL as the missing component in M. W. (Mac) Turner's studies on opsonin-deficient children who suffered repeated infections.[35] Continued studies on this protein after Steffen Thiel's return to Odense led to the realization that MBL had masqueraded as the proposed human conglutinin. Towbin, Staehelin and Gordon published their ground-breaking procedure of Western blotting,[36] and we persuaded Harry Towbin to come and show us how to perform it. The stage was set to identify the human proteins that react with the chicken antibovine conglutinin antibodies. Western blots of lectin preparations from human serum, developed with the chicken antibodies, revealed numerous bands besides MBL, most of which were also developed with normal chicken yolk IgG. Steffen and Jens moved to Aarhus in 1989, where Jens obtained a professorship in cancer immunology, and some years passed before he decided to concentrate on sorting out the bewildering results. Digging out old chicken antibodies, new immunizations of chickens (Figure 9.2), of rabbits and rats,

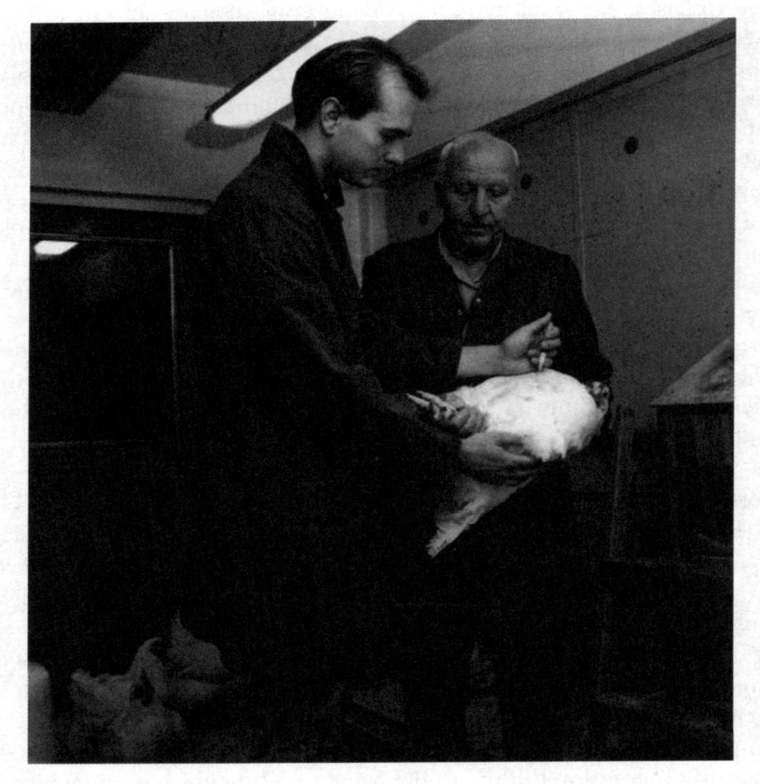

Figure 9.2 Immunization of chickens proved to be a determining strategy for making antibodies that specifically reacted with human MASP-2. The picture shows technician Hilmer Hald and Søren Hansen injecting a chicken with antigen. Søren Hansen contributed to the work on discovering MASP-2[1] and did parts of his PhD studies (1999) in the MRC Immunochemistry Unit in Ken Reid's laboratory working on SP-D.

purifications of cow and human lectins, and multiple new Western blots were produced. Interestingly, only chickens produced the cross-reactive antibodies. This seems to agree with the supposition that chickens, because of the large phylogenetic distance from mammalians, are good at producing species cross-reacting antibodies. The antibodies reacted with a 50 kDa protein, which looked interesting enough for us to solicit help to carry out N-terminal sequence analysis. In this respect A. C. (Tony) Willis at the MRC Immunochemistry Unit was well known as a master of the difficult art of deriving useful information from the minute samples offered to him by scientists hoping to learn what the bands on their Western blots consisted of. The limited sequence information produced by N-terminal sequencing showed some similarity with the C1r and C1s proteases, but the incompleteness of the databases of protein structures made claims over the nature of the protein in question difficult. Further progress could now only be made by devising some strategy to clone the cDNA that corresponded to the fragments of the protein sequence which had been determined by Tony Willis.

About this time Thomas was brought onboard the project with the intention that the cloning be part of his master thesis project, with Jens and Steffen Thiel as supervisors. Also in the group was Steen B. Laursen, who had played a role in the projects on conglutinin, which included the process that led to the mysterious new protein sequence. At the time he was working on the characterization of chicken MBL, which led to the cloning of the cDNA sequence for this protein.[37] A molecular biology program package for the Macintosh computer included an algorithm to calculate the phylogenetic trees of molecular ancestries.[38] When applied to the newly cloned DNA sequence for chicken MBL, as well as to other already known MBL sequences, the algorithm boldly predicted the time of the separation of the lineages that led to chickens and humans to 320 million years, a staggering precision considering the moderate amount of information fed into the program. As general enthusiasm for these calculations was at a high level in Aarhus, Thomas had already made several such phylogenetic inquiries into the molecular evolution of the group of proteins that we now know by the name of collectins.[39] None of this led to a separate publication, but one of these analyses has survived in a review paper by Jens and Uffe Holmskov.[40] By applying algorithms to test the phylogenetic relationship between proteins from the DNA sequence that encoded them it was possible to show that bovine conglutinin was sufficiently similar to the bovine lung surfactant protein SP-D that we could assume the most recent common ancestor of bovine conglutinin and bovine SP-D only existed at a later time point than the most recent common ancestor of the bovine and human SP-Ds. As the most recent common ancestor of bovine and human SP-D would likely coincide with the time at which the division of the evolutionary lineages that led to cows and humans, this finding suggested, more simply, that no human analogue of conglutinin was likely to exist. One would think that this should have posed a serious question about the reactivity of the chicken polyclonal antibodies and hence the rationale behind the project. But theoretical calculations in molecular biology have always had trouble outweighing experimental evidence. As is clear from the account below, this was not to be the only time that molecular phylogenetic algorithms would be more successful in forecasting the eventual outcome than the preconceptions of the investigators using them.

To start the project on cloning the most important decision was how to obtain the probe for screening libraries. Screenings had been performed successfully with short oligonucleotides. This approach requires the use of degenerate oligonucleotides to match the information provided by the protein sequence. With the aid and supervision of Knud Poulsen in the department at Aarhus, Thomas embarked on using such oligonucleotides for the polymerase chain reaction (PCR) amplification from a template of human genomic DNA. The idea was that a combination of primers derived from the peptides sequenced would amplify a DNA segment that contained more coding sequence. This strategy failed because of the location of intron sequences in the *MASP2* gene,[41] which we now know raised the length of the targeted PCR products above the maximum capacity of the PCR polymerases of the day. By happy coincidence, Jette Lovmand, a virologist in the department, made her considerable knowledge about PCR amplification of cDNA templates available to the project. With the application of this strategy, a PCR product of approximately 300 bp was finally amplified from human liver mRNA. This provided a piece of coding sequence that both strongly supported the similarity with the C1r and C1s proteases and identified the source of cDNA for a library that would contain the desired coding sequence.

As mentioned before, the cloning of cDNAs was not an entirely new thing in the Aarhus laboratory,[37] but also it was not a routine undertaking. Ken Reid kindly offered for the cloning work to be carried out in the Immunochemistry Unit by Thomas supervised by MRC scientists, who had plenty of experience in dealing with phage-based cDNA libraries. This approach quickly produced results in the form of phage clones with an inserted cDNA that encoded not only the sequence already known from the 300 bp probe, but also an additional sequence that placed the encoded protein as a close molecular relative of Matsushita and Fujita's MASP.

Although the cloned cDNA sequence clearly encoded a protein covering the domains found in sequences of C1r, C1s, and MASP-1, this was not contained in a single, open reading frame. At the time the simplest explanation was a technical artefact in the generation of the cDNA library. Wilhelm Schwaeble, at the University of Leicester and a long-time collaborator with the Unit, expressed a strong interest in the project, and Cordula Stover in his laboratory undertook the work, once again applying the cDNA probe to a human liver cDNA library. Through her meticulous investigation of the numerous clones generated not only was the final sequence of MASP-2 established, but it also generated an unexpected insight into the function of the *MASP2* gene as is discussed in more detail below.

With the cloning work completed at the end of 1995, the MASP-2 cDNA sequence arrived late, at least compared with other complement factors, and at a time when the number of known cDNAs was rapidly increasing. Novelty was no longer enough to carry through the publication of a cDNA sequence alone; a function of the encoded protein had to be demonstrated for publication in a high-impact journal. However, with the paradoxical circumstance that the protein found with antibovine conglutinin antibodies was not conglutinin-like at all, and with the growing suspicion that the earlier work on MASP had been carried out on preparations with more than one such molecule, it was not straightforward to suggest a strategy that would prove free of more contamination problems. It had already been shown in the Aarhus laboratory that the MASP-2 molecule was associated with MBL, but purification from MBL–MASPs complexes proved difficult. Contrary to the case for MASP-1, MASP-2 has not been purified from serum or plasma. A way to surmount this obstacle was devised by carrying out the separation on SDS-PAGE under gentle (non-reducing) conditions, followed by blotting onto a cellulose nitrate membrane and incubation of the membrane with C4 (provided by Alister Dodds). Activation of C4 would enable C4b to attach covalently to the membrane [this does not work with polyvinylidene fluoride (PVDF) membranes]. Deposited C4b could then be detected with antibodies to C4. The band developed coincided with the band developed with anti-MASP-2 antibodies, and not with the MASP-1 band. It was slightly perplexing that the mobility corresponded to a molecular weight of 50 kDa. On boiling the sample before SDS-PAGE, the band developed with anti-MASP-2 moved to an apparent molecular weight of 75 kDa, but no longer showed enzyme activity. This was the first evidence that MASP-2 was the C4-activating component of the MBL–MASP complexes, contrary to other evidence that attributed the activity to the original 'MASP', now redesignated as MASP-1.[10,42] Of course, it was a piece of good fortune that the MASP-2 enzyme was resilient enough to withstand the exposure to SDS and able to refold into an active protease on the Western blot membrane. The discovery of MASP-2 was presented at the British Society for Immunology meeting in December 1995,[43] and the complete sequence and the functional data in the summer of 1996[44] at the Complement Workshop in Boston. By then various suggestions had been offered on the enzymatic function of MASP (MASP-1), including cleavage of C2 and C4, as well as of C3.[2,10,45] In particular, the cleavage of C3 would nicely fit a view of the MBL–MASP complex as part of a primordial immune system in operation within life forms that predated the emergence of vertebrates. The discovery of MASP-2 suggested that complement activation through the MBL pathway could be similar to that of the classical pathway, in particular to the C1 complex that also had two proteases as part of the enzymatic machinery and was generally considered to originate with vertebrate evolution.

At the Boston conference we also listened to a brief historical account of the unfortunate ending of the life of Louis Pillemer (1908–1957) who, according to some people, suffered a nervous breakdown and killed himself, disappointed with the ridicule from colleagues on his postulate of the alternative pathway of complement activation. The MBL pathway was, instead, greeted with the excitement of the arrival of a new pet. In 1997 the finding was published in

Nature.[1] "How on earth did you get it into *Nature*" was heard from a few colleagues, while the much-admired Charles Janeway commented, "The paper is remarkably thorough – I will incorporate it in the new edition of my book". Jens had learned from Raymond Dwek in Oxford that nothing in life is more important than a *Nature* paper, and Raymond further stressed that you do not get a paper into *Nature* without a fight. Thus, the first direct rejection letter from Ursula Weiss was met by a much worked-on return letter. Reading it now, it seems to expose a slightly audacious edge, "Obviously, I understand all of your arguments. Seems fair enough. Only, I fail to see that any of them apply to our manuscript. Please, let me explain." Jens's arguments must have been received in good spirit – and, helped by friendly referees, the manuscript had a fair run.

9.5 Beyond Finding MASP-2

As mentioned above, Cordula Stover undertook a serious effort to characterize a number of cDNA clones identified with the MASP-2 probe and it became evident that the *MASP2* gene encoded more than one mRNA, a phenomenon referred to as alternative splicing (see Figure 9.1). Today, with the human genome sequenced, these processes have been used to explain how a mere 30 000–50 000 genes can meet the needs of a complicated multicellular organism, such as humans. However, the alternative splicing in the *MASP* gene was quite unexpected as it had no parallel among the C1 proteases. In the protein samples that contained MASP-2, low molecular weight protein species were visible following staining with antibodies to MASP-2. But more often than not in the world of protein purification, and in particular when one is dealing with proteases, this likely reflects protein degradation, which is a rather trivial observation. Thus the prominent band at 19 kDa that reacts with antibodies to MASP-2 was initially assumed to represent a degradation product.[43] However, the work of Cordula Stover and Wilhelm Schwaeble clearly showed that the *MASP2* gene gave rise to mRNA products which could encode such a polypeptide. MAp19, as the protein came to be known, only contains the two N-terminal domains of MASP-2 and not the serine protease domain.[13] A similar finding was published in the same year (1999) by Matsushita and Fujita's group.[46] Despite the biophysical characterization of the structure, as well as the kinetics of the interaction with MBL[25] we know nothing about the function of this product.

One important point about MAp19, as noted by Matsushita and Fujita, is that the existence of the truncated MASP-2 indicates that the MBL/MASP system was not as similar to the classical pathway as suggested by the many references made to the similarity between MASP and C1r and C1s.[46] However, the presence of two proteases in complex with MBL was conspicuously similar to the C1 complex, and on finding the ability of MASP-2 to cleave C4 it was natural to hypothesize that MASP-1 might serve as the activator of MASP-2, much as C1r activates C1s. However, phylogenetic analyses did not support this supposition. One would expect that the functional relationships proposed above would be reflected in a molecular ancestry with MASP-2 being closer to C1s than to C1r and *vice versa* for MASP-1. The first review paper on MASP-2 showed that this was not the case.[19] Nevertheless, such was the confidence in the symmetry between the classical and lectin pathways of complement activation that the failure of phylogenetic analysis was dismissed as the inability to infer the functional properties of proteins by comparison of their sequences alone. In the absence of methods to provide reasonably pure preparations of both MASP-1 and MASP-2 from plasma, recombinant synthesis methods were considered by the Aarhus laboratory. The high number of disulfide bridges in the native proteins, as well as the multidomain structure, clearly advocated the use of mammalian cell-line expression systems. After a brief spell in Wilhelm Schwaeble's laboratory working on generating a MASP-1 construct, Thomas returned to Oxford in the autumn of 1996, this time joining Simon J. Davis at his new laboratory at the Medical Science Division of the Nuffield Department of Clinical Medicine (NDM) at the John Radcliffe Hospital in Headington. With Simon Davis' expertise, which according to textbooks in the field,[47] included the world record in protein expression in CHO cell lines, appropriate constructs were made for the expression of MASP-1 and MASP-2. Compared with later, similar work it is an important observation that few ways to detect a recombinant protein existed and the only way forward required metabolic labelling of the recombinant proteins by including radioactive amino acids in the culture medium and subsequent capture of the protein through an engineered 'his tag'. This situation imposed numerous limitations on what was practicable.

Straining the situation even further was that it became gradually evident that the MASPs were not expressed. It was only the outstanding and committed supervision of Simon Davis and his insistence that every attempt to generate CHO cell clones be accompanied by a positive control to validate the methodology that made the data, if not immediately publishable, carry the scientific rigor that later allowed them to be a substantial part of a PhD thesis on this subject. As such, these experiments proved that good science may not always benefit humankind, but will benefit the person doing the experiments. However, even with this friendly philosophy as moral support, all projects meet their inescapable doomsday and laboratory notebooks must be opened to show how progress, if any, is made. A meeting was set up in the MRC Immunochemistry Unit one morning in the spring of 1997 between Jens, Thomas, Simon Davis and Ken Reid. It was clear from the beginning that at this stage there was little to report in terms of MASP expression, but as work was ongoing Thomas felt compelled to work up to literally the very last minute before this meeting to provide a glimpse of hope that the embarrassment always associated with things not working could be prevented. In the laboratories at NDM the man staring at autoradiographs absolutely devoid of bands (save the positive control, of course), but nevertheless claiming, with religious zest, to see weak shadows that confirmed his innermost wishes had, of course, already been the target for many remarks. However, with the arrival of Thomas's supervisor from Denmark, the high tension apparently called for an extraordinary poke of fun. The evening before the aforementioned meeting, Thomas completed one analysis of transfected CHO cell clones by placing a piece of photographic film on a dried SDS-PAGE with the radiolabelled CHO cell proteins. In the morning, in dizziness and with the rush to get from NDM to the MRC Immunochemistry Unit, the film was developed. Out of the machine came a rather confusing image with no resemblance to anything else these experiments had produced before, and at first entirely unintelligible. However, gradually and with the reluctant assistance of one other person present in laboratory, it became clear that this was the picture of the Virgin Mary carrying the infant Jesus. At this point there was no option but to bring the image to the meeting. Here it caused a good laugh, which can do a scientific project more good than even the strongest positive results. The carefully explored technique involves transfer of an image printed on an ordinary piece of paper onto photographic film commonly used for Western blots developed with chemiluminescence or autoradiography. By placing the printed image on top of the film, followed by a brief exposure to light and subsequent development of that film, the transfer would occur and add a remarkably artistic touch. The later-developed experimental film was as blank as any before it. The divine intervention marked a more realistic phase in the attempts to make MASP, now showing that even the 'belief' in MASP was more likely to produce miracles than MASP itself.

It is worth noting that Russell Wallis, working in the Glycobiology Institute, not far away from the John Radcliffe Hospital, observed that rat MASP-1 and -2 could not be made in CHO cells.[48] With methods and constructs at hand from Simon Davis' laboratory and improvements in the antibodies, thanks to the work of Steen Vang Petersen in Aarhus, we eventually found a way to make MASP-2 in a mammalian cell.[22] With recombinant MASP-2 available came a surprise: no MASP-1 was required for its activation. Recombinant MASP-2 in its proenzyme state was easily activated when exposed to ligand-bound MBL. Moreover, the MBL–MASP-2 complex was able to activate C4 and C2 to form the C3 convertase, C4b2b, otherwise associated with the classical pathway of complement activation.[22] The necessary tools for the job, plentiful supplies of freshly made complement components and the means to make them were found with Bob Sim and Alister Dodds from the Immunochemistry Unit. Similar findings were made later by Ce-Belle Chen and Russell Wallis with rat MASP-2,[49] and structural studies now seem to support this observation.[24] MASP-2 thus shared the properties of both C1r, which is the autoactivating and C1s-activating protease of the C1 complex, and C1s, which activates C4 and C2. This fits nicely with the phylogenetic analysis of the MASPs and other serine proteases of the complement system mentioned above.[19]

9.6 MASP-3: One Gene, Two Enzymes

If the presence of MAp19 and the apparent enzymatic autonomy of MASP-2 had not already finished off the C1 complex as a viable paradigm for the MBL–MASP complex, the nail in the coffin was provided by the discovery of MASP-3.[12] In some respects the discovery of this enzyme proceeded similarly to that of MASP-2 described above. A band reactive with antibodies in Western blotting provided enough protein for Tony Willis to sequence it. With this information it was possible to derive primers that, once again, amplified a segment of 174 bp from human liver cDNA. However, this time a screen of liver cDNA libraries was not necessary. As highlighted earlier, the human genome project was moving quickly forward and extensive data were already available. By entering the sequence determined by DNA sequencing of the amplified PCR product, Mads Dahl found that the segment aligned in a locus identical to the locus assigned for MASP-1. The exon–intron boundaries for *MASP1* were already established, so it was reasonably easy to work out that the amplified cDNA came from a segment of the *MASP1* gene that had not previously been thought to contribute to a mature mRNA product from this gene (Figure 9.1). The segment separates those exons already known for the serine protease domain of MASP-1 from the exons that encode the CUB–EGF–CUB–CCP–CCP part of MASP-1. The real surprise was that the presumed exon, which had been identified, could encode a complete serine protease domain, more similar to the domains of C1r, C1s, and MASP-2 than to the one encoding the MASP-1 serine protease domain. Experiment showed that, indeed, an enzyme was encoded by this segment of DNA, together with the upstream segments that encoded the A chain of MASP-1. We had an example of one gene that encoded two enzymes, and hence more than a semantic contrast to the famous concept in the early days of molecular biology of George Beadle and Edward Tatum of 'one gene, one enzyme'. Alternative splicing is one way in which organisms that require a large number of proteins can minimize the size of their genome. This is necessary according to postulates by J. B. S. Haldane in the early 1930s, who argued that a large number of genes increases the likelihood of lethal mutations. From data on mutational frequencies Haldane actually arrived at 30 000 genes as being the maximum load any organism could carry, in fair agreement with the most recent suggestion for the number of genes in the human genome. Although the generation of the MAp19 product by the *MASP2* gene is a similar process, it is certainly more striking that alternative splicing in the case of the *MASP1/3* gene exchange domains has a central functional significance. If, indeed, evolution has carefully compressed the gene load in a manner that would request such a sophisticated economy of making multiple enzymes, one would also expect these enzymes to carry out significant functions in the body. It quickly became clear, by identifying the MASP-3 orthologues in other species, that this serine protease domain is unusually well conserved. The first publication on MASP-3 reported a 95% conservation in primary structure between the B chain of the porcine and human analogues,[12] leading to the view that this enzyme fulfils an essential function that tolerates only little variation in the structure of the enzyme. Yet few clues exist today as to the function of the enzyme.

9.7 End Note

Despite all the MASPs having been known for years, significant problems remain unresolved and even unaddressed, with respect to both function and genetics. While MASP-2 appears to behave according to its now designated role, only a few researchers seem happy with MASP-1 playing a significant role in complement activation. MAp19 might have a controlling function (we could not reproduce this), and our own finding of inhibitory activity expressed by MASP-3 stretches the imagination. It seems fair to say that we await credible functions for these three molecules. The function of MASPs associated with the ficolins is still open to many questions, as is the function of the ficolins.

References

1. S. Thiel, T. Vorup-Jensen, C. M. Stover, W. Schwaeble, S. B. Laursen, K. Poulsen, A. C. Willis, P. Eggleton, S. Hansen, U. Holmskov, K. B. Reid and J. C. Jensenius, *Nature*, 1997, **386**, 506–510.
2. L. T. Mann, Jr., J. C. Jensenius, M. Simonsen and U. Abildgaard, *Science*, 1969, **166**, 517–518.
3. J. C. Jensenius, *Scand. J. Immunol.*, 1973, **2**, 541–550.
4. J. C. Jensenius, M. Crone and C. Koch, *Scand J. Immunol.*, 1975, **4**, 151–160.
5. J. C. Jensenius, *Immunology*, 1976, **30**, 145–155.
6. J. C. Jensenius, I. Andersen, J. Hau, M. Crone and C. Koch, *J. Immunol. Methods*, 1981, **46**, 63–68.
7. J. C. Jensenius, *Science*, 1995, **270**, 1104.
8. T. Kawasaki, R. Etoh and I. Yamashina, *Biochem. Biophys. Res. Commun.*, 1978, **81**, 1018–1024.
9. K. Ikeda, T. Sannoh, N. Kawasaki, T. Kawasaki and I. Yamashina, *J. Biol. Chem.*, 1987, **262**, 7451–7454.
10. M. Matsushita and T. Fujita, *J. Exp. Med.*, 1992, **176**, 1497–1502.
11. F. Takada, Y. Takayama, H. Hatsuse and M. Kawakami, *Biochem. Biophys. Res. Commun.*, 1993, **196**, 1003–1009.
12. M. R. Dahl, S. Thiel, M. Matsushita, T. Fujita, A. C. Willis, T. Christensen, T. Vorup-Jensen and J. C. Jensenius, *Immunity*, 2001, **15**, 127–135.
13. C. M. Stover, S. Thiel, M. Thelen, N. J. Lynch, T. Vorup-Jensen, J. C. Jensenius and W. J. Schwaeble, *J. Immunol.*, 1999, **162**, 3481–3490.
14. J. H. Lu, S. Thiel, H. Wiedemann, R. Timpl and K. B. M. Reid, *J. Immunol.*, 1990, **144**, 2287–2294.
15. M. Ohta, M. Okada, I. Yamashina and T. Kawasaki, *J. Biol. Chem.*, 1990, **265**, 1980–1984.
16. S. Thiel, S. V. Petersen, T. Vorup-Jensen, M. Matsushita, T. Fujita, C. M. Stover, W. J. Schwaeble and J. C. Jensenius, *J. Immunol.*, 2000, **165**, 878–887.
17. T. Fujita, *Nat. Rev.*, 2002, **2**, 346–353.
18. A. Krarup, S. Thiel, A. Hansen, T. Fujita and J. C. Jensenius, *J. Biol. Chem.*, 2004, **279**, 47513–47519.
19. T. Vorup-Jensen, J. C. Jensenius and S. Thiel, *Immunobiology*, 1998, **199**, 348–357.
20. S. Thiel, *Mol. Immunol.*, 2007, **44**, 3875–3888.

21. M. Matsushita, Y. Endo, M. Nonaka and T. Fujita, *Curr. Opin. Immunol.*, 1998, **10**, 29–35.
22. T. Vorup-Jensen, S. V. Petersen, A. G. Hansen, K. Poulsen, W. Schwaeble, R. B. Sim, K. B. M. Reid, S. J. Davis, S. Thiel and J. C. Jensenius, *J. Immunol.*, 2000, **165**, 2093–2100.
23. C. Gaboriaud, F. Teillet, L. A. Gregory, N. M. Thielens and G. J. Arlaud, *Immunobiology*, 2007, **212**, 279–288.
24. H. Feinberg, J. C. Uitdehaag, J. M. Davies, R. Wallis, K. Drickamer and W. I. Weis, *EMBO J.*, 2003, **22**, 2348–2359.
25. L. A. Gregory, N. M. Thielens, M. Matsushita, R. Sorensen, G. J. Arlaud, J. C. Fontecilla-Camps and C. Gaboriaud, *J. Biol. Chem.*, 2004, **279**, 29391–29397.
26. S. V. Petersen, S. Thiel, L. Jensen, R. Steffensen and J. C. Jensenius, *J. Immunol. Methods*, 2001, **257**, 107–116.
27. K. Stengaard-Pedersen, S. Thiel, M. Gadjeva, M. Moller-Kristensen, R. Sorensen, L. T. Jensen, A. G. Sjoholm, L. Fugger and J. C. Jensenius, *N. Engl. J. Med.*, 2003, **349**, 554–560.
28. J. C. Jensenius, H. C. Siersted and A. P. Johnstone, *J. Immunol. Methods*, 1983, **56**, 19–32.
29. P. J. Lachmann and H. J. Muller-Eberhard, *J. Immunol.*, 1968, **100**, 691–698.
30. S. B. Laursen, S. Thiel, B. Teisner, U. Holmskov, Y. Wang, R. B. Sim and J. C. Jensenius, *Immunology*, 1994, **81**, 648–654.
31. J. Bordet and O. Streng, *Ann. Inst. Pasteur*, 1906, **49**, 260–276.
32. M. A. Leon and R. Yokohari, *Science*, 1964, **143**, 1327–1328.
33. J. C. Jensenius, S. Thiel, G. Baatrup and U. Holmskov-Nielsen, *Biosci. Rep.*, 1985, **5**, 901–905.
34. P. Friis-Christiansen, S. Thiel, S. E. Svehag, R. Dessau, P. Svendsen, O. Andersen, S. B. Laursen and J. C. Jensenius, *Scand. J. Immunol.*, 1990, **31**, 453–460.
35. M. Super, S. Thiel, J. Lu, R. J. Levinsky and M. W. Turner, *Lancet*, 1989, **2**, 1236–1239.
36. H. Towbin, T. Staehelin and J. Gordon, *Proc. Natl. Acad. Sci. USA*, 1979, **76**, 4350–4354.
37. S. B. Laursen, T. S. Dalgaard, S. Thiel, B. L. Lim, T. V. Jensen, H. R. Juul-Madsen, A. Takahashi, T. Hamana, M. Kawakami and J. C. Jensenius, *Immunology*, 1998, **93**, 421–430.
38. J. Hein, *A unified appproach to alignment and phylogenies*, Academic Press, San Diego, 1990.
39. U. Holmskov, R. Malhotra, R. B. Sim and J. C. Jensenius, *Immunol. Today*, 1994, **15**, 67–74.
40. U. Holmskov and J. C. Jensenius, Chapter 3 in *Collectins and Innate Immunity*, R. A. B. Ezekowitz, K. Sastry and K.B.M. Reid (eds), R.G. Landes Company, Austin, 1996.
41. C. Stover, Y. Endo, M. Takahashi, N. J. Lynch, C. Constantinescu, T. Vorup-Jensen, S. Thiel, H. Friedl, T. Hankeln, R. Hall, S. Gregory, T. Fujita and W. Schwaeble, *Genes Immun.*, 2001, **2**, 119–127.

42. R. T. Ogata, P. J. Low and M. Kawakami, *J. Immunol.*, 1995, **154**, 2351–2357.
43. S. Thiel, T. V. Jensen, S. B. Laursen, A. Willis and J. C. Jensenius, *Immunology*, 1995, **86**, 101.
44. T. Vorup-Jensen, C. Stover, K. Poulsen, S. B. Laursen, P. Eggleton, K. B. M. Reid, A. C. Willis, W. Schwaeble, J. H. Lu, U. Holmskov, J. C. Jensenius and S. Thiel, *Mol. Immunol.*, 1996, **33**, 81.
45. M. Matsushita and T. Fujita, *Immunobiology*, 1995, **194**, 443–448.
46. M. Takahashi, Y. Endo, T. Fujita and M. Matsushita, *Int. Immunol.*, 1999, **11**, 859–863.
47. C. Bebbington, *Use of vectors based on gene amplification for the expression of cloned genes in mammalian cells*, Oxford University Press, Oxford, 1995.
48. R. Wallis and R. B. Dodd, *J. Biol. Chem.*, 2000, **275**, 30962–30969.
49. C. B. Chen and R. Wallis, *J. Biol. Chem.*, 2004, **279**, 26058–26065.
50. S. V. Petersen, S. Thiel and J. C. Jensenius, *Mol. Immunol.*, 2001, **38**, 133–149.
51. S. V. Petersen, S. Thiel, L. Jensen, T. Vorup-Jensen, C. Koch and J. C. Jensenius, *Mol. Immunol.*, 2000, **37**, 803–811.

CHAPTER 10

The Structure and Function of Ficolins, MBLs and MASPs

RUSSELL WALLIS,[a] ANDERS KRARUP[b] AND UMAKHANTH VENKATRAMAN GIRIJA[b]

[a] Department of Infection, Immunity and Inflammation, University of Leicester, Leicester, UK; [b] MRC Immunochemistry Unit, Department of Biochemistry, University of Oxford, Oxford, UK

10.1 Introduction

In this review we focus on the structure and function of mannose-binding lectin (MBL) and ficolins and their interactions with MBL-associated serine proteases (MASPs) to initiate activation of the lectin pathway of complement. Work carried out by Jinhua Lu and Steffen Thiel in 1990, in Ken Reid's group, made an important early contribution to this field, in a collaboration with Hanna Wiedemann and Rupert Timpl from the Max Plank Institute.[1] Lu *et al.* had shown in electron micrographs that MBL is composed of a mixture of oligomers that resemble bouquets. They were able to separate the different oligomeric forms of MBL and demonstrate that at least some of these oligomers (pentamers and hexamers) were able to bind and activate $C1r_2s_2$ complexes of the classical pathway of complement, *via* a mechanism that is independent of C1q. Subsequently, a different but related serine protease, called MASP-2, was shown to be the major complement-activating enzyme of MBL, through work that was pioneered by the group of Jen Jensenius from Aarhus, together with members of the Immunochemistry Unit and researchers from the University of Leicester.[2] Since these ground-breaking demonstrations of complement activation by MBL, much has been discovered about the 'new' pathway of complement, now

Molecular Aspects of Innate and Adaptive Immunity
Edited By Kenneth BM Reid and Robert B Sim

Published by the Royal Society of Chemistry, www.rsc.org

known as the lectin pathway, which paradoxically is probably the most ancient of all three complement pathways.[3] Numerous contributions have come from the Immunochemistry Unit, mainly from the laboratories of Ken Reid and Bob Sim. In addition, major contributions have been made by many international researchers, who have either worked within the Unit at some stage in their careers or who have close collaborations with members of the Unit. Here, we describe these findings and outline our current understanding of the molecular interactions that lead to complement activation *via* the lectin pathway.

10.2 MBL and Ficolins in the Innate Immune System

Ficolins and mannose-binding lectins (MBLs) are the first components of the lectin branch of the complement system, and provide a frontline defence within the immune system by neutralizing pathogens *via* antibody-independent mechanisms. They are both multidomain, oligomeric proteins assembled from multiple copies of a subunit, each of which is composed of three identical polypeptide chains.[4,5] They have characteristic bouquet-like shapes in electron micrographs, in which long stems joined to each other at one end are tipped by globular heads at the other. They bind directly to surface-exposed carbohydrates, or *N*-acetyl groups, on microorganisms *via* the globular heads, and activate the complement cascade *via* MBL-associated serine protease-2 (MASP-2), which binds to the stems. Complement activation leads to host-mediated lysis and phagocytosis of microorganisms and stimulates inflammatory and adaptive immune responses *via* arrays of complement receptors on host cells.[6]

The importance of the lectin pathway is highlighted by a common immunodeficiency, caused by mutations to the human MBL gene.[7] Affected individuals are susceptible to bacterial, viral and parasitic infections, particularly in early childhood, before the adaptive immune system is established,[8] or when adaptive immunity is compromised, for example during human immunodeficiency virus (HIV) infection or following chemotherapy.[9,10] The lectin pathway also plays a role in the pathogenesis of inflammatory disorders, such as cystic fibrosis and rheumatoid arthritis, in which variant MBL alleles are associated with more severe disease.[11] Although complement normally has a protective role, the lectin pathway has been directly implicated in causing significant complement-mediated damage to the host. For example, tissues (including heart and kidney) are subject to severe complement-mediated cytotoxicity upon reperfusion of oxygenated blood following transient ischaemia, although the underlying mechanism of this process is poorly understood.[12–14]

10.3 Genetics of MBLs and Ficolins

10.3.1 Genetics and Tissue Distributions of MBLs

All mammals investigated to date possess two MBL genes, although in humans, chimpanzees and gorillas, only one of these genes, designated *MBL2*, encodes a functional protein (Figure 10.1). The second gene, *MBL1P1*, is a transcribed pseudogene.[15–17] The two genes probably resulted from a duplication that occurred after the divergence of birds and mammals, because chickens and fish have just one MBL gene.[18–20] In humans, both MBL genes are clustered together with those of other collectins (SP-A and SP-D) on chromosome 10(q21-24).[21]

In rodents, syntenic regions on mouse and rat chromosomes 14 and 16 contain the genes for MBL-A, while the homologues of *MBL2* (*Mbl2* and *MAB-C_RAT*) are located on chromosomes 19 and 1, respectively.[22] Interestingly, the major serum MBL in rats and mice is MBL-A, whereas MBL-C has a relatively low complement activity.[23] MBL-C was originally identified as a liver protein, although more recent studies have shown that it is also present in serum.[24]

A common human immunodeficiency associated with variant *MBL2* alleles[25,26] is caused by a lack of functional MBL in the sera of affected individuals.[27] Three single point mutations were discovered in exon 1 of *MBL2*, which encodes the signal peptide, the cysteine-rich domain and the first seven Gly–Xaa–Yaa repeats of the collagen-like domain,[8,28,29] causing amino acid substitutions at codons 52, 54 and 57 (R52C, G54D and G57E), respectively. These variant alleles are known as D, B and C, respectively, and are collectively referred to as O, while the wild-type allele is called A. All three variants impair the function of MBL.

Plasma MBL levels are also influenced by three common promoter polymorphisms, which are given the allele symbols H/L, X/Y and P/Q.[29,30] Of these, the X allele, a single change at nucleotide position –221, has the greatest effect on MBL levels. XA/XA individuals have functional MBL concentrations comparable with those who are heterozygous for one of the exon 1 mutations, while XA/YO individuals have undetectable MBL levels. Individuals homozygous for the Q allele (at nucleotide position +4) have MBL levels approximately two-fold greater than heterozygotes and 2.5 times greater than P/P homozygotes, while those individuals with the H allele (at nucleotide position –550) have marginally higher MBL levels than those with the L allele. The three structural polymorphisms are in linkage disequilibrium with the promoter polymorphisms, and only seven of the possible haplotypes occur frequently: HYPA, LYPA, LYQA, LXPA, LYPB, LYQC and HYPD. For most disease-association studies the H/L and P/Q alleles are usually ignored, and the variant structural alleles grouped together, which gives just three common haplotypes: YA, XA and YO. The corresponding six genotypes can then be divided into three groups associated with low, medium and high levels of functional MBL. Individuals with the YO/YO and XA/YO genotypes fall into the low-level group and are essentially MBL deficient.[31]

Typical plasma MBL levels for Caucasians are 1.2–5.0 μg/ml for normal individuals, 0.05–1.0 μg/ml for heterozygotes (A/O) and undetectable (<20 ng/ml) for homozygote (O/O) mutants.[31] However, these estimates represent the total functional MBL concentrations rather than the total protein concentrations, because the antibody that is generally used to measure MBL (mAb 131-01, from Antibody Shop, Gentofte, Denmark) only detects the larger oligomers typically

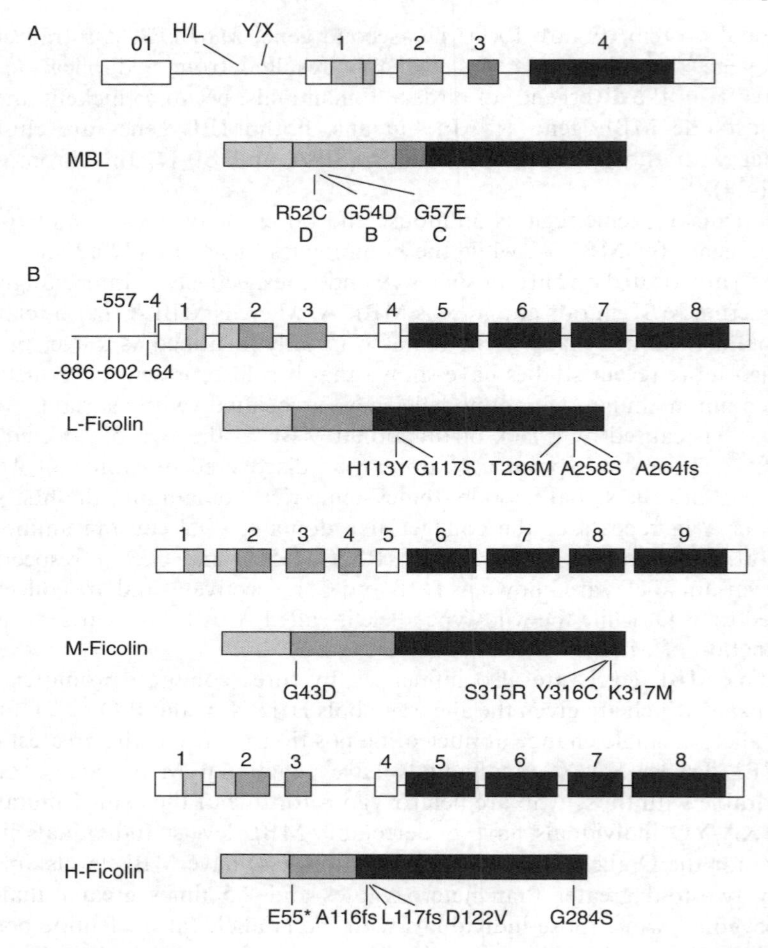

Figure 10.1 Gene organizations of human MBL and L-, M- and H-ficolins. In each case, the gene is shown on the *top* and the encoded polypeptide on the *bottom*. Numbering of polypeptide chains includes the leader sequences. Exons are represented as *boxes* and are connected by introns (not shown to scale). The cysteine-rich, collagen and target-recognition domains are shaded *light*, *medium* and *dark* grey, respectively, and exons that encode each domain are shaded correspondingly. (A) Organization of the human MBL gene. Positions of the common MBL mutations associated with immunodeficiency are marked on the polypeptide chain and the positions of the promoter mutations that affect serum concentrations are indicated on the gene. (B) Organization of human ficolin genes. Positions of polymorphisms are indicated.

found in A/A individuals. Recent measurements show that even sera from O/O homozygotes contain relatively high levels of MBL, but this MBL is composed of smaller oligomers (mainly monomers and dimers of subunits) that have low complement activities and are not recognized by mAb 131-01. Sera from A/O donors contain a mixture of larger and smaller oligomers and sera from O/O donors contain only smaller oligomers. Although they do not provide a true measure of total MBL serum concentrations, antibody-based assays are nevertheless useful for disease-association studies, because the amounts of MBL detected in plasma samples appear to correlate closely with their abilities to activate the lectin pathway.

10.3.2 Genetics and Tissue Distributions of Ficolins

Humans produce three ficolins called L-, H- and M- encoded by the *FCN2*, *FCN3* and *FCN1* genes located on chromosomes 9q34 (L- and M-) and 1p35.3 (H-ficolin)[5] (Figure 10.1). L-ficolin and M-ficolin are 79% identical,[32] and are believed to have arisen through a relatively recent gene duplication[33,34] based on their common location and high degree of sequence identity. H-ficolin is less closely related (45% identical to both L-ficolin and M-ficolin),[32] so probably diverged further back in time. L-ficolin is produced in the liver. The serum concentration has been estimated to range between 3.7 and 5.0 μg/ml in Caucasians and to be ~13.7 μg/ml in a Japanese population, with no more than a 10-fold concentration difference between the highest and the lowest concentrations measured.[35–38] Nevertheless, concentrations are affected by three polymorphisms in the promoter region and an additional polymorphism in the structural gene (exon 8). Some disease-association studies on L-ficolin have been published, although no clear links between L-ficolin concentrations and any infectious diseases have been reported so far.

Human H-ficolin is produced mainly in the liver by hepatocytes and bile-duct epithelial cells and in lungs by ciliated bronchial epithelial cells and type II alveolar epithelial cells. Serum concentrations are ~20 μg/ml. Several polymorphisms have been described in the FCN3 gene (Figure 10.1), including in one instance a stop codon within the region encoding the fibrinogen-like domain, which probably gives rise to a non-functional protein.

Human M-ficolin, unlike H- and L-ficolins, does not appear to circulate in serum. Rather, it is expressed in peripheral monocytes, lung and spleen. It is present in secretory granules in the cytoplasm of peripheral neutrophils and monocytes and in type II alveolar epithelial cells in the lung, which implies that it is a secretory protein released upon stimulation of these cells. M-ficolin has been found on the surface of monocytes where it is believed to function as an adhesion molecule promoting phagocytosis, but as the monocyte matures into a macrophage the M-ficolin gene is downregulated.[39,40]

Ficolins have been described in a number of mammals in addition to in humans, including rodents and pigs.[33] Mice and rats have only two ficolins, ficolin-A, which is a serum protein produced mainly in the liver with similar properties to human-L ficolin, and ficolin-B, which is a non-serum protein and is localized in the lysosomes of peripheral leukocytes and bone marrow and is probably the homologue of human M-ficolin. Interestingly, H-ficolin has only been identified in humans, although a pseudogene is present in mice.[41]

10.4 Structural Organization of MBLs and Ficolins

MBLs and ficolins have similar but not identical domain organizations.[5] Polypeptides comprise a short N-terminal domain that contains one or more cysteine residues, followed by a collagenous domain and a C-terminal target-recognition domain. In MBLs, target recognition is mediated through C-type (Ca^{2+}-dependent) carbohydrate-recognition domains (CRDs), which are linked to the collagenous domain through an α-helical coiled coil (Figure 10.2). Ficolins bind to microorganisms through fibrinogen-like domains that connect to the collagenous domain through a nine amino acid linker.

The collagenous domains of MBLs and ficolins resemble vertebrate collagens in their sequences and post-translational modifications. They consist of repeats of the tripeptide Gly–X–Y, where X and Y can be any residue other than glycine. Proline residues in the Y positions are usually hydroxylated after translation, but before folding, to form 4-hydroxyprolines.[23,42] This modification

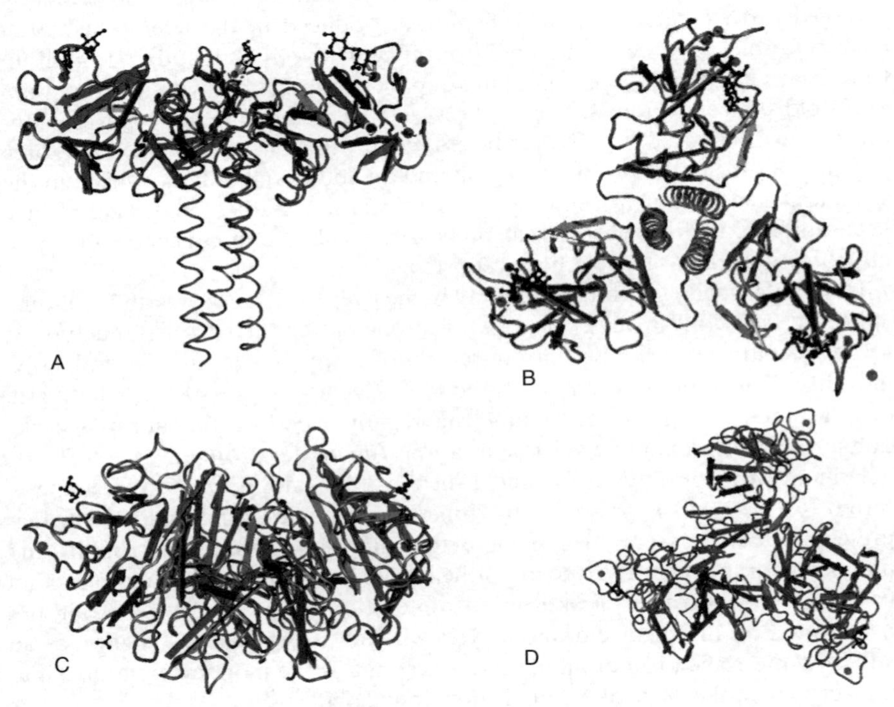

Figure 10.2 Pathogen recognition by MBL and ficolins. (A, B), Structure of a trimeric fragment of rat MBL-A, comprising the α-helical coiled coil and CRDs, in complex with a mannose-containing oligosaccharide.[94] The carbohydrate (Man-α1, 3-Man) is shown in *black*. Ca^{2+} are shown as *grey* spheres. (C, D), Fibrinogen-like domains of human H-ficolin in complex with D-fucose.[46] The fucose is shown in *black*. Ca^{2+} are shown as *grey* spheres. The images shown in (B) and (D) are rotated 90° with respect to those in (A) and (C).

facilitates folding and increases the stability of the collagenous domain. At least some of the lysine residues in the Y positions are hydroxylated and sometimes also glycosylated to form glucosyl-galactosyl- and galactosyl-5-hydroxylysines. The function of these modifications is unclear, but they might block non-specific interactions of the collagenous domain with other macromolecules and may also influence the interactions between the separate collagen helices.[43] Studies of rat MBL show that collectin polypeptides assemble into subunits soon after biosynthesis.[44] Each subunit consists of a long collagenous stalk attached *via* the neck to a rigid cluster of three CRDs. Assembly is probably initiated by association of the CRDs and neck regions, which form stable homotrimers even in the absence of the N-terminal domains[45] (Figure 10.2). By analogy, ficolins are also likely to fold in a C- to N-terminal direction. The fibrinogen-like domains form stable trimers, even in the absence of the N-terminal domains, and these structures probably drive the assembly of the collagen helices[46] (Figure 10.2).

During biosynthesis, multiple MBL or ficolin subunits associate to form supramolecular complexes, stabilized at the N-terminal ends of the polypeptides by disulfide bonds. Human MBL comprises dimers to hexamers of subunits, of which trimers and tetramers are the predominant species,[1,47] while rat MBL is composed mainly of dimers, trimers and tetramers of subunits.[48] Heterogeneity is a common feature of collectins and ficolins. It probably arises because of the way in which oligomers assemble and are stabilized by disulfide bond formation within the endoplasmic reticulum before secretion. Each polypeptide in human MBL and rodent MBL-As has three N-terminal cysteine residues, which potentially can form disulfide bonds. Certain bonding patterns allow covalent attachment to two additional MBL subunits and are thus compatible with further oligomerization, whereas other patterns permit linkage to only one subunit and thus probably serve as terminating patterns.[49] The extent of oligomerization is thus dependent on how many subunits associate before incorporation of terminators. Before secretion, any surface-exposed, unpaired cysteine residues are probably capped by disulfide bonding to cysteine, as occurs to many other secreted proteins, so only those oligomers with no exposed cysteines are secreted.[48,49]

Ficolins are often described as tetramers of subunits, but at least some heterogeneity probably occurs, as in MBLs. For example, rat ficolin-A consists of mixtures of oligomers ranging from monomers to tetramers of subunits, each assembled from three identical polypeptide chains.[50] Smaller oligomers have also been described in preparations of human L-ficolin from serum preparations.[51]

MBLs and ficolins probably contain multiple flexible regions (Figure 10.3), which permit the structural changes that induce activation of their associated MASPs upon binding to a target cell. The junction between the N-terminal domain and the collagen-like domain probably serves as a flexible hinge that causes the collagenous stems to splay apart to form the bouquet-like structures observed in electron micrographs. An interruption in the Gly–Xaa–Yaa collagen repeat, present in all mammalian MBLs and in some ficolins (L- and M-, but not H-ficolin) might also be flexible. While there is no direct evidence in MBLs, visualization of type IV collagen chains by electron microscopy shows

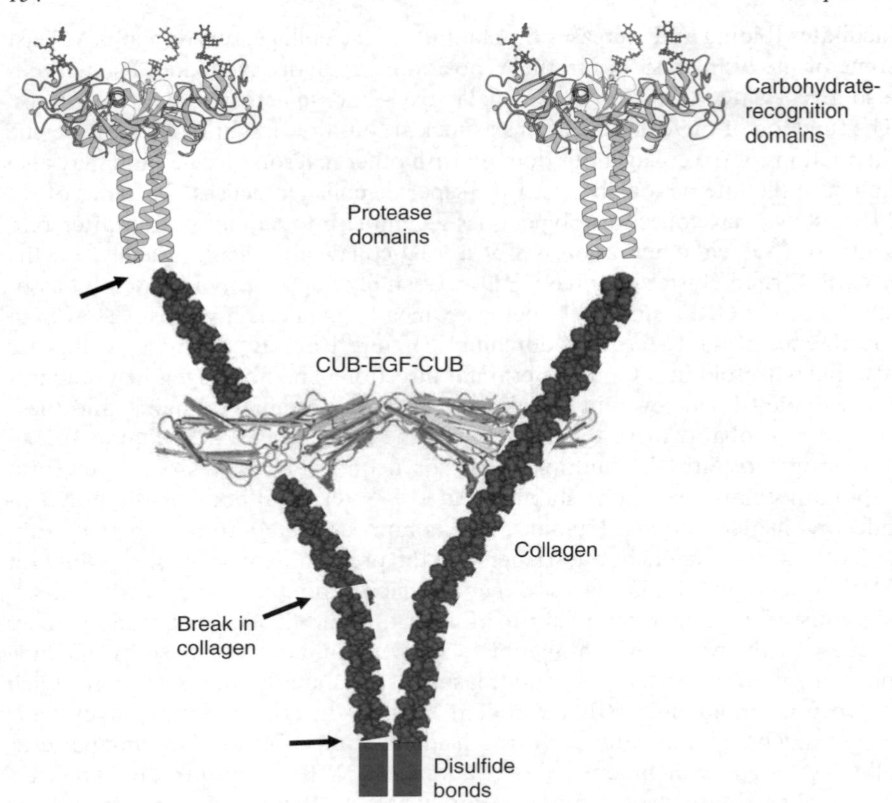

Figure 10.3 Model of the structural organization of a MBL–MASP-2 complex. Only the CUB1–EGF–CUB2 domains of MASP-2 are shown. The protease domains are probably located in the middle of the cone-shaped MBL dimer. Potentially flexible regions in MBL are indicated by arrows. Activation of the complex is probably induced by splaying apart of the MBL subunits.

some flexible sites, and the positions of these sites correlate with several sequence interruptions.[52] All human ficolins, including H-ficolin, can activate complement, so the interruption to the collagenous domain cannot be essential for complement activation, although it might modulate the activity in MBLs and L- and M-ficolins. The junction between the collagen triple helix and the α-helical neck region represents another potentially flexible region in MBL. At this point, the three polypeptides become aligned in the α-helical coiled coil, whereas they are staggered in the collagenous domain change. Corresponding regions in type I and II class A macrophage scavenger receptors are highly flexible.[53] Flexibility at this junction in MBL or at the collagen–fibrinogen junction in ficolins would enable the pathogen recognition domains to orient as the molecule docks onto a bacterial cell surface. The subsequent conformational changes induced at the N-terminal end of the molecule presumably cause a global conformational change, which probably provides the trigger for MASP activation (see Section 10.7).

10.5 Target Recognition by MBLs and Ficolins

10.5.1 Sugar Recognition by MBLs

MBLs recognize targets through their C-type CRDs.[4] CRDs belong to a larger domain family called C-type lectin-like domains (CTLDs),[54] only some of which are lectins. Each CRD has a single binding site for a terminal sugar moiety. Upon binding, the sugar forms a ternary complex with the CRD and a Ca^{2+} (Figure 10.2).[55] In rat MBL, asparagine and glutamate residues form hydrogen bonds to the 3- and 4-OH groups of the sugar and also make co-ordination bonds to the Ca^{2+}.[55] Binding specificity is determined largely by the configuration of the hydroxyl groups on the carbohydrate moiety, so that residues with equatorial 3- and 4-OH groups or their equivalent, such as mannose, *N*-acetylglucosamine and fucose, bind with similar affinities, whereas sugars such as galactose, where the 3- and 4-OH groups are axial, are poor ligands. The relatively small binding site and limited number of contacts between the sugar and the CRD, mean the affinity of a single interaction is weak ($\sim$1 mM). High affinity binding that initiates complement activation is achieved only through multiple CRD–sugar interactions.

The three CRDs of each MBL subunit are arranged in an ideal orientation to bind to ligands that project from a flat surface, such as a bacterial or fungal cell[56,57] (Figure 10.2). They are maintained in a fixed geometry through hydrophobic packing between the CRD of one polypeptide and the α-helix that forms the neck in an adjacent polypeptide. Most host sugars terminate in sialic acid residues, which are not ligands, so MBLs bind only transiently or not at all to host glycoproteins. Moreover, the spacing between the three binding sites on a MBL subunit is too far apart to allow multiple CRD–sugar interactions with a typical mammalian complex or hybrid sugar epitope, so even those host sugars that do terminate in mannose-like sugars generally interact only weakly.[57] Consequently, MBLs typically bind only transiently to viable host cells through their sugar-binding sites. However, there are exceptions. For example, human MBL binds to immobilised α2 macroglobulin and other thiolester-containing proteins *via* exposed oligomannose glycans. These interactions might lead to opsonization or activation of enzyme systems, including complement, and thus represent another amplification mechanism within the complement cascade.[58]

10.5.2 Ligand binding by Ficolins

Ficolins bind to *N*-acetylated structures, such as GlcNAc and GalN*A*c, through their trimeric fibrinogen-like domains[51] (Figure 10.2). These domains occur widely in both invertebrates as well as in vertebrates. *In vitro*, L-ficolin not only binds to sugars, but also to other acetylated structures, *e.g.* *N*-acetylglycine and *N*-acetylcysteine, implying that it is the N-acetyl group itself that is a key component of the interaction. L-Ficolin appears to have relatively wide ligand specificity compared to H- and M-ficolins. It binds to lipoteichoic acid,[59] a major component of the cell walls of Gram-positive bacteria, and β-D-glucan, a component of yeast cell surfaces. Interestingly, L-ficolin appears to have several different binding sites on each fibrinogen-like domain – a single outer binding pocket called the S1 site (which is also present in H-ficolin and in the invertebrate homologue tachylectin 5A) together with three additional unique sites, S2, S3 and S4, which appear to form a continuous binding surface for various acetylated and neutral carbohydrate ligands.[46] The S2 and S3 sites interact with a variety of monosaccharides, and the S3 and S4 sites make contacts with a four-residue 1,3-β-D-glucan structure. Thus, unlike MBL, which binds only to terminal sugar moieties, L-ficolin appears more suited to binding to extended oligosaccharides. The nature and linkage of the oligosaccharide ligand probably determines the specificity of the interaction, presumably favouring non-self structures over those present on host cell surfaces.

There are relatively few studies of the antimicrobial activity of L-ficolin. The purified protein has been shown to bind and deposit complement on the surface of *Salmonella typhimurium* TV119,[38,60,61] as well as increasing the opsonophagocytosis of group B streptococci (*Streptococcus agalactiae*).[62] The most comprehensive study of antimicrobial activity of L-ficolin was done with 20 *Str. pneumoniae* serotypes, as well as the 13 known *Staphylococcus aureus* serotypes.[36] It was found that L-ficolin bound selectively to the bacterial capsule constituents of some serotypes of both *Str. pneumoniae* and *Sta. aureus*, but far from all. Since the capsule composition of most of the *Str. pneumoniae* serotypes is known,[63] attempts have been made to identify an L-ficolin binding motif, but no consensus pattern has yet been identified.[36] In addition to its antimicrobial activities, L-ficolin has been shown to participate in the clearance of apoptotic and dead host cells.[64,65] Recently, L-ficolin was reported to bind to C-reactive protein (CRP), and possibly utilise CRP bound to microorganisms as a pattern of recognition.[66]

H- And M- ficolins appear to bind *N*-acetylated ligands through the S1 site and lack the additional binding sites seen in L-ficolin[46,67] (Figure 10.2). The fibrinogen-like domain of M-ficolin is able to accommodate sialic acid as well as the acetylated sugar residues that bind to L- and H-ficolins. This difference in binding specificity probably results from steric effects at the binding sites. In L-ficolin, access to the relatively bulky sialic acid residue is blocked by a phenylalanine residue at position 221 and a threonine residue at position 256, in place of the glycine and alanine residues at equivalent positions in M-ficolin. Another interesting feature of M-ficolin is that a conformational change occurs at lower pH and renders it unable to bind to its ligands.[67,68] This observation, together with the finding that M-ficolin appears to bind to the surface of peripheral leukocytes, is consistent with a role in opsonophagocytosis, in which M-ficolin traffics between the cell surface and the lysosome. Through this mechanism, M-ficolin secreted upon macrophage activation would bind to its target microorganism, and then become internalized as a complex with its ligand. The pH drop in the lysosomes would then trigger the conformational change switch, resulting in the release of ligand. Afterwards, M-ficolin may be recycled and deposited back on the cell membrane. While the main function of M-ficolin might be opsonophagocytosis, it is able to activate complement (together with L- and H-ficolin). Interestingly, mouse ficolin-B, the probable homologue of human M-ficolin, appears to have lost this activity.[69] Recombinant mouse ficolin-B is neither able to bind MASP-2 nor activate the complement cascade, probably as a result of a single amino acid substitution in the MASP binding site[50] (see Section 10.6).

Relatively few natural ligands for H- or M-ficolins have been described. M-Ficolin can bind to *Escherichia coli* and *Sta. aureus* and promote the uptake of these bacteria by phagocytosis.[40,70] H-Ficolin binds to *Aerococcus viridans*,[36,71] and polysaccharide derived from the same bacterium.[72] Recombinant H-ficolin was found to aggregate erythrocytes derivatized with lipopolysaccharides (LPSs) from *S. typhimurium*, *S. minnesota* and *E. coli* (O111), and this aggregation could be inhibited by GalNAc, GlcNAc and D-fucose.[73] These findings are in conflict with observations using serum-derived H-ficolin and the inhibition of H-ficolin binding to *A. viridans*,[51] where no inhibition by GlcNAc, GalNAc and ManNAc was observed. Furthermore, the fibrinogen-like domain could be co-crystallized with galactose and D-fucose, but with none of the *N*-acetylated sugars.[46] Thus, uncertainties still remain regarding the binding specificity of H-ficolin.

10.6 Interactions between MBLs, Ficolins and MASPs

MBL and most ficolins bind to three different MASPs (-1, -2 and -3) and a small non-enzymic protein called MAp19 or sMAP.[74] MASPs are homologues of C1r and C1s of the classical pathway of complement and comprise two complement Uegf–BMP-1 (CUB) domains, separated by a Ca^{2+}-binding epidermal growth factor (EGF)-like domain, followed by two complement control protein (CCP) modules and a serine protease domain. A short linker connects the CCP modules and the serine protease domains. MASP-1 and -3 are alternatively spliced products from a single gene, sharing the same N-terminal domains but with different linkers and serine protease domains. MAp19 is a truncated product of the MASP-2 gene and consists of the N-terminal CUB1-EGF domains of MASP-2.[75,76]

MASPs are synthesized as zymogens and circulate as complexes with MBL and ficolins. When these complexes bind to a target, MASP-1 and MASP-2 molecules autoactivate, which leads to cleavage of each MASP polypeptide within the short linker region at the N-terminal end of the protease domain. The protease domain of the MASP remains attached to the N-terminal domains through a single disulfide bond. While MASP-3 also becomes activated through an equivalent cleavage, this reaction is not autocatalyzed, but presumably occurs through an unknown protease or proteases in the serum.[77] Only MASP-2 has a clearly defined role within the lectin pathway of complement. Upon activation, it cleaves complement component C4, releasing the small anaphylotoxin C4a and the C4b fragment, which attaches covalently to the surface of a microorganism through its reactive thiolester bond.[2,78] C2 binds to C4b and is subsequently also cleaved by MASP-2 to generate the C3 convertase, C4b2a. *In vitro* studies show that MASP-1 cleaves C2 but not C4, so it might enhance complement activation triggered by lectin–MASP-2 complexes, but cannot initiate activation itself.[79] Relatively little is known about MASP-3, or its biological substrates.

Recently, work carried out in the laboratory of Bob Sim has revealed novel activities for MASPs within the clotting cascade. MASP-2 is able to promote fibrinogen turnover by cleavage of prothrombin, to generate thrombin.[80] The limited, localized thrombin activity is likely to deposit fibrin on the surface to which the MASP-2 is bound. The resulting release of fibrinopeptides and the deposition of fibrin will attract phagocytes that serve as adhesion points for immune system cells. MASP-1 might also initiate the clotting cascade. *In vitro* studies have shown that it has thrombin-like activity, and activates factor XIII and cleaves fibrinogen directly, so might also stimulate phagocytosis through fibrinogen activation.[81]

All three MASPs are homodimers arranged in an antiparallel configuration and stabilized through interactions between the CUB1 and EGF-like domains, so that the CCP and protease domains extend from either end of the dimer interface.[82,83] A single Ca^{2+} binds near the N-terminal end of each EGF-like domain and several residues, which lie near this Ca^{2+}, including a β-hydroxyasparagine residue, participate in interactions with the CUB1 domain of the adjacent protomer to stabilize the dimer. MASPs bind to the collagenous domain of MBL through interactions that involve the CUB1–EGF–CUB2 fragment of each subunit.[84] MAp19 (CUB1–EGF of MASP-2) also binds to MBL, but about five-fold more weakly than full-size MASP-2, which indicates that the CUB2 domain is important for stabilizing the MBL–MASP complex. The binding site for MBL is probably located near a second Ca^{2+} ion at the distal end of the CUB1 domain.[83] The head-to-tail arrangement of protomers in the MASP dimer means that each CUB1 domain is able to bind to a separate MBL subunit through equivalent interactions, so that each MASP dimer has binding sites for at least two MBL and/or ficolin subunits (Figure 10.3), and possibly up to four subunits (see below). MBL dimers thus bind to single MASP dimers, while MBL trimers and tetramers can bind to up to two MASPs, although 1:1 complexes are more stable than 1:2 complexes.[85]

Recently an additional binding site has been identified in the CUB2 domain of MASP-1. This extra site probably explains why full-size MASPs or CUB1–EGF–CUB2 fragments bind more tightly than CUB1–EGF fragments and suggests that each MASP dimer might interact with up to four MBL subunits.[86] However, the contribution of the CUB2–MBL interactions appears to be relatively small compared to the interactions that involve the CUB1–EGF domains, because, as noted above, MAp19 binds only about five-fold more weakly than full-size MASP-2.

While the stoichiometry of MBL oligomers binding to individual MASP-1 or MASP-2 proteins has been determined, the composition of circulating MBL–MASP complexes is still uncertain. Unlike C1r and C1s in the classical pathway of complement, which form $C1r_2s_2$ tetramers, MASPs-1 and -2 do not interact with each other in the absence of MBL.[85] However, the larger MBL oligomers can potentially bind more than one MASP, so *in vivo* these oligomers might circulate bound to two different MASPs or one MASP together with MAp19. Analysis of complexes separated from human serum tentatively supports this suggestion, because larger MBL oligomers were found to associate with MASP-2 and MASP-3, while smaller MBL oligomers bound to MAp19 and MASP-1, although the precise composition of these complexes was not determined.[87] However, more recent studies, which compared the MBL and MASP activities in the sera of over 100 individuals, found that MASP-1 activity on a mannan-coated surface is inversely correlated with MASP-2 activity and *vice versa*, suggesting that native MBL–MASP complexes do not have fixed MBL stoichiometries, but rather that there are separate populations of MBL–MASP-1 and MBL–MASP-2 complexes, the concentrations of which show wide inter-individual variation.[88]

MBLs and ficolins bind to MASPs *via* a short segment of the collagen-like domain C-terminal to the break in the collagenous domain in MBLs and in L- and M-ficolins.[43,50] Equivalent MASP-binding motifs are present in all MBLs and ficolins known to activate complement, and consist of the sequence OGKXGP, where X is generally an aliphatic amino acid or a methionine residue (Figure 10.4). Within this sequence, the lysine residue is essential for MASP-2 binding and complement activation in rat MBL and ficolin-A, as well as in human MBL.[89] The surrounding proline and hydroxyproline residues may not contribute directly to the interaction, but probably help to stabilize the MASP-binding site on the collagenous domain.[50] All three MASPs bind to the same region of the collagen-like domains, although mutagenesis studies suggest that the binding sites, while overlapping, are not identical.[43]

Not all MBL or ficolins contain the complete MASP-binding motif. Notably, mouse ficolin-B naturally contains a glutamic acid residue in the position equivalent to the aliphatic–methionine residue. Interestingly, the mouse protein does not bind to MASP-2 or activate complement, which suggests that the acidic residue at this position prevents MASP-2 binding. In addition, porcine MBL-C and ficolin-A both lack the lysine residue within the binding motif that is known to be essential for MASP binding. Porcine MBL-A contains the standard binding motif, whereas ficolin-B contains a similar sequence, VGKAGP. It is not known whether these proteins activate complement.

It is of interest that C1qs also contain sequences similar to the MASP-binding motif, at equivalent positions in all three polypeptide chains. For example, the A, B and C chains of human C1q contain the sequences OGKVGY, OGKVGP and OGKNGP, respectively. These sites probably form part of the binding sites for the MASP homologues, C1r and C1s. *In vivo*, there is no apparent cross-reactivity between C1q and MASPs or between MBLs and/or ficolins and C1r and/or C1s, although interactions between MBL and C1r or C1s have been

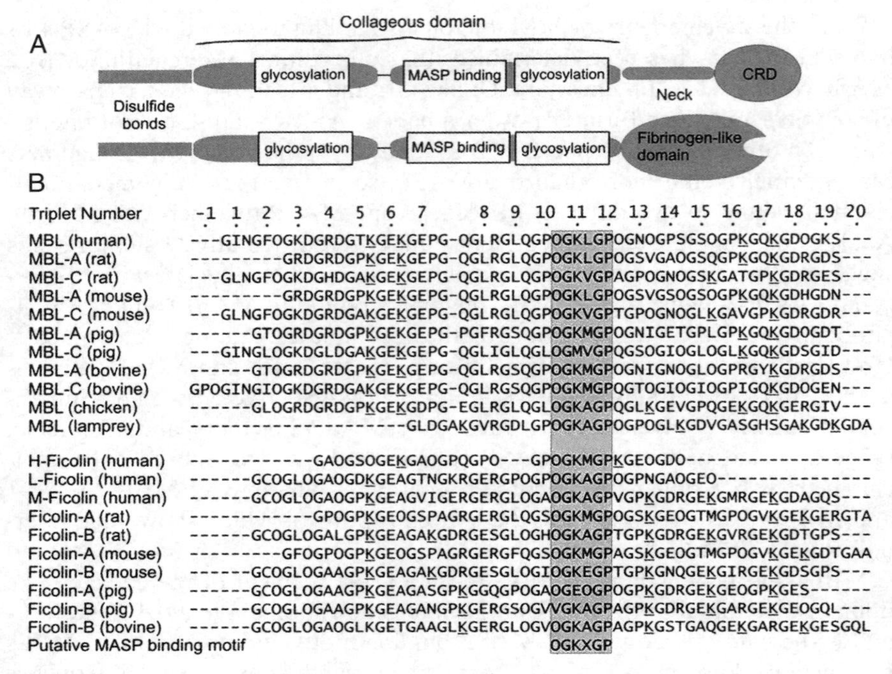

Figure 10.4 Sequence alignment of the collagenous domains of MBLs and ficolins highlighting the putative MASP-binding sites. (A) Domain organizations of MBL (top) and ficolins (bottom). MASP-binding sites are located between two glycosylated regions, which probably prevent non-specific interactions. (B) Aligned sequences of the collagen-like domains of MBLs and ficolins. Numbering of triplets is based on the sequence of human MBL. The putative MASP-binding motif is shaded. Lysine residues within the Y position of the Gly–X–Y repeat are underlined. All such residues are at least partially hydroxylated and glycosylated in rat MBL-A and MBL-C.[23,48] Most of the proline residues in the Y positions are represented as hydroxyproline (O) based on the sequences of rat MBL-A and MBL-C, in which all such residues are at least partially derivatized, except for the proline residue that immediately precedes the interruption to the Gly–X–Y collagen consensus repeat, which is unmodified in each case.

described *in vitro*.[1] It will be of interest to determine how binding specificity is achieved by the initiating complexes of the classical and lectin pathways.

In conclusion, the likely organization of an MBL–MASP complex is where each MASP dimer bridges two or more MBL subunits by interacting with a portion of the collagenous domain, just C-terminal to the interruption in the Gly–Xaa–Yaa repeat (Figure 10.3). This binding arrangement would allow the CCP and serine protease domains of the MASP to occupy the space between the MBL subunits in the middle of the cone-shaped MBL oligomer, with enough room to permit the conformational changes required to activate the protease upon binding of MBL to a cell surface.[82]

10.7 Mechanism of Complement Activation in the Lectin Pathway

Kinetic studies have shown that MBL activates MASP-2 by increasing the rate of autocatalysis when MBL–MASP-2 complexes bind to an activating surface, such as a bacterial cell wall.[79] The MASP bound to the collagenous domains of MBL is far removed from the sugar-binding sites on the CRDs (Figure 10.3), so the 'activation signal' must be transmitted from the C-terminal to the N-terminal ends of the molecule. In theory, a conformational change in each MBL subunit could induce activation. However, no significant changes occur when isolated CRDs or trimeric fragments that comprise the neck and CRDs bind to sugars.[55,90] The most likely activation mechanism is where binding of lectin-MASP complexes to a bacterial surface induces a global change in the arrangement of MBL subunits. Insights into the type of conformational changes that might occur have come from recent studies using atomic force microscopy, which show that the CRDs appear to splay apart upon binding to a sugar-coated surface.[91] Each MASP spans two or more MBL subunits, so it would be sensitive to movement of the collagenous stems. The resulting conformational change in the MASP presumably causes the serine protease domains to move so that the catalytic site of one subunit of the dimer can access the linker region of its partner. Reciprocal cleavage of MASP polypeptides would then fix the protease domains in the active conformation, enabling recognition and activation of downstream substrates to initiate the reaction cascade. A major challenge will be to gain structural insight into this activation mechanism.

Structural insights into zymogen activation and substrate recognition have emerged from structural studies of the CCP2–serine protease and CCP1–CCP2–serine protease domain of human MASP-2.[92,93] The larger fragment forms an extended 'club-' or 'mace-like' structure, in which the ellipsoid CCP domains project in an extended conformation from the globular serine protease domain. In the inactive enzyme, the serine protease domain shows typical features of zymogen structures of the chymotrypsin family. The catalytic triad is present in the active conformation, but the oxyanion hole and the substrate specificity pocket are missing. Upon activation, significant conformational changes occur to the activation domain, as well as to other loops of the serine protease domain. Interestingly, although MASP-2 and C1s of the classical pathway of complement have very similar substrate specificities, the enzyme–substrate interactions that mediate these activities are different in the two serine protease domains.[93]

10.8 Conclusions

In conclusion, the molecular mechanisms that lead to complement activation *via* the lectin pathway are beginning to emerge, and considerable progress has been made over the past few decades in defining the components of the pathway and characterizing the way in which they interact with each other to initiate activation. It is clear that the lectin pathway plays important roles in preventing infectious disease, particularly when the adaptive immune system is compromised, but also in causing damage under certain circumstances. Further characterization of the mechanisms of complement activation *via* the lectin pathway will not only help us to understand how the immune system fights infection, but will also provide a sound basis on which to design novel therapeutics to control complement activation and prevent self damage.

References

1. J. Lu, S. Thiel, H. Wiedemann, R. Timpl and K. B. M. Reid, *J. Immunol.*, 1990, **144**, 2287–2294.
2. S. Thiel, T. Vorup-Jensen, C. M. Stover, W. Schwaeble, S. B. Laursen, K. Poulsen, A. C. Willis, P. Eggleton, S. Hansen, U. Holmskov, K. B. M. Reid and J. C. Jensenius, *Nature*, 1997, **386**, 506–510.
3. A. W. Dodds, *Immunobiology*, 2002, **205**, 340–354.
4. K. Drickamer and M. E. Taylor, *Ann. Rev. Cell Biol.*, 1993, **9**, 237–264.
5. T. Fujita, M. Matsushita and Y. Endo, *Immunol. Rev.*, 2004, **198**, 185–202.
6. M. C. Carroll, *Nat. Immunol.*, 2004, **5**, 981–986.
7. M. W. Turner, *Immunology Today*, 1996, **17**, 532–540.
8. M. Sumiya, M. Super, P. Tabona, R. J. Levinsky, T. Arai, M. W. Turner and J. A. Summerfield, *Lancet*, 1991, **337**, 1569–1570.
9. P. Garred, H. O. Madsen, U. Balslev, B. Hofmann, C. Pedersen, J. Gerstoft and A. Svejgaard, *Lancet*, 1997, **349**, 236–240.
10. O. Neth, I. Hann, M. W. Turner and N. J. Klein, *Lancet*, 2001, **358**, 614–618.
11. D. C. Kilpatrick, *Biochim. Biophys. Acta.*, 2002, **1572**, 401–413.
12. M. L. Hart, K. A. Ceonzo, L. A. Shaffer, K. Takahashi, R. P. Rother, W. R. Reenstra, J. A. Buras and G. L. Stahl, *J. Immunol.*, 2005, **174**, 6373–6380.
13. M. Moller-Kristensen, W. Wang, M. Ruseva, S. Thiel, S. Nielsen, K. Takahashi, L. Shi, A. Ezekowitz, J. C. Jensenius and M. Gadjeva, *Scand. J. Immunol.*, 2005, **61**, 426–434.
14. M. C. Walsh, T. Bourcier, K. Takahashi, L. Shi, M. N. Busche, R. P. Rother, S. D. Solomon, R. A. Ezekowitz and G. L. Stahl, *J. Immunol.*, 2005, **175**, 541–546.
15. N. Guo, T. Mogues, S. Weremowicz, C. C. Morton and K. N. Sastry, *Mamm. Genome*, 1998, **9**, 246–249.
16. J. Seyfarth, P. Garred and H. O. Madsen, *Hum. Mol. Genet.*, 2005, **14**, 2859–2869.
17. M. V. Verga Falzacappa, L. Segat, B. Puppini, A. Amoroso and S. Crovella, *Genes Immun*, 2004, **5**, 653–661.
18. S. B. Laursen, T. S. Dalgaard, S. Thiel, B. L. Lim, T. V. Jensen, H. R. Juul-Madsen, A. Takahashi, T. Hamana, M. Kawakami and J. C. Jensenius, *Immunology*, 1998, **93**, 421–430.

19. R. Sastry, J. S. Wang, D. C. Brown, R. A. Ezekowitz, A. I. Tauber and K. N. Sastry, *Mamm. Genome*, 1995, **6**, 103–110.
20. L. Vitved, U. Holmskov, C. Koch, B. Teisner, S. Hansen, J. Salomonsen and K. Skjodt, *Immunogenetics*, 2000, **51**, 955–964.
21. K. Sastry, G. A. Herman, L. Day, E. Deignan, G. Bruns, C. C. Morton and R. A. Ezekowitz, *J. Exp. Med.*, 1989, **170**, 1175–1189.
22. R. A. White, L. L. Dowler, L. R. Adkison, R. A. Ezekowitz and K. N. Sastry, *Mamm. Genome*, 1994, **5**, 807–809.
23. R. Wallis and K. Drickamer, *Biochem. J.*, 1997, **325**(Pt 2), 391–400.
24. S. Hansen, S. Thiel, A. Willis, U. Holmskov and J. C. Jensenius, *J. Immunol.*, 2000, **164**, 2610–2618.
25. J. Miller and R. N. Germain, *J. Exp. Med.*, 1986, **164**, 1478–1489.
26. M. W. Turner, J. F. Mowbray and D. R. Roberton, *Clin. Exp. Immunol.*, 1981, **46**, 412–419.
27. M. Super, S. Thiel, J. Lu, R. J. Levinsky and M. W. Turner, *Lancet*, 1989, **2**, 1236–1239.
28. R. J. Lipscombe, M. Sumiya, A. V. Hill, Y. L. Lau, R. J. Levinsky, J. A. Summerfield and M. W. Turner, *Hum. Mol. Genet.*, 1992, **1**, 709–715.
29. H. O. Madsen, P. Garred, J. A. Kurtzhals, L. U. Lamm, L. P. Ryder, S. Thiel and A. Svejgaard, *Immunogenetics*, 1994, **40**, 37–44.
30. H. O. Madsen, P. Garred, S. Thiel, J. A. Kurtzhals, L. U. Lamm, L. P. Ryder and A. Svejgaard, *J. Immunol.*, 1995, **155**, 3013–3020.
31. P. Garred, F. Larsen, H. O. Madsen and C. Koch, *Mol. Immunol.*, 2003, **40**, 73–84.
32. S. Thiel, *Mol. Immunol.*, 2007, **44**, 3875–3888.
33. Y. Endo, M. Matsushita and T. Fujita, *Immunobiology*, 2007, **212**, 371–379.
34. Y. Endo, M. Takahashi and T. Fujita, *Immunobiology*, 2006, **211**, 283–293.
35. D. C. Kilpatrick, T. Fujita and M. Matsushita, *Immunol. Lett.*, 1999, **67**, 109–112.
36. A. Krarup, U. B. Sorensen, M. Matsushita, J. C. Jensenius and S. Thiel, *Infect. Immun.*, 2005, **73**, 1052–1060.
37. Y. Le, S. H. Lee, O. L. Kon and J. Lu, *FEBS Lett.*, 1998, **425**, 367–370.
38. S. Taira, N. Kodama, M. Matsushita and T. Fujita, *Fukushima J. Med. Sci.*, 2000, **46**, 13–23.
39. P. D. Frederiksen, S. Thiel, C. B. Larsen and J. C. Jensenius, *Scand. J. Immunol.*, 2005, **62**, 462–473.
40. C. Teh, Y. Le, S. H. Lee and J. Lu, *Immunology*, 2000, **101**, 225–232.
41. Y. Endo, Y. Liu, K. Kanno, M. Takahashi, M. Matsushita and T. Fujita, *Genomics*, 2004, **84**, 737–744.
42. K. J. Colley and J. U. Baenziger, *J. Biol. Chem.*, 1987, **262**, 10290–10295.
43. R. Wallis, J. M. Shaw, J. Uitdehaag, C. B. Chen, D. Torgersen and K. Drickamer, *J. Biol. Chem.*, 2004, **279**, 14065–14073.
44. C. T. Heise, J. R. Nicholls, C. E. Leamy and R. Wallis, *J. Immunol.*, 2000, **165**, 1403–1409.

45. H.-J. Hoppe, P. N. Barlow and K. B. M. Reid, *FEBS Letters*, 1994, **344**, 191–195.
46. V. Garlatti, N. Belloy, L. Martin, M. Lacroix, M. Matsushita, Y. Endo, T. Fujita, J. C. Fontecilla-Camps, G. J. Arlaud, N. M. Thielens and C. Gaboriaud, *EMBO J.*, 2007, **26**, 623–633.
47. F. Teillet, B. Dublet, J. P. Andrieu, C. Gaboriaud, G. J. Arlaud and N. M. Thielens, *J. Immunol.*, 2005, **174**, 2870–2877.
48. R. Wallis and K. Drickamer, *J. Biol. Chem.*, 1999, **274**, 3580–3589.
49. P. H. Jensen, D. Weilguny, F. Matthiesen, K. A. McGuire, L. Shi and P. Hojrup, *J. Biol. Chem.*, 2005, **280**, 11043–11051.
50. U. V. Girija, A. W. Dodds, S. Roscher, K. B. Reid and R. Wallis, *J. Immunol.*, 2007, **179**, 455–462.
51. A. Krarup, S. Thiel, A. Hansen, T. Fujita and J. C. Jensenius, *J. Biol. Chem.*, 2004, **279**, 47513–47519.
52. A. Mohs, M. Popiel, Y. Li, J. Baum and B. Brodsky, *J. Biol. Chem.*, 2006, **281**, 17197–17202.
53. D. Resnick, J. E. Chatterton, K. Schwartz, H. Slayter and M. Krieger, *J. Biol. Chem.*, 1996, **271**, 26924–26930.
54. K. Drickamer, *Curr. Opin. Struct. Biol.*, 1999, **9**, 585–590.
55. W. I. Weis, K. Drickamer and W. A. Hendrickson, *Nature*, 1992, **360**, 127–134.
56. S. Sheriff, C. Y. Y. Chang and R. A. B. Ezekowitz, *Nat. Struct. Biol.*, 1994, **1**, 789–794.
57. W. I. Weis and K. Drickamer, *Structure*, 1994, **2**, 1227–1240.
58. J. N. Arnold, R. Wallis, A. C. Willis, D. J. Harvey, L. Royle, R. A. Dwek, P. M. Rudd and R. B. Sim, *J. Biol. Chem.*, 2006, **281**, 6955–6963.
59. N. J. Lynch, S. Roscher, T. Hartung, S. Morath, M. Matsushita, D. N. Maennel, M. Kuraya, T. Fujita and W. J. Schwaeble, *J. Immunol.*, 2004, **172**, 1198–1202.
60. M. Matsushita, Y. Endo and T. Fujita, *J. Immunol.*, 2000, **164**, 2281–2284.
61. M. Matsushita and T. Fujita, *J. Immunol.*, 1996, **33**, 44.
62. Y. Aoyagi, E. E. Adderson, J. G. Min, M. Matsushita, T. Fujita, S. Takahashi, Y. Okuwaki and J. F. Bohnsack, *J. Immunol.*, 2005, **174**, 418–425.
63. J. P. Kamerling, Pneumococcal Polysaccharides: A Chemical View, In: *Streptococcus pneumoniae: molecular biology and mechanisms of disease*, ed. A. Thomasz, Mary Ann Liebert, New York, 2000, pp. 81–114.
64. M. L. Jensen, C. Honore, T. Hummelshoj, B. E. Hansen, H. O. Madsen and P. Garred, *Mol. Immunol.*, 2007, **44**, 856–865.
65. M. Kuraya, Z. Ming, X. Liu, M. Matsushita and T. Fujita, *Immunobiology*, 2005, **209**, 689–697.
66. P. M. Ng, A. Le Saux, C. M. Lee, N. S. Tan, J. Lu, S. Thiel, B. Ho and J. L. Ding, *EMBO J.*, 2007, **26**, 3431–3440.
67. V. Garlatti, L. Martin, E. Gout, J. Reiser, T. Fujita, G. J. Arlaud, N. M. Thielens and C. Gaboriaud, *Mol. Immunol.*, 2007, **44**, 3928–3929.

68. M. Tanio, S. Kondo, S. Sugio and T. Kohno, *J. Biol. Chem.*, 2007, **282**, 3889–3895.
69. Y. Endo, N. Nakazawa, Y. Liu, D. Iwaki, M. Takahashi, T. Fujita, M. Nakata and M. Matsushita, *Immunogenetics*, 2005, **57**, 837–844.
70. Y. Liu, Y. Endo, D. Iwaki, M. Nakata, M. Matsushita, I. Wada, K. Inoue, M. Munakata and T. Fujita, *J. Immunol.*, 2005, **175**, 3150–3156.
71. M. Tsujimura, T. Miyazaki, E. Kojima, Y. Sagara, H. Shiraki, K. Okochi and Y. Maeda, *Clin. Chim. Acta*, 2002, **325**, 139–146.
72. M. Tsujimura, C. Ishida, Y. Sagara, T. Miyazaki, K. Murakami, H. Shiraki, K. Okochi and Y. Maeda, *Clin. Diagn. Lab. Immunol.*, 2001, **8**, 454–459.
73. R. Sugimoto, Y. Yae, M. Akaiwa, S. Kitajima, Y. Shibata, H. Sato, J. Hirata, K. Okochi, K. Izuhara and N. Hamasaki, *J. Biol. Chem.*, 1998, **273**, 20721–20727.
74. W. Schwaeble, M. Dahl, S. Thiel, C. Stover and J. Jensenius, *Immunobiology*, 2002, **205**, 455–466.
75. C. M. Stover, S. Thiel, N. J. Lynch, T. Vorup-Jensen, J. C. Jensenius and W. J. Schwaeble, *J. Immunol.*, 1999, **162**, 3481–3490.
76. M. Takahashi, Y. Endo, T. Fujita and M. Matsushita, *Int. Immunol.*, 1999, **11**, 859–863.
77. S. Zundel, S. Cseh, M. Lacroix, M. R. Dahl, M. Matsushita, J. P. Andrieu, W. J. Schwaeble, J. C. Jensenius, T. Fujita, G. J. Arlaud and N. M. Thielens, *J. Immunol.*, 2004, **172**, 4342–4350.
78. V. Rossi, S. Cseh, I. Bally, N. M. Thielens, J. C. Jensenius and G. J. Arlaud, *J. Biol. Chem.*, 2001, **276**, 40880–40887.
79. C. B. Chen and R. Wallis, *J. Biol. Chem.*, 2004, **279**, 26058–26065.
80. A. Krarup, R. Wallis, J. S. Presanis, P. Gal and R. B. Sim, *PLoS ONE*, 2007, **2**, e623.
81. K. Hajela, M. Kojima, G. Ambrus, K. H. Wong, B. E. Moffatt, J. Ferluga, S. Hajela, P. Gal and R. B. Sim, *Immunobiology*, 2002, **205**, 467–475.
82. H. Feinberg, J. C. Uitdehaag, J. M. Davies, R. Wallis, K. Drickamer and W. I. Weis, *EMBO J.*, 2003, **22**, 2348–2359.
83. L. A. Gregory, N. M. Thielens, M. Matsushita, R. Sorensen, G. J. Arlaud, J. C. Fontecilla-Camps and C. Gaboriaud, *J. Biol. Chem.*, 2004, **279**, 29391–29397.
84. R. Wallis and R. B. Dodd, *J. Biol. Chem.*, 2000, **275**, 30962–30969.
85. C. B. Chen and R. Wallis, *J. Biol. Chem.*, 2001, **276**, 25894–25902.
86. C. Gaboriaud, F. Teillet, L. A. Gregory, N. M. Thielens and G. J. Arlaud, *Immunobiology*, 2007, **212**, 279–288.
87. M. D. Dahl, S. Thiel, M. Matsushita, T. Fujita, A. C. Willis, T. Christensen, T. Vorup-Jensen and J. C. Jensenius, *Immunity*, 2001, **15**, 127–135.
88. K. R. Mayilyan, J. S. Presanis, J. N. Arnold and R. B. Sim, *Int. J. Immunopathol. Pharmacol.*, 2006, **19**, 567–580.
89. F. Teillet, M. Lacroix, S. Thiel, D. Weilguny, T. Agger, G. J. Arlaud and N. M. Thielens, *J. Immunol.*, 2007, **178**, 5710–5716.

90. W. I. Weis, M. E. Taylor and K. Drickamer, *Immunol. Rev.*, 1998, **163**, 19–34.
91. M. Dong, S. Xu, C. L. Oliveira, J. S. Pedersen, S. Thiel, F. Besenbacher and T. Vorup-Jensen, *J. Immunol.*, 2007, **178**, 3016–3022.
92. P. Gal, V. Harmat, A. Kocsis, T. Bian, L. Barna, G. Ambrus, B. Vegh, J. Balczer, R. B. Sim, G. Naray-Szabo and P. Zavodszky, *J. Biol. Chem.*, 2005, **280**, 33435–33444.
93. V. Harmat, P. Gal, J. Kardos, K. Szilagyi, G. Ambrus, B. Vegh, G. Naray-Szabo and P. Zavodszky, *J. Mol. Biol.*, 2004, **342**, 1533–1546.
94. K. K. Ng, A. R. Kolatkar, S. Park-Snyder, H. Feinberg, D. A. Clark, K. Drickamer and W. I. Weis, *J. Biol. Chem.*, 2002, **277**, 16088–16095.

CHAPTER 11

Surfactant Protein D and Glycoprotein 340

JENS MADSEN[a] AND UFFE HOLMSKOV[b]

[a] Infection, Inflammation and Repair Division, University of Southampton, UK; [b] Medical Biotechnology Centre, University of Southern Denmark, Denmark

11.1 Introduction

Glycoprotein 340 (DMBT1$^{gp\text{-}340}$) was discovered at the MRC Immunochemistry Unit in Oxford as an impurity during a sodium dodecyl sulfate polyacrylamide gel electrophoresis (SDS-PAGE) analysis of a surfactant protein D (SP-D) preparation that had been purified from the 10 000×*g* supernatant from bronchiolar alveolar lavage (BAL) of a patient suffering from alveolar proteinosis.[1] When the supernatant was applied to a maltose-TSK column, in the presence of calcium, the component was bound and could be eluted with maltose together with SP-D and anti-carbohydrate antibodies. The component could then be separated from SP-D and immunoglobulins by gel permeation chromatography in the presence of ethylenediaminetetraacetic acid (EDTA). Rechromatography of the recovered component on a maltose-TSK column, in the presence of calcium, showed no binding, which indicates that the initial retention of the component on the column was mediated by one of the other proteins bound to the maltose. Further characterization showed a calcium-dependent binding of gp-340 to SP-D and the binding was not inhibited by maltose, which showed the interaction was not being mediated through the lectin activity of SP-D, but indicated a protein–protein interaction.[1] Amino acid sequence analysis of gp-340 showed that the protein contained a scavenger cysteine-rich domain (SRCR) and thereby belonged to the

Molecular Aspects of Innate and Adaptive Immunity
Edited By Kenneth BM Reid and Robert B Sim

Published by the Royal Society of Chemistry, www.rsc.org

SRCR superfamily.[1] The number and spacing of the cysteine residues showed that gp-340 belongs to group B of the SRCR superfamily, in which the SRCR domain has eight cysteine residues, whereas group A members have six cysteine residues.[2] Most members of the group B SRCR are transmembrane proteins and many of them are located on immunocompetent cells such as B-lymphocytes, T-lymphocytes or macrophages. It is now well-established that many of these molecules acts as pattern-recognition receptors.[2] At the same time Mollenhauer and colleagues, at the German Cancer Centre in Heidelberg, were working on the cloning of a gene localized on chromosome 10, in a region that was deleted in malignant brain tumours (DMBT1). Sequence analysis showed that DMBT1 was the gene encoding DMBT1$^{gp\text{-}340}$.[3,4] Furthermore, saliva agglutinin (DMBT1^{SAG}), which had been identified in 1983 as a 300–400 kDa glycoprotein that could be isolated by affinity adsorption of saliva to *Streptococcus mutans*,[5] was shown to share identical peptide sequences with DMBT1$^{gp\text{-}340}$ and it was also shown that DMBT1^{SAG} bound SP-D and that DMBT1$^{gp\text{-}340}$ agglutinated *S. mutans*.[6,7] Furthermore, monoclonal antibodies raised against DMBT1$^{gp\text{-}340}$ recognized DMBT1^{SAG} and *vice versa* and, immunohistochemically, the distribution of gp-340 in the submandibular saliva gland was identical to the localization of DMBT1^{SAG}.[6,7] These results showed that DMBT1$^{gp\text{-}340}$ and DMBT1^{SAG} are proteins isolated from different tissues but with similar characteristics and encoded by the same gene, DMBT1.

11.2 Domain Organization and Expression

The gene encoding DMBT1 contains putative 55 exons and spans more than 80 kb of genomic DNA. Of the 55 putative exons, 54 have been confirmed by their presence in various alternative spliced mRNAs.[8] The first six exons encode the signal peptide and a motif of approximately 90 amino acids (aa) of unknown function. This is followed by a repeated pattern, like pearls on a string, of SRCR domains separated by scavenger interspersed domains (SIDs). Each SRCR domain is encoded by a single exon while most SIDs are encoded by two exons.[8] The SRCR domains are followed by a complement Uegf–BMP-1 (CUB) domain, another SRCR domain, a second CUB domain and finally a zona pellucida (ZP) domain (Figure 11.1A). Preliminary electron microscopy (EM) pictures of DMBT1$^{gp\text{-}340}$ show the molecule as having a central core, from which spider-like extrusions appear. It is tempting to speculate that CUB and ZP domains form the central core and that the extensions stem from the repeating SRCR domains (Figure 11.1B).

DMBT1 undergoes extensive alternative splicing in the SRCR and SID region on the mRNA level, which gives rise to mRNAs of several different sizes, of which three have been characterized. The longest form of 8 kb encodes DMBT1$^{gp\text{-}340}$ with 13 SRCR domains in a row, while the shortest DMBT1 form of 6 kb has eight SRCR domains.[3,4] The extensive post-translational modifications, such as N- and O-linked glycosylations, take place and differ between tissues all add to the complexity of the DMBT1 proteins.[1,9–11] DMBT1 has also been identified and cloned in several animals such as mouse (CRP-ductin, Vomeroglandin,

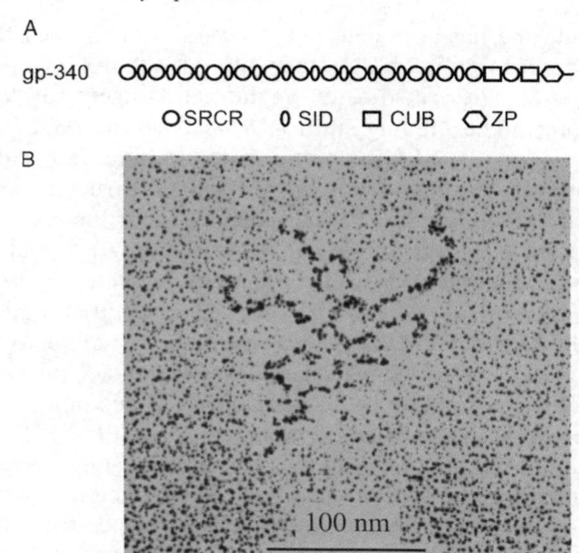

Figure 11.1 Structure of DMBT1$^{gp\text{-}340}$. (A) Simplified structure of DMBT1$^{gp\text{-}340}$ based on cDNA sequence. (B) Electron microscopy of purified DMBT1$^{gp\text{-}340}$ from bronchoalveolar lavage (courtesy of Rupert Timpl and Hanna Widermann).

Muclin),[12–14] rat (Ebnerin),[15] rabbit (Hensin),[16] porcine DMBT1,[17] bovine DMBT1 (bovine gall bladder mucin),[18] and monkey (H3).[19] These show different numbers of SRCR domains from animal to animal, but all have a domain organization similar to that of human DMBT1. Furthermore, an exon encoding for a possible transmembrane region has been identified in rodent and porcine DMBT1, where mice express two alternative spliced forms of DMBT1^{mouse}, one form with and one form without the transmembrane region, while in rats only a DMBT1^{rat} mRNA encoding the transmembrane region has been identified.[12,15] A similar transmembrane region is encoded by exon 55 in the human gene, but it has not been identified on the mRNA level despite extensive investigation.[1,4,8]

Reverse transcriptase–polymerase chain reaction and Northern blotting have shown major sites of expression, the respiratory and gastrointestinal tracts.[3,4] Immunohistochemistry showed localization mainly to mucosal epithelial cells in these organs.[1,20] DMBT1 is, in general, expressed and localized to epithelial cells on surfaces that have contact with the outside world, such as skin and mucosal tissues, for example the respiratory tract, gastrointestinal channel and associated organs, and the urogenital organs.[1,20,21]

11.3 Gp-340, Agglutinin, DMBT1 and Immune Functions

The localization of DMBT1 in skin and mucosal surfaces is ideal for a protein with immune functions in the first line of defence. Furthermore, gp-340 is a secreted molecule and has been detected in several human body fluids, such as BAL, saliva, pancreatic juice and tear film.[1,5,11,22] As mentioned above, DMBT1^{SAG} was first described as an agglutinating agent for *S. mutans* and other streptococci.[5] The binding site for *S. mutans* has been identified to a specific region within the SRCR domain and the binding is calcium dependent.[23] Computer modelling based on the crystal structure of the SRCR domain from the Mac-2 binding protein indicates that the region is a surface loop structure with a β-turn.[23] This region, furthermore, bound to the gram-positive bacterium *S. gordonii* as well as the gram-negative bacteria *Escherichia coli* and *Helicobacter pylori*.[24] *S. mutans* was also bound and agglutinated by the mouse orthologue CRP-ductin and, in addition, bound to *Haemophilus influenza*, *Klebsiella oxytoca*, *Staphylococcus aureus* and *S. pneumoniae*.[25] The region responsible for binding to bacteria in CRP-ductin was shown to be identical to the region in the SRCR domains of DMBT1^{SAG}.[24]

Besides having the capability to bind potentially harmful bacteria, DMBT1 also binds to and inhibits infectivity of viruses like influenza A virus (IAV) and human immunodeficiency virus type I (HIV-I). As opposed to the binding of bacteria, the binding to IAV is not calcium-dependent and the anti-IAV effect takes place by the virus binding to sialic acid-bearing carbohydrates on DMBT1$^{gp\text{-}340}$.[26] The sialylation of DMBT1$^{gp\text{-}340}$ varies between individuals and this is also observed for the anti-IAV activity of DMBT1$^{gp\text{-}340}$.[10] Human saliva inhibits HIV-I infection *in vitro* and DMBT1^{SAG} was shown to be one of the inhibitory components.[27,28] DMBT1^{SAG} interacts with gp-120 from the envelope of HIV-1 and, unlike the interaction with IAV, the binding to gp-120 is calcium-dependent.[29] However, the binding can be inhibited by antibodies raised against carbohydrates, which implies that the interaction between DMBT1^{SAG} and HIV-1 is similar to the interaction characteristic of that between DMBT1$^{gp\text{-}340}$ or DMBT1^{SAG} and IAV.[29]

Besides binding to bacteria and viruses, DMBT1^{SAG} and DMBT1$^{gp\text{-}340}$ bind to several endogenous protein ligands, all of which show an involvement in innate immunity, such as: SP-D[1] and SP-A,[30] secretory immunoglobulin A (sIgA),[31] trefoil factors (TFFs),[32] MUC5B,[33] complement factor C1q[34] and lactoferrin.[35]

11.4 DMBT1$^{gp\text{-}340}$ and its Interaction with SP-D

DMBT1$^{gp\text{-}340}$ binds to SP-D *via* a calcium-dependent protein–protein interaction.[1] DMBT1$^{gp\text{-}340}$ also binds to SP-A and this binding is also calcium dependent and not mediated through the lectin activity of SP-A.[30] Both SP-A and SP-D have important roles in innate immunity by binding to and aggregating bacteria and viruses.[36] DMBT1$^{gp\text{-}340}$ binds several bacteria and viruses and it is therefore obvious to look at the cooperative effect of these molecules on invading microorganisms. Hartshorn and colleagues showed a cooperative effect between these molecules and that they inhibited IAV in the following order: SP-D > DMBT1$^{gp\text{-}340}$ > SP-A.[37] The cooperative effect was most evident in viral aggregation, but was also observed in haemagglutinin inhibition and viral neutralization assays.[37] However, the cooperative effect between SP-D and DMBT1$^{gp\text{-}340}$ was not mediated through SP-D binding to DMBT1$^{gp\text{-}340}$, but more likely through independent aggregation activities of the individual proteins.[37] The binding sites on SP-D for carbohydrates and DMBT1$^{gp\text{-}340}$ are both located in the CRD region of SP-D, but they do not overlap.[1,37] However, if the affinity for SP-D and DMBT1$^{gp\text{-}340}$ becomes higher than the affinity between SP-D and IAV or between DMBT1$^{gp\text{-}340}$ and IAV, respectively, the two proteins will reciprocally inhibit each other's antiviral activities by binding to each other. Thereby they block binding to IAV because of the steric hindrance of the non-overlapping binding sites in the CRD of SP-D.[10] Sialylation of DMBT1$^{gp\text{-}340}$ varies from donor to donor, with a higher degree of sialylation of DMBT1$^{gp\text{-}340}$ showing higher antiviral activity against avian-like IAV strains. Differences were also seen with DMBT1$^{gp\text{-}340}$ from different donors and their interaction with SP-D.[10] This shows that a strong interaction between components in the immune system might not always be a beneficial factor, since these might end up binding to each other instead of binding to an invading micro-organism, and thereby not facilitating clearance.

The cooperative effect of SP-D and DMBT1$^{gp\text{-}340}$ also influences the interaction between IAV and neutrophils. Preincubation of IAV with SP-D strongly increases neutrophil respiratory-burst response to the virus *in vitro*.[38] However, when DMBT1$^{gp\text{-}340}$ was added a significant reduction in the neutrophil respiratory-burst response was observed.[38] This shows that the interaction of these proteins increased the neutrophil uptake of IAV while reducing the respiratory burst to the virus, thereby limiting the potential harmful effect of this burst.

11.5 DMBT1 and its Interaction with Other Host Molecules

DMBT1 binds to a variety of other host proteins, including serum and sIgA, C1q, lactoferrin, MUC5B and albumin. DMBT1^{SAG} is naturally found associated with sIgA.[31,39] The interaction between DMBT1^{SAG} and sIgA is calcium-dependent.[31,39] The binding can be inhibited by high concentrations of salt, which indicates that electrostatic interactions are involved.[40] The interaction with IgA is destroyed after the reduction of DMBT1^{SAG}, which suggests that a protein moiety that contains disulfide bridges was involved in the binding.[40] Experiments using antibodies against IgA or DMBT1^{SAG} showed that it is mainly DMBT1 that agglutinates *S. mutans* and *Salmonella typhimurium*, not sIgA.[31] The interaction between DMBT1^{SAG} and IgA in agglutinating *S. mutans* was found to be additive and the calcium-dependent properties of the DMBT1^{SAG}–sIgA complex favoured the enhancement of their respective activities.[31] The same effect was found when only using the surface protein antigen (Pac) of *S. mutans*.[41] A number of consensus-based peptides of the SRCR domains and SRCR interspersed domains were designed and synthesized to further pinpoint the binding domain for IgA on DMBT1^{SAG}. Enzyme-linked immunosorbent assay (ELISA) binding studies with IgA indicated that only one of the peptides tested, comprising amino acids 18–33 (QGRVEVLYRGSWGTVC) of the 109 amino-acid SRCR domain, exhibited binding to IgA.[40] This domain is identical to the domain of DMBT1^{SAG} that is involved in binding to bacteria.[23] Despite this similar binding site, IgA did not inhibit binding of *S. mutans* to SAG or peptide.[40]

DMBT1 also binds to bovine lactoferrin and this binding inhibits the binding of *S. mutans* to DMBT1^{SAG}.[35,42] Lactoferrin is a non-haem iron-binding protein widely localized in external fluids, like milk, and in mucosal secretions with a role in host protection against microbial infections (reviewed by Ward and Conneely[43]). The peptide domain of bovine lactoferrin, which inhibits the interaction between DMBT1^{SAG} and *S. mutans*, was shown to bind to the same surface protein antigen on *S. mutans* as does DMBT1^{SAG}.[35]

DMBT1^{SAG} has been shown to bind to the C1q globular heads and activate the classic complement pathway through native C1 in freshly isolated normal human serum *in vitro*.[34,44] Although DMBT1^{SAG} and DMBT1$^{gp\text{-}340}$ are present mainly on mucosal surfaces and C1q is a serum protein, these proteins could come into contact with each other during local inflammatory reactions and thereby provide an additional way of local complement activation.

DMBT1$^{gp\text{-}340}$ is associated with the mucin MUC5B *in vivo*.[33,45] Mucins are large, oligomeric gel-forming glycoproteins and, when cross-linking through their cysteine residues, they make viscous mucus gel. Mucus gel performs a critical function in defending every mucosal surface against pathogenic and environmental challenges. Using respiratory mucus or whole saliva, DMBT1$^{SAG/gp\text{-}340}$ was found to be associated with MUC5B in both secretions.[33,45]

Porcine DMBT1 was found to be bind to porcine TFF2.[32] There are three known TFF proteins, TFF1, TFF2 and TFF3 and they are – like DMBT1 – associated with mucosal surfaces,[46] where they are involved in tissue homeostasis and maintenance (reviewed by Taupin and Podolsky[47]).

11.6 Conclusion

It is now clear that DMBT1 alone and through its interactions with other molecules plays an important role as an innate immune defence molecule. However, it is also well-documented that DMBT1 also plays an important role in cell differentiation and cancer, and aspects of DMBT1 were recently reviewed by Kang and Reid.[48]

Direct evidence indicating that DMBT1 also plays a role in the protection and prevention of inflammation recently came from Renner *et al.* who showed that DMBT1$^{(-/-)}$ mice display enhanced susceptibility to dextran sulfate sodium-induced colitis and elevated tumour necrosis factor (TNF), interleukin 6 (IL6), and nucleotide-binding oligomerization domain containing 2 (NOD2) expression levels during inflammation.[49] Furthermore, they showed that DMBT1 is upregulated in inflammatory bowel diseases (IBDs) in humans.[49] The DMBT1$^{(-/-)}$

mice now make it possible to further answer many of the remaining questions in relation to the physiological relevance of DMBT1. For example, what is the relative importance of DMBT1 in relation to different bacterial and viral infections? Will DMBT1$^{(-/-)}$/SP-D$^{(-/-)}$ double knock-out mice be more susceptible to infection or inflammatory damage than the corresponding single gene deficient mice? Are the DMBT1$^{(-/-)}$ mice more susceptible to cancer than the corresponding wild-type mice? Mollenhauer and colleagues have generated a vector system to express recombinant full-length DMBT1 with properties similar to those of native DMBT1^{SAG}.[50] As DMBT1 purified from different natural sources is very heterogeneous, this new tool will help to explore more precisely the full spectrum of microbial molecules that bind to DMBT1.

References

1. U. Holmskov, P. Lawson, B. Teisner, I. Tornoe, A. C. Willis, C. Morgan, C. Koch and K. B. Reid, Isolation and characterization of a new member of the scavenger receptor superfamily, glycoprotein-340 (gp-340), as a lung surfactant protein-D binding molecule, *J. Biol. Chem.*, 1997, **272**(21), 13743–13749.
2. D. Resnick, A. Pearson and M. Krieger, The SRCR superfamily: a family reminiscent of the Ig superfamily, *Trends Biochem. Sci.*, 1994, **19**(1), 5–8.
3. U. Holmskov, J. Mollenhauer, J. Madsen, L. Vitved, J. Gronlund, I. Tornoe, A. Kliem, K. B. Reid, A. Poustka and K. Skjodt, Cloning of gp-340, a putative opsonin receptor for lung surfactant protein D, *Proc. Natl. Acad. Sci. U.S.A.*, 1999, **96**(19), 10794–10799.
4. J. Mollenhauer, S. Wiemann, W. Scheurlen, B. Korn, Y. Hayashi, K. K. Wilgenbus, A. von Deimling and A. Poustka, DMBT1, a new member of the SRCR superfamily, on chromosome 10q25.3-26.1 is deleted in malignant brain tumours, *Nat. Genet.*, 1997, **17**(1), 32–39.
5. T. Ericson and J. Rundegren, Characterization of a salivary agglutinin reacting with a serotype c strain of *Streptococcus mutans*, *Eur. J. Biochem.*, 1983, **133**(2), 255–261.
6. T. J. Ligtenberg, F. J. Bikker, J. Groenink, I. Tornoe, R. Leth-Larsen, E. C. Veerman, A. V. Nieuw Amerongen and U. Holmskov, Human salivary agglutinin binds to lung surfactant protein-D and is identical with scavenger receptor protein gp-340, *Biochem. J.*, 2001, **359**(Pt 1), 243–248.
7. A. Prakobphol, F. Xu, V. M. Hoang, T. Larsson, J. Bergstrom, I. Johansson, L. Frangsmyr, U. Holmskov, H. Leffler, C. Nilsson, T. Boren, J. R. Wright, N. Stromberg and S. J. Fisher, Salivary agglutinin, which binds *Streptococcus mutans* and *Helicobacter pylori*, is the lung scavenger receptor cysteine-rich protein gp-340, *J. Biol. Chem.*, 2000, **275**(51), 39860–39866.
8. J. Mollenhauer, U. Holmskov, S. Wiemann, I. Krebs, S. Herbertz, J. Madsen, P. Kioschis, J. F. Coy and A. Poustka, The genomic structure of the DMBT1 gene: evidence for a region with susceptibility to genomic instability, *Oncogene*, 1999, **18**(46), 6233–6240.

9. C. Eriksson, L. Frangsmyr, L. Danielsson Niemi, V. Loimaranta, U. Holmskov, T. Bergman, H. Leffler, H. F. Jenkinson and N. Stromberg, Variant size- and glycoforms of the scavenger receptor cysteine-rich protein gp-340 with differential bacterial aggregation, *Glycoconj. J.*, 2007, **24**(2-3), 131–142.
10. K. L. Hartshorn, A. Ligtenberg, M. R. White, M. Van Eijk, M. Hartshorn, L. Pemberton, U. Holmskov and E. Crouch, Salivary agglutinin and lung scavenger receptor cysteine-rich glycoprotein 340 have broad anti-influenza activities and interactions with surfactant protein D that vary according to donor source and sialylation, *Biochem. J.*, 2006, **393**(Pt 2), 545–553.
11. B. L. Schulz, D. Oxley, N. H. Packer and N. G. Karlsson, Identification of two highly sialylated human tear-fluid DMBT1 isoforms: the major high-molecular-mass glycoproteins in human tears, *Biochem. J.*, 2002, **366**(Pt 2), 511–520.
12. H. Cheng, M. Bjerknes and H. Chen, CRP-ductin: a gene expressed in intestinal crypts and in pancreatic and hepatic ducts, *Anat. Rec.*, 1996, **244**(3), 327–343.
13. R. C. De Lisle, Characterization of the major sulfated protein of mouse pancreatic acinar cells: a high molecular weight peripheral membrane glycoprotein of zymogen granules, *J. Cell. Biochem.*, 1994, **56**(3), 385–396.
14. F. Matsushita, A. Miyawaki and K. Mikoshiba, Vomeroglandin/CRP-ductin is strongly expressed in the glands associated with the mouse vomeronasal organ: identification and characterization of mouse vomeroglandin, *Biochem. Biophys. Res. Commun.*, 2000, **268**(2), 275–281.
15. X. J. Li and S. H. Snyder, Molecular cloning of Ebnerin, a von Ebner's gland protein associated with taste buds, *J. Biol. Chem.*, 1995, **270**(30), 17674–17679.
16. J. Takito, L. Yan, J. Ma, C. Hikita, S. Vijayakumar, D. Warburton and Q. Al-Awqati, Hensin, the polarity reversal protein, is encoded by DMBT1, a gene frequently deleted in malignant gliomas, *Am. J. Physiol.*, 1999, **277**(2 Pt 2), F277–F289.
17. B. Haase, S. J. Humphray, S. Lyer, M. Renner, A. Poustka, J. Mollenhauer and T. Leeb, Molecular characterization of the porcine deleted in malignant brain tumors 1 gene (DMBT1), *Gene*, 2006.
18. D. P. Nunes, A. C. Keates, N. H. Afdhal and G. D. Offner, Bovine gallbladder mucin contains two distinct tandem repeating sequences: evidence for scavenger receptor cysteine-rich repeats, *Biochem. J.*, 1995, **310**(Pt 1), 41–48.
19. C. I. Ace and W. C. Okulicz, A progesterone-induced endometrial homolog of a new candidate tumor suppressor, DMBT1, *J. Clin. Endocrinol. Metab.*, 1998, **83**(10), 3569–3573.
20. J. Mollenhauer, S. Herbertz, U. Holmskov, M. Tolnay, I. Krebs, A. Merlo, H. D. Schroder, D. Maier, F. Breitling, S. Wiemann, H. J. Grone and A. Poustka, DMBT1 encodes a protein involved in the immune defense and in epithelial differentiation and is highly unstable in cancer, *Cancer Res.*, 2000, **60**(6), 1704–1710.

21. E. Stoddard, G. Cannon, H. Ni, K. Kariko, J. Capodici, D. Malamud and D. Weissman, gp340 expressed on human genital epithelia binds HIV-1 envelope protein and facilitates viral transmission, *J. Immunol.*, 2007, **179**(5), 3126–3132.
22. M. Gronborg, J. Bunkenborg, T. Z. Kristiansen, O. N. Jensen, C. J. Yeo, R. H. Hruban, A. Maitra, M. G. Goggins and A. Pandey, Comprehensive proteomic analysis of human pancreatic juice, *J. Proteome. Res.*, 2004, **3**(5), 1042–1055.
23. F. J. Bikker, A. J. M. Ligtenberg, K. Nazmi, E. C. I. Veerman, W. van't Hof, J. G. M. Bolscher, A. Poustka, A. V. Nieuw Amerongen and J. Mollenhauer, Identification of the bacteria-binding peptide domain on salivary agglutinin (gp-340/DMBT1), a member of the scavenger receptor cysteine-rich superfamily, *J. Biol. Chem.*, 2002, **277**(35), 32109–32115.
24. F. J. Bikker, A. J. Ligtenberg, C. End, M. Renner, S. Blaich, S. Lyer, R. Wittig, W. van't Hof, E. C. Veerman, K. Nazmi, J. M. de Blieck-Hogervorst, P. Kioschis, A. V. Nieuw Amerongen, A. Poustka and J. Mollenhauer, Bacteria binding by DMBT1/SAG/gp-340 is confined to the VEVLXXXXW motif in its scavenger receptor cysteine-rich domains, *J. Biol. Chem.*, 2004, **279**(46), 47699–47703.
25. J. Madsen, I. Tornoe, O. Nielsen, M. Lausen, I. Krebs, J. Mollenhauer, G. Kollender, A. Poustka, K. Skjodt and U. Holmskov, CRP-ductin, the mouse homologue of gp-340/deleted in malignant brain tumors 1 (DMBT1), binds Gram-positive and Gram-negative bacteria and interacts with lung surfactant protein D, *Eur. J. Immunol.*, 2003, **33**(8), 2327–2336.
26. K. L. Hartshorn, M. R. White, T. Mogues, T. Ligtenberg, E. Crouch and U. Holmskov, Lung and salivary scavenger receptor glycoprotein-340 contribute to the host defense against influenza A viruses, *Am. J. Physiol. Lung Cell Mol. Physiol.*, 2003, **285**(5), L1066–L1076.
27. T. Nagashunmugam, D. Malamud, C. Davis, W. R. Abrams and H. M. Friedman, Human submandibular saliva inhibits human immunodeficiency virus type 1 infection by displacing envelope glycoprotein gp120 from the virus, *J. Infect. Dis.*, 1998, **178**(6), 1635–1641.
28. D. C. Shugars, S. P. Sweet, D. Malamud, S. H. Kazmi, K. Page-Shafer and S. J. Challacombe, Saliva and inhibition of HIV-1 infection: molecular mechanisms, *Oral. Dis.*, 2002, **8**(Suppl 2), 169–175.
29. Z. Wu, D. Van Ryk, C. Davis, W. R. Abrams, I. Chaiken, J. Magnani and D. Malamud, Salivary agglutinin inhibits HIV type 1 infectivity through interaction with viral glycoprotein 120, *AIDS Res. Hum. Retroviruses*, 2003, **19**(3), 201–209.
30. M. J. Tino and J. R. Wright, Glycoprotein-340 binds surfactant protein-A (SP-A) and stimulates alveolar macrophage migration in an SP-A-independent manner, *Am. J. Respir. Cell Mol. Biol.*, 1999, **20**(4), 759–768.
31. J. Rundegren and R. R. Arnold, Differentiation and interaction of secretory immunoglobulin A and a calcium-dependent parotid agglutinin for several bacterial strains, *Infect. Immun.*, 1987, **55**(2), 288–292.

32. L. Thim and E. Mortz, Isolation and characterization of putative trefoil peptide receptors, *Regul. Pept.*, 2000, **90**(1-3), 61–68.
33. D. J. Thornton, J. R. Davies, S. Kirkham, A. Gautrey, N. Khan, P. S. Richardson and J. K. Sheehan, Identification of a nonmucin glycoprotein (gp-340) from a purified respiratory mucin preparation: evidence for an association involving the MUC5B mucin, *Glycobiology*, 2001, **11**(11), 969–977.
34. R. J. Boackle, M. H. Connor and J. Vesely, High molecular weight non-immunoglobulin salivary agglutinins (NIA) bind C1Q globular heads and have the potential to activate the first complement component, *Mol. Immunol.*, 1993, **30**(3), 309–319.
35. T. Oho, F. J. Bikker, A. V. Nieuw Amerongen and J. Groenink, A peptide domain of bovine milk lactoferrin inhibits the interaction between streptococcal surface protein antigen and a salivary agglutinin peptide domain, *Infect. Immun.*, 2004, **72**(10), 6181–6184.
36. U. Holmskov, Lung surfactant proteins (SP-A and SP-D) in non-adaptive host responses to infection, *J. Leukoc. Biol.*, 1999, **66**(5), 747–752.
37. M. R. White, E. Crouch, M. van Eijk, M. Hartshorn, L. Pemberton, I. Tornoe, U. Holmskov and K. L. Hartshorn, Cooperative anti-influenza activities of respiratory innate immune proteins and neuraminidase inhibitor, *Am. J. Physiol. Lung Cell Mol. Physiol.*, 2005, **288**(5), L831–L840.
38. M. R. White, E. Crouch, J. Vesona, P. J. Tacken, J. J. Batenburg, R. Leth-Larsen, U. Holmskov and K. L. Hartshorn, Respiratory innate immune proteins differentially modulate the neutrophil respiratory burst response to influenza A virus, *Am. J. Physiol. Lung Cell Mol. Physiol.*, 2005, **289**(4), L606–L616.
39. J. L. Rundegren and R. R. Arnold, Bacteria-agglutinating characteristics of secretory IgA and a salivary agglutinin, *Adv. Exp. Med. Biol.*, 1987, **216B**, 1005–1013.
40. A. J. Ligtenberg, F. J. Bikker, J. M. De Blieck-Hogervorst, E. C. Veerman and A. V. Nieuw Amerongen, Binding of salivary agglutinin to IgA, *Biochem. J.*, 2004, **383**(Pt 1), 159–164.
41. T. Oho, H. Yu, Y. Yamashita and T. Koga, Binding of salivary glycoprotein-secretory immunoglobulin A complex to the surface protein antigen of *Streptococcus mutans*, *Infect. Immun.*, 1998, **66**(1), 115–121.
42. M. Mitoma, T. Oho, Y. Shimazaki and T. Koga, Inhibitory effect of bovine milk lactoferrin on the interaction between a streptococcal surface protein antigen and human salivary agglutinin, *J. Biol. Chem.*, 2001, **276**(21), 18060–18065.
43. P. P. Ward and O. M. Conneely, Lactoferrin: role in iron homeostasis and host defense against microbial infection, *BioMetals*, 2004, **17**(3), 203–208.
44. M. S. Kojouharova, I. G. Tsacheva, M. I. Tchorbadjieva, K. B. Reid and U. Kishore, Localization of ligand-binding sites on human C1q globular head region using recombinant globular head fragments and single-chain antibodies, *Biochim. Biophys. Acta*, 2003, **1652**(1), 64–74.

45. C. Wickstrom, C. Christersson, J. R. Davies and I. Carlstedt, Macromolecular organization of saliva: identification of 'insoluble' MUC5B assemblies and non-mucin proteins in the gel phase, *Biochem. J.*, 2000, **351**(Pt 2), 421–428.
46. J. Madsen, O. Nielsen, I. Tornoe, L. Thim and U. Holmskov, Tissue localization of human trefoil factors 1, 2, and 3, *J. Histochem. Cytochem.*, 2007, **55**(5), 505–513.
47. D. Taupin and D. K. Podolsky, Trefoil factors: initiators of mucosal healing, *Nat. Rev. Mol. Cell Biol.*, 2003, **4**(9), 721–732.
48. W. Kang and K. B. Reid, DMBT1, a regulator of mucosal homeostasis through the linking of mucosal defense and regeneration? *FEBS Lett.*, 2003, **540**(1-3), 21–25.
49. M. Renner, G. Bergmann, I. Krebs, C. End, S. Lyer, F. Hilberg, B. Helmke, N. Gassler, F. Autschbach, F. Bikker, O. Strobel-Freidekind, S. Gronert-Sum, A. Benner, S. Blaich, R. Wittig, M. Hudler, A. J. Ligtenberg, J. Madsen, U. Holmskov, V. Annese, A. Latiano, P. Schirmacher, A. V. Amerongen, M. D'Amato, P. Kioschis, M. Hafner, A. Poustka and J. Mollenhauer, DMBT1 confers mucosal protection *in vivo* and a deletion variant is associated with Crohn's disease, *Gastroenterology*, 2007, **133**(5), 1499–1509.
50. C. End, S. Lyer, M. Renner, C. Stahl, J. Ditzer, A. Holloschi, H. M. Kuhn, H. T. Flammann, A. Poustka, M. Hafner, J. Mollenhauer and P. Kioschis, Generation of a vector system facilitating cloning of DMBT1 variants and recombinant expression of functional full-length DMBT1, *Protein Expr. Purif.*, 2005, **41**(2), 275–286.

Section 4
Cell Surface Proteins – Immunoglobulin Superfamily and Integrins

CHAPTER 12

Leukocyte Surface Proteins – Purification and Characterization

A. NEIL BARCLAY

Sir William Dunn School of Pathology, University of Oxford, Oxford OX1 3RE

12.1 Background

In 40 years the analysis of the lymphocyte cell surface has been transformed with major leaps following technological advances. However, considerable progress was made in the early 1970s by the group of Alan Williams (Figure 12.1A) who joined the MRC Immunochemistry Unit (ICU) in 1970 and turned to studying the lymphocyte cell surface. While one long-term goal was to identify the T cell receptor (TCR) it soon became clear that good methods needed to be developed to quantitate cell surface proteins and, later, these provided excellent techniques when monoclonal antibodies (mAbs) were introduced to revolutionize the field. One of the features of the ICU was the high regard for good quantitation and this critical approach was driven by Rod Porter and then carried on by Ken Reid and Alan Williams [from 1978 in the sister Unit, the MRC Cellular Immunology Unit (CIU)]. I worked in the ICU from 1973 to 1976 and the CIU from 1978 onwards and in this chapter I outline some of the key problems, findings and how these impacted on later studies after the group moved to the CIU in 1978 and in the development of concepts about the lymphocyte cell surface.

12.1.1 Quantitation of Cell Surface Proteins

In the 1960s and 1970s many new cell surface proteins were defined using alloantisera – *i.e.* antibodies raised by immunizing one strain of mouse or rat

Molecular Aspects of Innate and Adaptive Immunity
Edited By Kenneth BM Reid and Robert B Sim

Published by the Royal Society of Chemistry, www.rsc.org

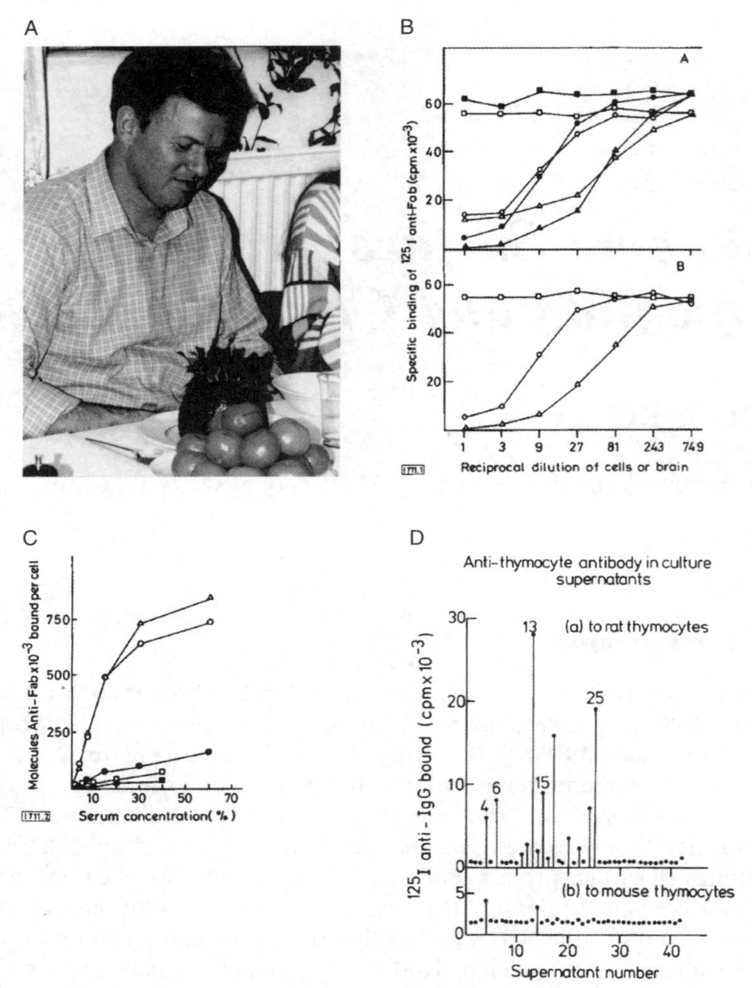

Figure 12.1 Alan Williams and the indirect radioactive binding assay. (A) Alan Williams. (B) Specificity of the assay for Thy-1.1 shown by inhibition with thymocytes and brain from appropriate congenic mouse strains. The C3H and AKT are mouse strains differing in thy-1 allele. A/thy-1.1 and A/Jax are congenic strains for Thy-1.1 and Thy-1.2, respectively. BALB/c anti-WRB is a xenogeneic mouse anti-rat brain serum. (C) Binding assay using saturating amounts of purified labelled $F(ab')_2$ anti-mouse Ig to measure maximum binding of anti-Thy-1.1 antibodies and estimate the number of antigenic sites. (D) Screening of individual clones of hybridomas from the W3 fusion. Good binding was obtained for W3/13 and W3/25, which were later shown to recognize what were later called CD43 and CD4, respectively. Data are from Acton *et al.*[2] and Williams *et al.*[24] and reproduced with permission from Wiley and Cell Press, respectively.

with cells of another strain. One problem with these sera was they tended to be of low affinity, low titre and not necessarily specific. The introduction of congenic mouse strains allowed the specificities to be defined more clearly, although early characterization was hampered by the presence of antibodies against natural mouse viruses. A knowledge of the amount of a particular surface protein is of key importance in being able to identify and biochemically characterize a cell surface protein. The majority of studies in the early 1970s had used cytotoxicity assays. These are very sensitive, so small amounts of precious sera could be used, and were widely used to type histocompatibility antigens. However, they involved lysing cells, which was dependent on the site density, and the efficiency of lysis could vary between batches of complement. To quantitate the presence of an antigen, good data could be obtained by inhibition assays – a method exploited by the early classical studies of Reif and Allen on Thy-1.[1] A few studies had used radioactive assays, but the big advance by Jensenius and Williams was to combine mild radioiodination methods with purified anti-immunoglobulin reagents; using this reagent under trace conditions provided a sensitive assay economical with reagents and, under saturating conditions, the number of antigens being recognized could be determined (Figures 12.1B and 12.1C).[2,3] One disadvantage of using antibodies that had been purified by affinity chromatography of immunoglobulin G (IgG) bound to Sepharose™ and acid elution was that they tended to aggregate, which gave non-saturatable binding, but this could be overcome by using $F(ab')_2$. With a knowledge of the specific activity the precise number of antigenic determinants could be determined and the percentage of cells labelled measured by autoradiography.[3] The values obtained for Thy-1 of around 600 000 sites per cell[2] (Figure 12.1C) were very close to those determined using mAb several years later.[4] The purified high-quality reagents were of value in other studies, such as establishing that the cells carrying cytoplasmic IgA in the thoracic duct also expressed surface IgA using immunofluorescence[5] and the origin of Ig on T cells (see below).

12.1.2 Was there a Receptor on T Cells that Incorporated Ig Variable Regions?

The early quantitative analysis of Ig on B cells revealed that there was very little Ig on T cells or thymocytes.[3] At this time there was a great willingness for the TCR to use the same variable region gene repertoire as Ig's although there was also good evidence that the binding specificity of the TCR differed from that of Ig on B cells. Hunt and Williams were able to demonstrate, using autoradiography with purified anti-Ig reagents and the manipulation of T cells in rats differing in the allotype of their Ig's, that the small amount of Ig on T cells was acquired from the host in transfer experiments and not synthesized by these cells.[6] Much of the analysis in the literature lacked quantitative rigour and, in an excellent and critical review, Jensenius and Williams[7] concluded that there was no reliable evidence that the TCR used Ig variable regions. As it turned out, of course, they did use another family of T cell variable regions with many similarities to Ig variable domains. One feature of the research at that time was the prevalence of studies on a variety of factors, many of which were claimed to be antigen specific. In a critical quantitative environment, such as the ICU, they failed to pass further scrutiny and it was not surprising to us that they faded away, particularly after the I-J region in the genome was shown not to contain a corresponding gene.[8]

12.1.3 Assays that Worked in Detergent

Some of the early studies on membrane proteins used detergents and, from early attempts to do immunoprecipitations, these seemed likely to be the way forward for lymphocyte surface proteins. Binding assays using live cells or cytotoxicity assays had a disadvantage in that the detergent would lyse the target cell, although this could be minimized to some extent by using large quantities of protein, such as bovine serum albumin, to mop up the detergent. The trick here was to fix the target cells with enough glutaraldehyde to make them resistant to lysis by detergent, but not sufficient to destroy the antigenic activity of the target antigen.[9] One extra advantage was that these cells could be prepared as a large batch and stored frozen.

12.1.4 Thy-1 – The Pioneer for Characterizing a Lymphocyte Membrane Protein

Why Thy-1? The distribution of Thy-1 had been determined in detail in the classical studies of Reif and Allen using quantitative inhibition of cytotoxicity assays.[1] It was present on thymocytes and brain cells in the mouse and later it was found in the rat, and all rat strains tested had the equivalent of the mouse Thy-1.1 allele.[10] Thy-1 was abundant and its presence in rat brain provided a much larger tissue from which to attempt biochemical studies than did thymus and, in addition, xenogeneic (cross-species) antisera could be used to detect Thy-1. In this assay an anti-rat brain Thy-1 serum was assayed back on mouse thymocytes (Thy-1.1) to produce a specific assay for Thy-1.1, and these xenogeneic antibodies were of higher affinity than alloantisera (Figure 12.1B).[2] This was important in that if antigen was solubilized to a monomeric form it had to out-compete the antibodies that bind divalently to the polymeric target antigen (cells) – see the discussion in Williams *et al.*[11] This is not the case with weak antisera, which gives the observation that the detergent destroys antigenic activity, whereas it is usual that the well-solubilized monomeric material cannot compete in this assay.

12.1.5 Solubilization of Surface Proteins

A systematic study of all the available detergents showed that, while non-ionic detergents such as Triton X-100 and Lubrol PX were good at solubilizing major histocompatibility complex (MHC) antigens they were not good for Thy-1. With Thy-1 the antigenic activity was apparently solubilized with Lubrol PX in that it did not precipitate when centrifuged at high speed. However, hydrodynamic analysis by gel filtration and sucrose gradient centrifugation showed that it formed a large low-density complex. In contrast, Thy-1 was efficiently solubilized by sodium deoxycholate, which turned out to be most effective detergent for a range of surface proteins and was used extensively in later studies.[12] Deoxycholate had the advantage that it could be removed easily by dialysis or precipitation, unlike the non-ionic detergents, but it had to be kept at pH above 8 and with low salt concentrations, which precluded purification methods such as ion-exchange chromatography.

12.1.6 Purification Using Lectin Affinity Columns

The key to the purification of Thy-1 was the introduction of a method developed by Crumpton and colleagues at the National Institute for Medical Research at Mill Hill using a lectin from lentils bound to Sepharose™ beads to form an affinity column.[13] This lectin could be purified in large amounts, bound to most of the Thy-1 (50% in thymus and 100% in brain) and worked efficiently as an affinity column in which the bound glycoproteins could be eluted with high concentrations of monosaccharides.[14,15] The knowledge of the number of antigenic sites and the assumption the antigen was an average-sized protein made it possible to estimate that around a 1000-fold purification would yield pure antigen. The big worry was that a major protein might be purified but the real antigen would be a small proportion of it (as, indeed, happened in other studies on Thy-1 where the purification factor of 60–120 fold was insufficient).[16] Thus, when a single band was obtained it was likely that this was the protein of interest. This argument was strengthened by running a sodium dodecyl sulfate polyacrylamide gel electrophoresis (SDS PAGE) gel with a low SDS concentration and without boiling or reducing the sample, cutting into slices and showing that antigenic activity co-migrated with the protein band.[14]

Although deoxycholate had to be used at low salt and high pH it did have the advantage that it could be removed by dialysis or by precipitating the glycoprotein with ethanol at −20 °C. Thy-1 precipitated in this way could be redissolved in water and analysis by ultracentrifugation showed that it formed multimers that contained about 16 molecules, presumably in the form of a micelle. In detergent it was monomeric and an accurate *M*r of 18K could be determined, overcoming inaccuracies by SDS PAGE.[17]

12.1.7 Antibody Affinity Columns

Affinity chromatography worked well in purifying proteins, such as the anti-immunoglobulin reagents and lectins discussed above. The problem with antibody affinity chromatography was that antisera generally contain only a small content of specific high-affinity antibodies unless raised against sufficient quantities of purified antigen. The Thy-1 project saw two early applications of the technique. Firstly, the availability of reasonable quantities of Thy-1 purified from rat brain[15] allowed a high-affinity antisera to be produced in rabbits with very high titre. With the quantitative techniques discussed above it was estimated that it contained about 1 mg of specific antibody per millilitre of serum and thus using purified IgG in the column meant it contained about 5–10% specific antibody (note alloantisera would be 10–100 fold less and also of much lower affinity). This column was used to purify thymocyte Thy-1. This was of particular value as lentil lectin only bound to 50% of the Thy-1 in thymocytes due to differential glycosylation (see below).[14] It also confirmed that the protein band identified on thymocytes carried antigen activity found in the purified brain protein used for the immunization. The second application used rabbit anti-lentil lectin prepared as an affinity column to remove contaminant lentil lectin in Thy-1 purifications in which the lectin had leached from the affinity column.[14]

12.1.8 Biochemical Analysis of Thy-1

The amino acid analysis indicated that Thy-1 did not have a particularly high content of amino acids with hydrophobic side chains and the compositions of the thymocyte and brain forms were very similar.[18] Carbohydrate analysis indicated that the Thy-1 was highly glycosylated with different amounts and compositions of carbohydrate in the two tissues.[18] With studies using a variety of antibodies, it was not possible to distinguish the Thy-1 from different tissues antigenically and it seemed likely that we were looking at tissue-specific glycosylation of the same polypeptide with the difference in apparent *M*r on SDS PAGE being due to differences in glycosylation. The carbohydrate analysis was carried out by Ralph Faulkes in Mike Crumpton's laboratory and he had extracted the samples with chloroform–methanol to remove any glycolipds. This fraction, obtained after hydrolysis, contained stearic acid in amounts roughly equimolar to the amount of Thy-1.[18] It was intriguing that this might be covalently attached to Thy-1, but at that time it was not possible to rule out that it was a contaminant. Later studies on Thy-1 established that it was anchored to the membrane by a novel mechanism – the glycophosphatidylinositol (GPI) anchor.[19] Thy-1 also provided a paradigm for membrane protein glycosylation in that detailed analysis of the individual peptides on brain and thymocyte Thy-1 established that all three potential N-linked sites were occupied, but there was site-specific glycosylation as well as tissue-specific glycosylation.[20] Much later it was interesting to note that recombinant Thy-1 lacking the GPI anchor signal sequence was well-expressed but, unlike the membrane form, did not show full occupancy of N-linked glycosylation sites suggested that selection of fully glycosylated protein was affected by the anchor.[21]

12.1.9 Amino Acid Sequence Analysis

The ICU and, in particular L Mole and J Gagnon, had made a major effort to set up protein sequencing with major projects on the sequence of immunoglobulins and later complement components. Although only a few milligrams of Thy-1 was available, it seemed practical to attempt, particularly as it was a small protein of 25 KD on an SDS gel and contained about 25% by weight carbohydrate. Thus, in a collaboration between Ken Reid and Alan Williams, peptides were prepared and sequenced. These peptides showed sequence similarities to immunoglobulins.[22] This striking result suggested that Ig-like domains were used for functions other than in antibodies and might be general recognition proteins (see below).

12.1.10 Monoclonal Antibodies to Recognize New Cell Surface Proteins

When the first paper by Kohler and Milstein in Cambridge on mAbs was published in 1975,[23] it was not immediately obvious how widely its applicability would be. Indeed, many of the first applications were to make better reagents to proteins already known. Alan was the first to see and test their applicability in shotgun screening for new specificities. In this collaboration with Cesar Milstein, supernatants from the fusion were sent from Cambridge to Oxford, assayed using the radioactive binding assay and the results sent back. The availability of sensitive quantitative assays was central to picking the clones and then cloning them (Figure 12.1D). Out of the first shotgun fusion two mAbs, W3/25 and W3/13, went on to be extensively characterized as what we now call CD4 and CD43.[24] W3/25 mAb was instrumental in determining some of the early features of CD4, such as a marker for helper T cells,[25,26] and blocking *in vitro*[27] and *in vivo* immune responses.[28] The aim was to work on experimental systems. but the principle was established for human cells and one of the most used human mAb W6/32 was produced.[29] In subsequent years mAbs were produced regularly and called MRC OX1, 2, *etc.* Currently the series has reached OX129. The availability of mAbs also made simpler the purification of antigens using the methods developed for the polyclonal sera, and early examples were the membrane protein CD45 or, as it was then known, the leukocyte common antigen.[30] mAbs also allowed the simple purification of soluble proteins, such as the less-abundant complement components Factor I or C3b inactivator.[31]

12.1.11 The Immunoglobulin Superfamily Concept

The partial amino acid sequence of Thy-1 immediately suggested that it might be related to Ig, breaking the link between Ig-related sequences and immune recognition, as Thy-1 was primarily a brain antigen with variable expression in lymphoid cells according to species. By now Alan Williams was established in the CIU, but the collaboration with Ken Reid and Jean Gagnon continued to establish this relatedness, including the sequence of mouse Thy-1 from mouse strains that expressed different alleles (Thy-1.1 and Thy-1.2).[32–34] The difference between these alleles was only a single amino acid substitution Arg/Gln,[34]

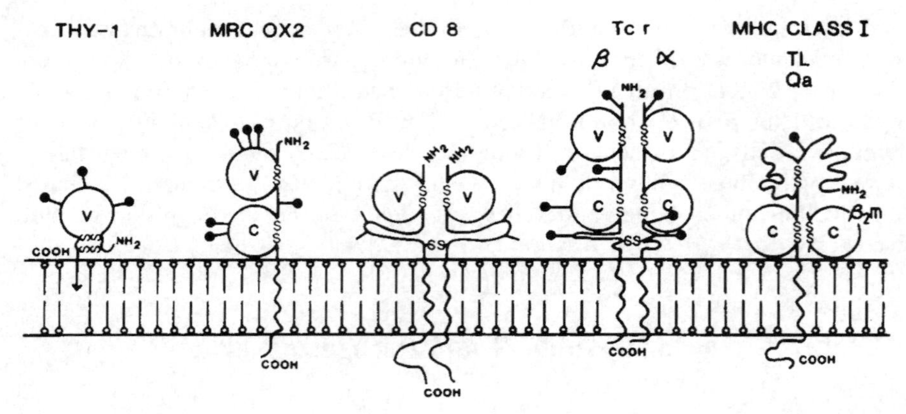

Figure 12.2 The immunoglobulin superfamily begins; cartoon to show proteins that demonstrated significant similarity to Ig sequences in 1985. The finding of two proteins, Thy-1 and OX2, present mainly in brain, and leucocytes clearly established the use of this domain type outwith the immune system recognition. Reproduced from Williams[65] with permission from Nature Publishing Group.

but this had been enough to be detected by antibodies and led to its identification and purification.

Another membrane protein for which sequencing was attempted at the protein level was OX2 (now CD200); this was chosen as we had predicted that it might be Ig-related as it had biochemical properties similar to Thy-1.[35] It was also present in thymus and brain, but at lower levels than Thy-1. Although a partial sequence was obtained, cDNA cloning was introduced and this was the first of many cell surface proteins to be sequenced in this way. OX2 turned out to have two Ig-like domains and a single transmembrane region (Figure 12.2).[36] The development of cDNA cloning led to an explosion in sequences for surface proteins and by 1988 around 30 Ig-related sequences had been defined, which established the Ig superfamily concept (IgSF).[37] The hypothesis was that IgSF domains had evolved as cell recognition molecules and that those associated with immune diversity (antibodies and the TcR) were a late addition to a very large family.[37] In a major review of the leukocyte cell surface the position of the IgSF was established in that almost one-third of the cell surface proteins contained IgSF domains and of these, half had just two IgSF domains.[38,39] Clearly, this topology is vital and can be explained by many of these proteins being involved in cell–cell interactions, whereby they span around 14 nm and probably many give productive engagement in synapse-like patches of interaction between cells.[40]

12.1.12 The Follow-up

From the above summary, it is apparent that the methods established in the ICU established a rigorous quantitative approach to the study of the cell surface. Of course the introduction of mAbs ensured that this progressed fast, but the techniques to quantify the cell surface proteins and to follow them quantitatively continued to be essential to their study. The main legacy, continued in the CIU under Alan Williams and Neil Barclay, saw the study of lymphocyte surface proteins blossom with three main foci.

Firstly, setting up cDNA cloning saw several cell surface proteins being cloned for the first time, including OX40, CD45, CD43, CD48, CD147, CD200R and CD8β, together with species homologues of CD2, CD4, CD8α and TcR. Most of these were Ig-related and gave an indication of the repertoire of the lymphocyte cell surface.[41] An important aspect of this was the concept of the size of the proteins and their abundance. So the biochemistry, including electron microscopy[42–44] and sequencing, showed that CD45 and CD43[39,45] were very large proteins in addition to being very abundant, with about 100 000 molecules per cell. In contrast, proteins such as CD2, CD200 and the TcR were much less abundant in the range 10 000 to 20 000 sites per cell (see Barclay *et al.*[38]). Early cartoons[46] illustrated the contrast between the small proteins involved in signalling and the abundant proteins, like CD43 and CD45, that others developed with imaging techniques to build up the concept of the immunological synapse (reviewed in Bromley *et al.*[40]). Figure 12.3 illustrates some of the proteins in an updated version of the original cartoon.

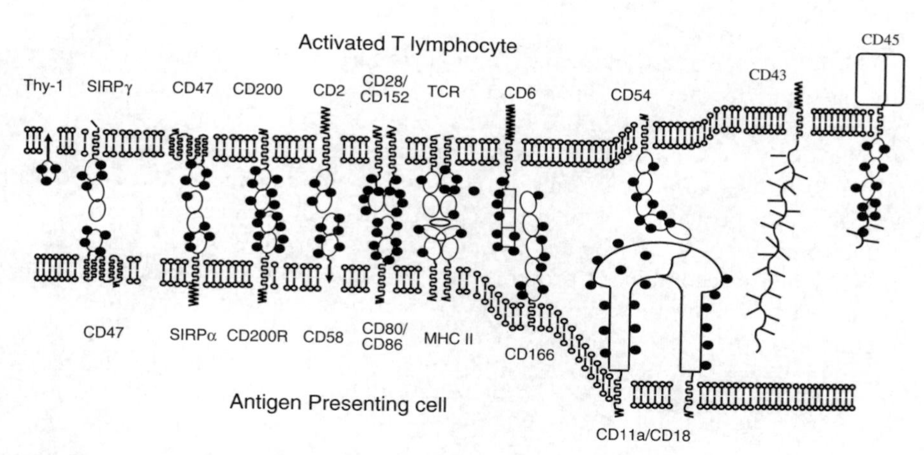

Figure 12.3 Cartoon to indicate the approximate size of some of the proteins at the surface of T cells and an antigen-presenting cell. The IgSF and fibronectin type III (in CD45) domains are indicated by ovals, N-linked glycosylation sites by blobs and regions with extensive O-linked glycosylation by short dashes. CD6 contains scavenger receptor cysteine-rich domains and CD43 contains no recognizable domains. The cytoplasmic region of CD45 contains two regions with phosphatase homology. Note some proteins are much more abundant than others, with proteins like CD2 and CD4 around 10 000–20 000 molecules per cell compared to 100 000 for the large proteins CD43 and CD45. Adapted from Figure 10 in reference Barclay *et al.*[38]

The cDNA cloning also established that the reason that so many proteins could be identified with alloantisera was that the extracellular regions of lymphocyte surface proteins were diverging in evolution much more rapidly than intracellular proteins and, indeed, from many membrane proteins from other tissues. For example, the cytoplasmic region of CD45 is about 90% identical between human and rodents, compared to around 45% for the extracellular regions of CD4, CD8 and CD45 between these species. A typical membrane protein expressed mainly in brain, such as neural cell adhesion molecule (NCAM), is more than 90% conserved throughout between these species.[38] The polymorphisms may result from a single amino acid substitution, as in Thy-1 (see above), or from several residues, *e.g.* in CD45.[47]

Secondly, the cDNA and recombinant DNA technology provided a powerful combination and, in particular, provided the means to express large amounts of proteins. The breakthrough here was the introduction of the glutamine synthetase expression system developed by Celltech Ltd, which allowed routine expression of recombinant protein corresponding to the extracellular regions of the membrane proteins in the 10–100 mg/L range.[48] At last large amounts of proteins were available for structural analysis, with the X-ray crystal structures of CD2, CD4 and SIRPα being determined.[49–52] Bacterial expression of these types of protein was less successful, but a nuclear magnetic resonance (NMR) structure of rat CD2 domain one was obtained that was important in definitively showing that this domain had an Ig-like fold,[53] a prediction that had been of some controversy.[54]

Thirdly, the big question for the lymphocyte surface proteins was, and indeed still is, what do they do? There were very few enzymes in the extracellular regions[38] and, given that IgSF domains were so good at being recognized, it seemed likely that most of the surface proteins were there to recognize other proteins, either soluble or cell surface, to regulate immune reactions. The focus was on those proteins that interacted with other cell surfaces that were predicted to be important in the fine-tuning of immune responses. The availability of high-quality recombinant proteins and the introduction of surface plasmon resonance to follow these interactions in real time allowed good kinetics of interactions to be determined. What was immediately clear was that these interactions were of much lower affinity than had previously been indicated – with K_D in the range 1–100 μM.[55,56] This fitted with the transient nature of lymphocyte interactions, but also made identification of new interactions difficult. Many groups used Fc fusion proteins that comprised the extracellular domains fused to Fc chains from IgG to produce a dimeric protein. We devised a method to couple recombinant monomeric extracellular regions to beads to provide a highly avid reagent for ligand identification.[57,58] This was successful in identifying ligands for CD244, SIRPα and SIRPγ, and CD200.[57,59,60] The availability of monomeric proteins allowed for kinetic analysis to be carried out once the interaction had been identified. The recombinant proteins themselves and, of course, mAb remain powerful tools for dissecting the roles of these cell surface proteins. The quantitative analysis allows one to consider hierarchies of interactions, which is particularly important in intracellular interactions where some proteins can interact with many targets.[61,62] Considerable effort has been made to localize various cytoplasmic proteins, but what is still lacking is good quantitation of the proteins.

12.2 The Legacy

The past 40 years has seen immense progress from the early days when just obtaining good recognition of the cell surface proteins was a challenge. Now we have an idea of the complexity of the problem with one good estimate by Simon Davis and colleagues of 400 different membrane proteins on a cytotoxic T cell.[63] The quantitation can tell us about the likelihood of an interaction, but what is apparent is that the topology of the proteins at the surface is vital and one cannot assume that just because two cells express interacting membrane proteins they are in productive engagement. Although the synapse has clearly been shown to be involved in leukocyte signalling, smaller areas of contact are also likely to of importance, especially in early recognition events.[64]

Acknowledgements

I would like to acknowledge the critical training and friendships of Rod Porter, Ken Reid, Alan Williams and colleagues in the ICU from 1972 to 1976 and the contribution of colleagues in the ICU and since 1978 in the CIU. The majority of the work described was funded by the Medical Research Council.

References

1. A. E. Reif and J. M. V. Allen, *J. Exp. Med.*, 1964, **120**, 413–433.
2. R. T. Acton, R. J. Morris and A. F. Williams, *Eur. J. Immunol.*, 1974, **4**, 598–602.
3. J. C. Jensenius and A. F. Williams, *Eur. J. Immunol.*, 1974, **4**, 91–97.
4. D. W. Mason and A. F. Williams, *Biochem. J.*, 1980, **187**, 1–20.
5. A. F. Williams and J. L. Gowans, *J. Exp. Med.*, 1975, **141**, 335–345.
6. S. V. Hunt and A. F. Williams, *J. Exp. Med.*, 1974, **139**, 479–496.
7. J. C. Jensenius and A. F. Williams, *Nature*, 1982, **300**, 583–588.
8. M. Kronenberg, M. Steinmetz, J. Kobori, E. Kraig, J. A. Kapp, C. W. Pierce, C. M. Sorensen, G. Suzuki, T. Tada and L. Hood, *PNAS*, 1983, **80**, 5704–5708.
9. A. F. Williams, *Eur. J. Immunol.*, 1973, **3**, 628–632.
10. T. C. Douglas, *J. Exp. Med.*, 1972, **136**, 1054–1062.
11. A. F. Williams, A. N. Barclay, M. Letarte-Muirhead and R. J. Morris, *Cold Spring Harb Symp. Quant. Biol.*, 1977, **41**, 51–61.
12. M. L. Muirhead, R. T. Action and A. F. Williams, *Biochem. J.*, 1974, **143**, 51–61.
13. D. Allan, J. Auger and M. J. Crumpton, *Nat. New. Biol.*, 1972, **236**, 23–25.

14. M. Letarte-Muirhead, A. N. Barclay and A. F. Williams, *Biochem. J.*, 1975, **151**, 685–697.
15. A. N. Barclay, M. Letarte-Muirhead and A. F. Williams, *Biochem. J.*, 1975, **151**, 699–706.
16. D. Sauser, C. Anckers and C. Bron, *J. Immunol.*, 1974, **113**, 617–624.
17. P. W. Kuchel, D. G. Campbell, A. N. Barclay and A. F. Williams, *Biochem. J.*, 1978, **169**, 411–417.
18. A. N. Barclay, M. Letarte-Muirhead, A. F. Williams and R. A. Faulkes, *Nature*, 1976, **263**, 563–567.
19. A. G. Tse, A. N. Barclay, A. Watts and A. F. Williams, *Science*, 1985, **230**, 1003–1008.
20. R. B. Parekh, A. G. Tse, R. A. Dwek, A. F. Williams and T. W. Rademacher, *EMBO J.*, 1987, **6**, 1233–1244.
21. M. Devasahayam, P. D. Catalino, P. M. Rudd, R. A. Dwek and A. N. Barclay, *Glycobiology*, 1999, **9**, 1381–1387.
22. D. G. Campbell, A. F. Williams, P. M. Bayley and K. B. Reid, *Nature*, 1979, **282**, 341–342.
23. G. Kohler and C. Milstein, *Nature*, 1975, **256**, 495–497.
24. A. F. Williams, G. Galfre and C. Milstein, *Cell*, 1977, **12**, 663–673.
25. A. N. Barclay, *Immunology*, 1981, **42**, 593–600.
26. R. A. White, D. W. Mason, A. F. Williams, G. Galfre and C. Milstein, *J. Exp. Med.*, 1978, **148**, 664–673.
27. M. Webb, D. W. Mason and A. F. Williams, *Nature*, 1979, **282**, 841–843.
28. S. W. Brostoff and D. W. Mason, *J. Immunol.*, 1984, **133**, 1938–1942.
29. C. J. Barnstable, W. F. Bodmer, G. Brown, G. Galfre, C. Milstein, A. F. Williams and A. Ziegler, *Cell*, 1978, **14**, 9–20.
30. C. A. Sunderland, W. R. McMaster and A. F. Williams, *Eur. J. Immunol.*, 1979, **9**, 155–159.
31. L. Hsiung, A. N. Barclay, M. R. Brandon, E. Sim and R. R. Porter, *Biochem. J.*, 1982, **203**, 293–298.
32. D. G. Campbell, J. Gagnon, K. B. Reid and A. F. Williams, *Biochem. J.*, 1981, **195**, 15–30.
33. F. E. Cohen, J. Novotny, M. J. Sternberg, D. G. Campbell and A. F. Williams, *Biochem. J.*, 1981, **195**, 31–40.
34. A. F. Williams and J. Gagnon, *Science*, 1982, **216**, 696–703.
35. A. N. Barclay and H. A. Ward, *Eur. J. Biochem.*, 1982, **129**, 447–458.
36. M. J. Clark, J. Gagnon, A. F. Williams and A. N. Barclay, *EMBO J.*, 1985, **4**, 113–118.
37. A. F. Williams and A. N. Barclay, *Annu. Rev. Immunol.*, 1988, **6**, 381–405.
38. A. N. Barclay, M. H. Brown, S. K. A. Law, A. J. McKnight, M. G. Tomlinson and P. A. van der Merwe, *Leucocyte Antigens Factsbook - second edition*, Academic Press, London, 1997.
39. A. N. Barclay, D. I. Jackson, A. C. Willis and A. F. Williams, *EMBO J.*, 1987, **6**, 1259–1264.

40. S. K. Bromley, W. R. Burack, K. G. Johnson, K. Somersalo, T. N. Sims, C. Sumen, M. M. Davis, A. S. Shaw, P. M. Allen and M. L. Dustin, *Annu. Rev. Immunol.*, 2001, **19**, 375–396.
41. A. N. Barclay, *Semin. Immunol.*, 2003, **15**, 215–223.
42. J. G. Cyster, D. M. Shotton and A. F. Williams, *EMBO J.*, 1991, **10**, 893–902.
43. M. N. McCall, D. M. Shotton and A. N. Barclay, *Immunology*, 1992, **76**, 310–317.
44. G. R. Woollett, A. F. Williams and D. M. Shotton, *EMBO J.*, 1985, **4**, 2827–2830.
45. N. Killeen, A. N. Barclay, A. C. Willis and A. F. Williams, *EMBO J.*, 1987, **6**, 4029–4034.
46. A. F. Williams and A. N. Barclay, in *Handbook of Experimental Immunology*, Blackwell Scientific Publications, Editon edn., 1986, Vol. **1**, pp. 22.21–22.24.
47. A. Symons and A. N. Barclay, *Immunogenetics*, 2000, **51**, 747–750.
48. S. J. Davis, H. A. Ward, M. J. Puklavec, A. C. Willis, A. F. Williams and A. N. Barclay, *J. Biol. Chem.*, 1990, **265**, 10410–10418.
49. R. L. Brady, E. J. Dodson, G. G. Dodson, G. Lange, S. J. Davis, A. F. Williams and A. N. Barclay, *Science*, 1993, **260**, 979–983.
50. E. Y. Jones, S. J. Davis, A. F. Williams, K. Harlos and D. I. Stuart, *Nature*, 1992, **360**, 232–239.
51. D. Hatherley, K. Harlos, D. C. Dunlop, D. I. Stuart and A. N. Barclay, *J. Biol. Chem.*, 2007, **282**, 14567–14575.
52. A. J. Murray, S. J. Lewis, A. N. Barclay and R. L. Brady, *Proc. Natl. Acad. Sci. USA*, 1995, **92**, 7337–7341.
53. P. C. Driscoll, J. G. Cyster, I. D. Campbell and A. F. Williams, *Nature*, 1991, **353**, 762–765.
54. L. K. Clayton, P. H. Sayre, J. Novotny and E. L. Reinherz, *Eur. J. Immunol.*, 1987, **17**, 1367–1370.
55. P. A. van der Merwe and A. N. Barclay, *Trends Biochem. Sci.*, 1994, **19**, 354–358.
56. P. A. van der Merwe, D. C. McPherson, M. H. Brown, A. N. Barclay, J. G. Cyster, A. F. Williams and S. J. Davis, *Eur. J. Immunol.*, 1993, **23**, 1373–1377.
57. M. H. Brown, K. Boles, P. A. van der Merwe, V. Kumar, P. A. Mathew and A. N. Barclay, *J. Exp. Med.*, 1998, **188**, 2083–2090.
58. S. Preston, G. J. Wright, K. Starr, A. N. Barclay and M. H. Brown, *Eur. J. Immunol.*, 1997, **27**, 1911–1918.
59. E. F. Vernon-Wilson, W. J. Kee, A. C. Willis, A. N. Barclay, D. L. Simmons and M. H. Brown, *Eur. J. Immunol.*, 2000, **30**, 2130–2137.
60. G. J. Wright, M. J. Puklavec, A. C. Willis, R. M. Hoek, J. D. Sedgwick, M. H. Brown and A. N. Barclay, *Immunity*, 2000, **13**, 233–242.
61. N. J. Hassan, S. J. Simmonds, N. G. Clarkson, S. Hanrahan, M. J. Puklavec, M. Bomb, A. N. Barclay and M. H. Brown, *Mol. Cell. Biol.*, 2006, **26**, 6727–6738.

62. N. J. Hutchings, N. Clarkson, R. Chalkley, A. N. Barclay and M. H. Brown, *J. Biol. Chem.*, 2003, **278**, 22396–22403.
63. E. J. Evans, L. Hene, L. M. Sparks, T. Dong, C. Retiere, J. A. Fennelly, R. Manso-Sancho, J. Powell, V. M. Braud, S. L. Rowland-Jones, A. J. McMichael and S. J. Davis, *Immunity*, 2003, **19**, 213–223.
64. A. D. Douglass and R. D. Vale, *Cell*, 2005, **121**, 937–950.
65. A. F. Williams, *Nature*, 1985, **314**, 579–580.

CHAPTER 13

Cell Surface Integrins

SUET-MIEN TAN AND S. K. ALEX LAW

School of Biological Sciences, Nanyang Technological University, 60 Nanyang Drive, Singapore 637551, Singapore

13.1 Introduction

I have written this chapter and S. K. Alex Law provided background information on complement and earlier studies on integrin from the MRC Immunochemistry Unit (the Unit). My stay at the MRC Immunochemistry Unit, Department of Biochemistry, University of Oxford (1998–2001) was memorable. To commemorate the last year of the Unit after 41 years since its inception in 1967, this short chapter provides a personal perspective of my experience in the Unit. It reviews the structure–function relationship of integrins based on studies made by us in Oxford, as well as on subsequent work after we moved to Singapore in 2001 to 2002.

13.2 From Complement Proteins to Integrins

The integrins are type I heterodimer transmembrane proteins formed by an α and a β subunit via non-covalent association (Figure 13.1A). Many integrins are present on the surface of leukocytes, and these integrins can be named based on the cluster of differentiation (CD) nomenclature.[1] For ease of reference, however, the earlier nomenclatures of some of the integrins discussed herein will be used for historical reasons. Subsequently, these will be based entirely on the type of α and β subunits that form the integrin heterodimer. In the early 1980s there was a huge interest in the characterization of complement receptors. The third component of the complement C3 binds covalently to cell

Molecular Aspects of Innate and Adaptive Immunity
Edited By Kenneth BM Reid and Robert B Sim

Published by the Royal Society of Chemistry, www.rsc.org

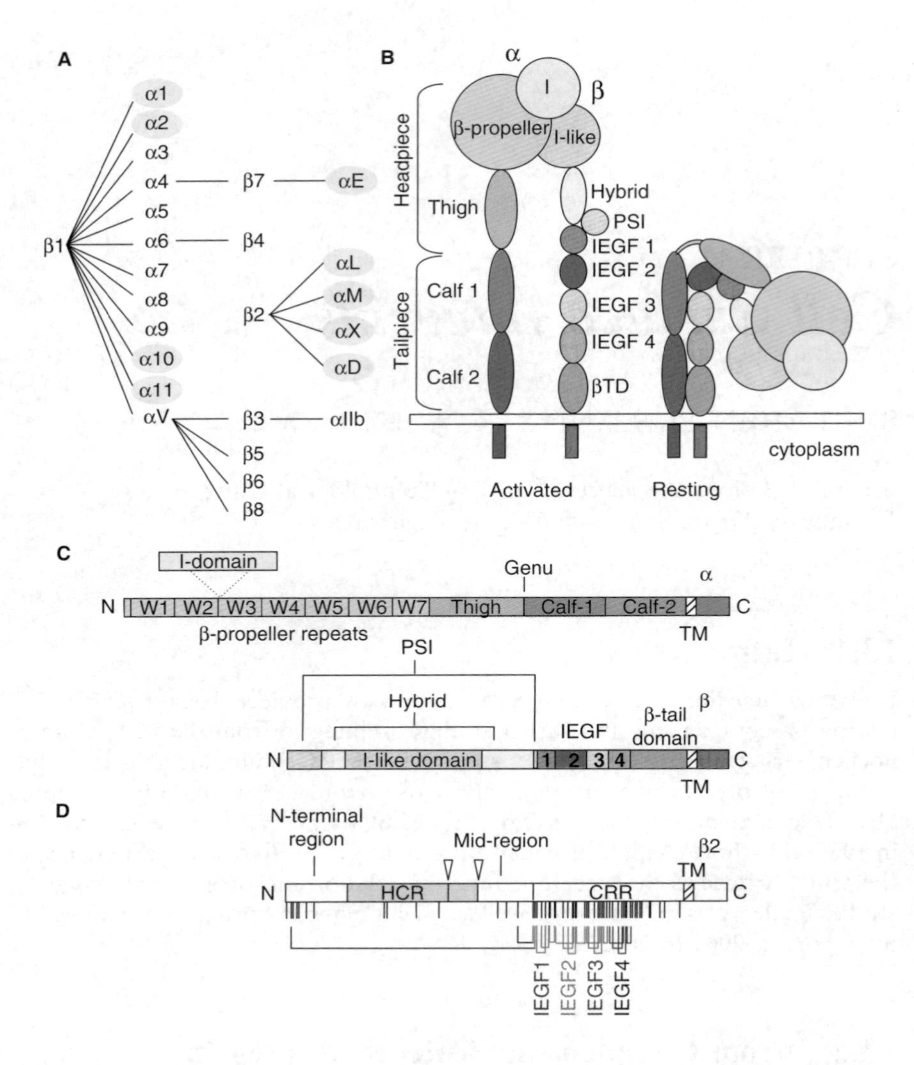

Figure 13.1 The domain organization of the integrins. (A) The human integrins. The integrin α subunit that contains the I domain is highlighted in cyan. (B) Cartoon representation of the bent and extended integrin that contains the I domain. The designations of headpiece and tailpiece are shown. I (inserted) domain; PSI, plexin-semaphorin-integrin domain; I-EGF, integrin epidermal growth factor fold; βTD, β tail domain. (C) Linear domain organization of the integrin subunits. The I domain, found in nine of the human integrins, is inserted between blades 2 and 3 of the β-propeller. (D) Schematic illustrating the integrin β2 subunit highlighting the high content of cysteines (vertical lines), and the initial demarcation of the boundaries N-terminal region, HCR (highly conserved region), mid-region and CRR (cysteine-rich region). TM is the transmembrane domain. The two boundaries of the HCR (I-like domain) based on the crystal structure of integrin αM I domain (front arrowhead)[55] and structure prediction[77] (back arrowhead) are shown. The disulfide bridges formed by the eight cysteines in each of the four I-EGFs are shown. Note that the first two cysteines of I-EGF1 form disulfide bonds with cysteines found in other regions.

surfaces in the form of C3b,[2] which is rapidly cleaved into C3b-inactivator (iC3b).[3] iC3b is very stable on the cell surface, and its conversion to the fluid phase C3c and surface-bound C3d requires the presence of exogenous proteases, such as trypsin, in the absence of serum, which is a source of protease inhibitors. Thus, it was postulated that in addition to the C3b-receptor (CR1, CD35) and the C3d-receptor (CR2, CD21), there is also a receptor for iC3b.[3] The receptor for iC3b was identical to the Mac-1 antigen[4] (also known as Mo1)[5] based on the characterization of a rat monoclonal antibody M1/70 to an antigen found on mouse and human myeloid cells.[6] The following year, Mac-1 [also known as complement receptor 3 (CR3)], together with the lymphocyte function-associated antigen (LFA-1) and p150,95, were reported as leukocyte differentiation antigens.[7] Mac-1, LFA-1 and p150,95 have different α subunits, but share a common β subunit. These are αMβ2, αLβ2 and αXβ2 (Figure 13.1A).[7] Subsequently, it was found that p150,95 also binds iC3b, and was referred to as CR4.[8] The importance of these antigens in the immune system was recognized when patients with a rare inherited autosomal disease, leukocyte adhesion deficiency (LAD) type I, suffered from recurrent opportunistic bacterial infections. Their leukocytes were found to be deficient in the expression of Mac-1,[9] LFA-1[10] and p150,95.[11]

Three decades earlier, it was proposed that an integral membrane protein maintains the link between fibronectin and the actin cytoskeleton of a cell because the extracellular matrix (ECM) molecule fibronectin had a major impact on adherent cell morphology.[12] In 1986 several independent investigations led to the discovery of a new family of molecules that mediate cell–cell and cell–ECM interactions. The platelet glycoprotein gpIIb/IIIa is related to the fibronectin and vitronectin receptors because it binds synthetic RGD peptide,[13] and the designation ‘cytoadhesins’ was proposed for this family of adhesion receptors.[14] The β subunit of the chicken fibronectin receptor was cloned, and the name integrin was coined.[15] The cloning and N-terminal amino acid sequencing of the α subunit of vitronectin receptor[16] revealed not only sequence homology with the α subunit of the fibronectin receptor, but also that of the Mac-1 and LFA-1.[17] It was apparent that Mac-1, LFA-1 and p150,95 have identical β subunit based on co-precipitation studies with relevant monoclonal antibodies, but its sequence identity was unclear.[7] The fact that the N-terminus of this β subunit is blocked adds to the challenge in defining its sequence.[17] In 1987, two independent groups successfully determined the primary sequence of this β subunit – one led by T. A. Springer at the Dana-Faber Cancer Institute (Boston, MA), and the other led by S. K. Alex Law at the MRC Immunochemistry Unit (Oxford, UK).[18,19] Alex, who was with the Unit from 1981 to 2002 and my DPhil supervisor (1998–2001), was working on the complement proteins C3 and C4. His interest in the characterization of Mac-1 (CR3) led him to clone the β subunit of Mac-1, which is common to LFA-1 and p150,95.[20] By then, what emerged from the characterization of the β subunit of the fibronectin receptor in the very late antigens (VLA) family of molecules,[2,21,22] the platelet glycoprotein gpIIb/IIIa[23,24] and the Mac-1, LFA-1 and p150,95 was the concept of three subfamilies of the integrins based on distinct β subunits, namely β1, β3 and β2, respectively.[18,19] Subsequently, five other β subunits were reported in humans.[25–29] The 18 human integrin α subunits were also reported.[30,31] The β2 integrins in humans, which are the focus of our research, comprise Mac-1 (αMβ2), LFA-1 (αLβ2), p150,95 (αXβ2), with the later addition of αDβ2.

After cloning the integrin β2 subunit, there was a period of transition from complement to integrin research in Alex's laboratory. From 1987 to 1998, much work was still focussed on complement research that culminated in a seminal report on the molecular mechanism of C4 internal thioester in 1996.[32] During this period, the polymorphism,[33] and the genomic organization of human β2 subunit[34] were addressed by his team. There was also a growing interest in the molecular basis of LAD type-1 (LAD-1).[35–37] These set the stage for later research focus on β2 integrins. My final year undergraduate project, with Jin-Hua Lu at the National University of Singapore, involved the purification of the complement protein mannan-binding lectin (MBL).[38] My MSc. project turned to investigate the biosynthesis and expression of various integrins during the differentiation of monocytes to macrophages because I was fascinated by their roles in adhesion and migration. In 1998 I came to Oxford as a DPhil student under the supervision of Alex Law to study the molecular mechanism of integrin function. A year before, Aymen Al-Shamkhani, a postdoctoral fellow with Alex, observed that under certain conditions the conformations of the common β2 subunit in association with the respective α subunits of Mac-1, LFA-1 and p150,95 may be different.[39] Following-on from that work, I started to characterize a panel of β2 integrin-specific monoclonal antibodies, and found distinct reactivity of these antibodies to the β2 integrins. This study was largely unpublished, but fuelled my interest in examining integrin conformation and regulation.

13.3 Integrins as Modular Proteins

The domain organization of the integrin α and β subunits is shown (Figures 13.1B and 13.1C). Generally, the integrins contain a large ectodomain, two transmembrane domains and two relatively short cytoplasmic tails, with the exception of the β4 subunit. The α subunit consists of a seven-blade β-propeller fold, thigh, calf-1 and calf-2 domains. Nine of the human α subunits have an inserted (I) domain found between blades two and three of the β-propeller. The β subunit consists of a plexin–semaphorin–integrin (PSI) domain, an I-like domain because of its structural similarity with the α subunit I domain, a hybrid domain, four integrin–epidermal growth factor (I-EGF) folds and a β tail domain (β-TD). The integrin ectodomain can be segregated into two collective regions referred to as the headpiece and tailpiece.[40] Under resting conditions, the integrin adopts an obtuse bent conformation, as revealed by the partial structure of αVβ3 (Figure 13.2A).[41] The conversion of an integrin from a bent conformation to an extended conformation is a widely accepted hallmark of integrin activation (Figure 13.1B).[42]

When the integrin β2 subunit was cloned in 1987,[18,20] the domain organization of the integrins was poorly defined, although the sequences of the β1 and β3, along with a number of α subunits, many of which were partial, were reported.[15–17,23,24,43] This includes the *Drosophila* α position-specific (PS)-2 antigen, which shares sequence homology with the integrin α subunits.[44] Nonetheless, a highly conserved region (HCR) and a cysteine-rich region (CRR) were identified when the sequence of the human β2 subunit was compared with that of the chicken fibronectin receptor (β1 subunit; Figure 13.1D). Interestingly, the subsequent definition of the HCR, which is now referred to as the I-like domain, β A or β I domain,[42–45] was driven largely by the characterization of the I domain of the α subunit.

When the cDNA clones for human αX,[46] murine αM,[47] and human αM[48,49] were isolated a 180-amino acid sequence was identified in these subunits that was absent from the α subunits of the fibronectin (α5),[43] platelet gpIIb/IIIa (αIIb)[50] and vitronectin (αV)[16] receptors. This sequence bears significant homology with the A domain (A1–A3) of the von Willebrand factor that mediates platelet aggregation.[51,52] The A domain is also found in a large number of proteins, including the complement protein C2 and Factor B, the cartilage matrix protein, and the α chains of type VI collagen.[53] Thus, the sequence was referred to as the integrin A domain. As this domain is 'inserted' into the β-propeller of some integrin α subunits, the integrin I domain was also used. In fact, Alex had managed to express and purify a large quantity of the αM I domain with the aim to solve its structure, but the study was set aside when the first I domain structure, also from αM, was reported in 1995[54,55] This was followed by the I domain structures of αL,[56] α1,[57] and α2.[58] These structures present a domain with a dinucleotide-binding fold that is found in the dinucleotide-binding proteins and the G proteins. Also known as the Rossmann fold, they contain a central β sheet surround by amphipathic α helices (Figure 13.2B). In line with Colombatti and Bonaldo's (1991)[53] proposal that the A domain serves a ligand-recognition function in the A domain-containing proteins, the I domain was later shown to be the primary ligand-binding domain in the I domain-containing integrins.[59–65] On the top face of the I domain lies a metal ion dependent adhesion site (MIDAS) with a signature Asp–X–Ser–X–Ser (X is another amino acid) sequence. Together with two non-contiguous residues located in the I domain, they form a divalent metal cation coordination sphere that is essential for ligand binding.[59,66] Apart from these, other residues in the proximity of MIDAS are found also to be important.[65,67] Structural information on other domains of the integrins was still lacking during this period. The N-terminal β-propeller of the α subunit was modelled based on the β subunit of the trimeric G protein.[68] The C-terminal half of the α subunit was shown to be fold independent of the β-propeller, and was predicted to fold into two-layer, antiparallel β-sheet structures.[69] In fact, my DPhil Thesis, which was submitted in 2000, a year before the resolution of the integrin αVβ3 structure (Fig 13.2A),[41] contains 'crude' drawings, in blocks and spheres, of these integrin domains!

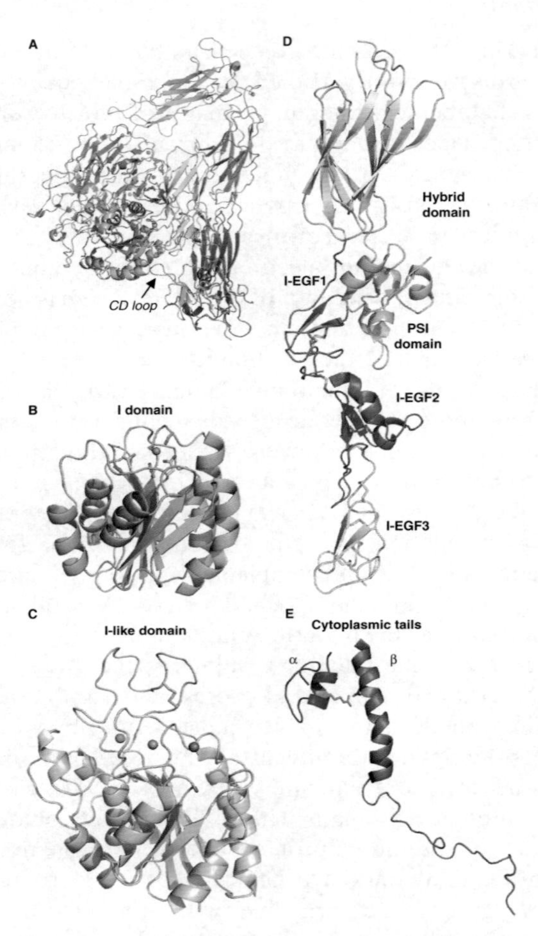

Figure 13.2 Structures of different integrin domains. (A) The bent structure of integrin αVβ3. Coordinates taken from 1JV2. The calcium cation is shown as a green sphere.[41] The PSI and the I-EGF1-3 are not resolved in this structure. αV is presented in different shades of blue and β3 is in other colours. (B) The structure of integrin αM I domain. Structure coordinates 1IDO[55] were used. The MIDAS coordinating residues are shown in sticks. DxSxS is illustrated as magenta sticks, and the two non-contiguous amino acids that form the MIDAS are shown as yellow sticks. The divalent cation is shown as a sphere. (C) Structure of the β3 I-like domain. Structure coordinates 1I5G[90] were used. The divalent cations in the MIDAS, ADMIDAS and LIMBS are shown as gold, red and green spheres, respectively. The specificity loop is in cyan. The disulfide bond in the specificity loop is shown (yellow). (D) Structure of integrin β2 hybrid domain, PSI, and I-EGF1-3. Structure coordinates 2P28 were used. The disulfide bonds (sticks) are shown. (E) Structure of the integrin αIIbβ3 cytoplasmic tails. Structure coordinates 1M80 were used.[154] The salt-bridge forming residues are shown in sticks. The figures were generated using PyMOL (DeLano WL 2002).

While structural information for the I domain in the integrin α subunit began to evolve, when compared with the domain organization of the integrin β subunit it was poorly defined, despite the demarcations of HCR and CRR. However, it was apparent that the folding of the HCR of the β2 subunit is dependent on its association with the αL subunit,[70] which perhaps explains why the structure of an isolated β I domain was not reported, because it may not adopt a proper fold in the absence of the α partner. The CRR was initially suggested to comprise three or four repeating elements that contained eight cysteines in each repeat.[19,71] The rigidity of the CRR, because of the disulfide linkages, could affect the function of the integrin (Figure 13.1D).[18] Indeed, the work of Wendy A. Douglass, a former DPhil student of Alex Law from 1991 to 1995, showed by analyses of β subunit chimeras that the CRR apparently constrain the integrin in a default resting state,[72] which corroborated well with the observations that several of the monoclonal antibodies that activate β2 integrins have epitopes residing in this region.[70,73–75] These prompted us to investigate the requirement of different regions of the β2 in the biosynthesis and function of the β2 integrins.[76] This study relied on the systematic deletion of the β2 subunit from its C-terminus. The challenge was to choose the 'correct' boundaries to maintain correctly folded domains when these deletions were made. At that time, HCR was known to share sequence homology with the I domain. However, the I-like domain, as it was referred to subsequently, was predicted to encompass a larger region than that based on the α I domain[77] (Figure 13.1D). A structural model of the β2 I-like domain was also sculpted.[78] In this model, the I-like domain has an overall fold similar to that of the I domain, but with two additional loops, one of which is referred to as the specificity-determining loop reported to be critical in determining ligand-binding specificity (Figure 13.2C).[79] The I-like domain boundaries and its conformation were finally substantiated by the integrin αVβ3 crystal structure.[41]

We were also intrigued by the possible number of repeating folds found in the CRR. Each of these should contain eight cysteines. However, whether there are three or four repeats depends on how the cluster of eight cysteines is assigned. By testing the reactivity of several conformational sensitive monoclonal antibodies that have epitopes residing in the CRR with a series of β2 CRR truncated mutants, we defined four repeating units, each having the XC–C–C–CxCxxCxC–Cx (C is cysteine, X represents other amino acids, and – represents a stretch of 4 to 14 amino acids) arrangement (Figure 13.1D).[80] These repeats are now referred to as the I-EGF folds because they bear structural similarity to the EGF domain (Figure 13.2D). While working on this, the structure of I domains, and the model of the I-like domain, showed that the N- and C-termini of these domains converge at the bottom face of the domain. In our initial assignment of the domain boundaries for β2,[72,76] there are two other regions apart from HCR and CRR. The N-terminal region (NTR) found before the HCR, and the mid-region because it lies in the middle of the molecule that connects HCR and CRR (Figure 13.1D). The proximity of the N- and C-termini of the I-like domain should allow NTR and the mid-region to be juxtaposed. Indeed, we were able to demonstrate that these two regions are in close proximity to each other, perhaps interacting and forming a complex fold.[81] This study was made by using chimeras of β2 in which different regions of the β2, including NTR and the mid-region, were replaced with the corresponding segments of the integrin β7 subunit. A criticism from the reviewer of our work suggested that human–mouse chimeric β2 should be used instead.[81] However, in our subsequent study we found that the recommended strategy does not serve well in demonstrating the complex formations of NTR and the mid-region.[82] Nonetheless, our data on NTR and the mid-region complex formation corroborated well with the crystal structure of αVβ3[41] reported a month later. It is clear a domain that is similar to an I-set Ig-like domain is formed by two discontinuous segments – one from the NTR and the other from the mid-region, and it was aptly named the hybrid domain.[41]

In late 2001, I took up a faculty position at the School of Biological Sciences, Nanyang Technological University (SBS-NTU), Singapore. With a series of email exchanges, Alex and I were pleasantly surprised by the structure of αVβ3 because our proposal on the NTR–mid-region complex was correct and, importantly, we found a possible explanation for the inhibitory effect of a monoclonal antibody 7E4 on the ligand-binding function of αLβ2 reported in the same NTR–mid-region manuscript.[81] This led us eventually to define the epitope of 7E4, which lies in the hybrid domain, and to demonstrate that the movement of the hybrid domain is essential for the transmission of integrin-activating signal,[82] which is in line with that reported of α5β1[83] and αIIbβ3.[84] In 2002, Alex Law shifted his laboratory to SBS-NTU, Singapore.

13.4 The Flow of Conformational Changes in Integrin During its Activation

In 1985 a molecule with a large globular head and two tails was observed when electron microscopy (EM) studies of integrin αIIbβ3 were made.[85,86] The susceptibility of αIIbβ3 to dissociate when divalent cations were depleted, together with the images collected from these samples, prompted the authors to propose that the globular portion of the molecule is formed by αIIb, whereas the two tails are contributed by β3.[85] Subsequent EM studies on α5β1 revealed otherwise.[87] Each subunit contains a stalk and contributes to the formation of the globular head. What emerged from the sketch of the α5β1 'structure' and its composition was an extended molecule that presents its head at least 12 nm from the plasma membrane. While working on the integrin β2 truncation mutants, we came across a study that reported a truncated GPI-anchored β2 mutant, which contained only the top half of the β2 subunit. It could form a heterodimer with αM, and the resultant integrin retained its C3bi-binding capacity.[88] The ability of truncated β2 mutants to form functional αMβ2 corroborates well with our data. The use of a GPI anchor to attach the top half of the β2 subunit to the cell membrane is also reasonable, but we were intrigued by the observation that the

GPI-anchored β2 mutant, presumably proximal to the cell membrane, could interact with the distal αM globular head to form a heterodimer. The bent structure of an integrin, with its headpiece orientated towards the plasma membrane, provides the explanation (Figure 13.2A).[41]

The molecular basis of integrin activation is an area of intensive research because information obtained from these studies is valuable for the design of integrin-specific therapeutics.[89] When the structure of the bent αVβ3 in complex with a ligand mimetic was resolved, it was proposed that an obtuse bent integrin is able to bind ligand.[90] This was supported by EM studies of ligand-bound αVβ3 that showed a compact and triangular shape similar to that of an unliganded αVβ3.[91] A 'deadbolt' model was proposed for integrin activation.[92] An elongated CD loop of the β–TD interacts with the I-like domain, and its displacement would allow I-like domain activation, and consequently integrin activation (Figure 13.2A). However, in a cell-based system, the bent integrin has its headpiece, which contains the ligand-binding site, pointed towards the plasma membrane. This would prevent favourable accessibility to its ligands, many of which are macromolecules. Thus, it was proposed that the bent integrin may adopt an 'angle-poise' conformation that is facilitated by changes in the length of the transmembrane to allow better accessibility to ligands.[93] Nonetheless, the unbending of an integrin, in a 'switchblade' manner, into an extended structure is becoming a widely accepted hallmark of integrin activation.[94,95] Whether a fully extended integrin is required for ligand binding remains to be clarified.[96–99] Our recent studies demonstrate that αLβ2 engineered in a half-bent conformation could bind its native ligand[100] (and unpublished data).

Integrins are devoid of enzymatic activities, but they are *bona fide* signalling receptors that mediate bidirectional signalling. The recruitment of cytosolic protein(s), many of which are signalling proteins, to integrin cytoplasmic tails are essential for integrin signalling and its activation. Integrin activation is dependent on the flow of conformational changes from one region of the molecule to the next.

Three regions of the integrin are discussed here in the context of integrin activation by cytosolic molecules, referred to as inside-out activation – the integrin cytoplasmic tails, joint regions and headpiece. The membrane proximal regions of the integrin cytoplasmic tails are clasped by a salt bridge that constrains the integrin in a low-affinity ligand-binding state (Figure 13.2E).[101] The release of constraint by mutation of the residues involved in salt-bridge formation activates the receptor.[101] Phosphorylation of the cytoplasmic tails is reported to modulate the activity of the integrins.[102–105] The cytosolic protein talin is also shown to bind the β subunit cytoplasmic tail of the integrin.[106–108] As a result, the integrin cytoplasmic tails are separated, which leads to integrin activation.[109] We also showed that the integrin adopts an extended conformation when talin associates with its cytoplasmic tails.[110] An increasing number of cytosolic molecules have been found to activate integrin. These include cytohesin-1,[111] RapL,[112] Rap1[113] and RIAM.[114] It was shown that the effect of Rap1 and RIAM on integrin αIIbβ3 activation is mediated by talin.[115] The proteins SLAP-130, Rac-1 and Vav-1 are reported to enhance integrin-mediated adhesion by promoting integrin clustering rather than affinity changes.[116–118] Interestingly, talin can also promote αLβ2 clustering, which apparently predominates over talin-mediated αLβ2 affinity upregulation during immune synapse formation.[119] The conflict between the importance of affinity over that of valency regulation or vice-versa may be reconciled by considering the temporal events of cell adhesion.[120–123] Affinity upregulation of integrins will allow the initial formation of tethering of cell–cell or cell–ECM contact that would eventually lead to stable adhesion promoted by receptor clustering.

When integrin unbends, the region that undergoes a marked change in conformation is at its flexible joint (Figure 13.3A). On the α subunit, the joint is formed by residues in the thigh and calf-1, which are connected by a genu.[41] It was proposed that the extension of an integrin is effected by structural rearrangements at the joint.[124] The structure of the joint in the β subunit was not resolved in the αVβ3 crystal data.[41] This structure will provide key information on the switchblade mechanism of integrin activation. We have resolved the structure of I-EGF-1 linked to the PSI and hybrid domain.[125] The entire structure is held in a rigid orientation because of the extensive interactions of the entities. Thus, structural perturbation of the I-EGF-1 is translated into movement of the hybrid domain. Recently, we determined the structure of I-EGF-1, -2 and -3 with the PSI and hybrid domain (Figure 13.2D).[100] The structures not only substantiated the boundaries of the I-EGFs but also shed light on the joint of the integrin β subunit (Figure 13.3A). The joint is formed by I-EGF-1 and -2, which was further verified by mutagenesis and functional assays. Different degrees of integrin unbending were observed from EM images of recombinant αLβ2 and αXβ2.[40] It was proposed that a dynamic equilibrium between bent and extended integrins is maintained on the cell surface. Under resting conditions there is a larger population of bent integrins compared to the extended forms. In the presence of an activation signal, the equilibrium is shifted towards the extended forms. The observations that αLβ2 could adopt different bent conformations lent support to the hypothesis (unpublished data).[99]

The final outcome of integrin unbending is the projection of the headpiece away from the membrane. From the EM and crystal structures of ligand-bound α5β1 and αIIbβ3, respectively, the headpieces of these integrins underwent a significant change in shape, attributed to the swing-out of the hybrid domain tethered to the bottom face of the I-like domain (Figures 13.3B to 13.3E).[84,126] In integrins that do not contain the I domain, the I-like domain together with the β-propeller participate directly in ligand binding (Figures 13.3B and 13.3D). In integrins that contain the I domain, the I-like domain in the β subunit regulates the function of the I domain in the α subunit by allostery (Figures 13.3C and 13.3E). The importance of the I-like domain in integrin function is well exemplified in the large number of mutations found in the I-like domain of integrin β2 and β3 subunits, many of which have deleterious effects on the integrins that predispose affected individuals to LAD-1 and Glanzmann thrombasthenia, respectively.[127]

Unlike the α subunit I domain, which has only one MIDAS, the β subunit I-like domain contains three metal ion coordinating sites – the MIDAS, the adjacent to MIDAS (ADMIDAS) and the ligand-associated metal-binding site (LIMBS; Figure 13.2C).[90] The MIDAS of the β subunit serve to bind ligand, and the ADMIDAS and LIMBS serve as negative and positive regulatory sites, respectively.[128–130] Point mutations of residues that form these sites, identified from LAD-1 and Glanzmann thrombasthenia, disrupt the ligand-binding capacity of the I-like domain in the respective integrins.[127] In a series of LAD-1 characterization studies made by Alex Law and co-workers, a number of these LAD-1 mutations provide insights into the functional regulation of the β2 integrins.[35,36,131–136] Recently, we described a gain-of-function LAD-1 mutation that does not involve the residues of the metal ion coordinating sites, but is found in the last helix of the β2 subunit I-like domain.[137] The last helix of the I-like domain, which is connected to the hybrid domain, is shown to undergo downward displacement in the ligand-mimetic bound αIIbβ3.[84] In line with this, the aforementioned mutation activates not only αLβ2, but also αIIbβ3 when the corresponding residue was mutated.[137]

As mentioned above, there is a difference between integrin that contains an I domain and those that do not in terms of ligand binding. In integrins that lack the I domain, the I-like domain of the β subunit is directly involved in ligand binding. By contrast, in integrins that contain an I domain, the I domain is the primary ligand-binding domain, and the I-like domain of the β subunit regulates the activity of the I domain by allostery. It was shown that the I-like domain MIDAS of the β subunit binds to an intrinsic ligand, which is a conserved glutamate located in the last helix of the α subunit I domain (Figure 13.3C).[138] The downward displacement of this helix is required for the I domain to convert from a closed (low-affinity ligand-binding) to an open (high-affinity ligand-binding) conformation based on evidence gathered from structural studies of recombinant I domains, and functional studies with engineered I domain conformers.[54,55,58,139] It was proposed that the activation of I domain by the I-like domain is by a 'pull-spring' or 'bell-rope' mechanism in which the I-like domain when activated 'pulls' on the last helix of the I domain in a downward motion.[95,138] The displacement of the last helix breaks the socket for isoleucine (SILEN) contact that maintains an I domain in its closed conformation (Figures 13.3C and 13.3E). The SILEN contact is formed by an isoleucine located in the last helix and a hydrophobic socket in the I domain.[140] However, whether SILEN is a universal regulatory switch of the I domains remains to be clarified because crystal structures of integrin α2, αM and αX I domains in their closed conformations show well-defined SILEN contacts, but not in αL I domain.[55,56,141,142] Indeed, we showed that the intrinsic regulation of αL and αM I domains is different, and SILEN serves a primary role in αM but not in αL I domain regulation.[143] An isolated αL I domain that was expressed on the cell surface with a transmembrane anchor showed high affinity for its ligand. By contrast, disruption of the SILEN was required to promote upregulation of αM I domain affinity.

A relay of activation signals from the integrin cytoplasmic tails to its joint, and finally to its headpiece, is envisaged. The regulation of integrin activity is analogous to that of a rheostat rather than a simple 'on–off' switch. The integrins αIIbβ3, α4β1 and αLβ2 are known to exist in different affinity states under

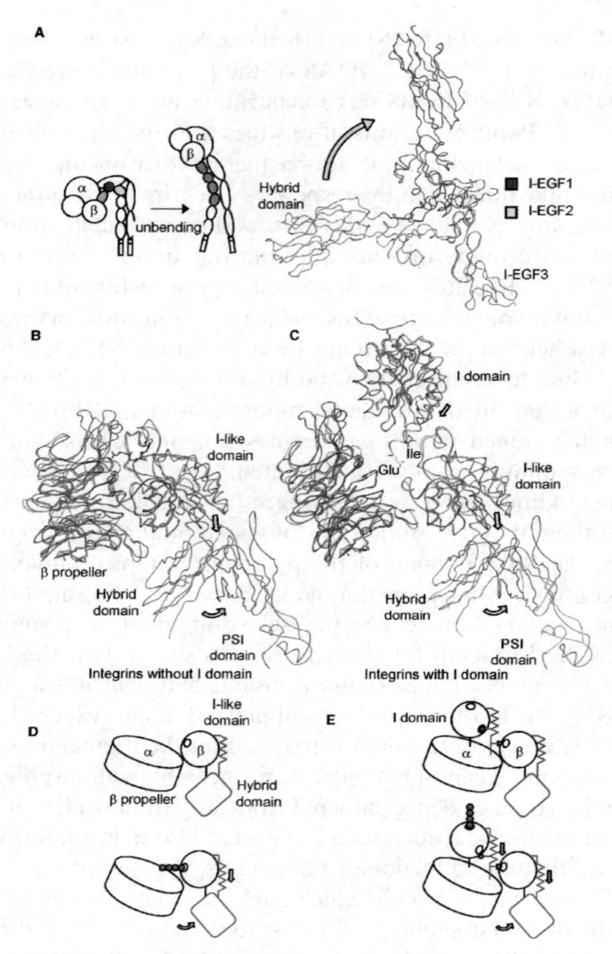

Figure 13.3 Molecular mechanisms of integrin unbending and the opening of the integrin headpiece. (A) A proposed model of integrin unbending at the β2 genu involving I-EGF1 and I-EGF2. The cartoon illustrates the unbending process. Structure coordinates used are 2P26 and 2P28.[100] (B–E) Illustrations of shape changes in the headpiece of an integrin without I domain and integrin with I domain. The closed headpiece structure uses the αVβ3 coordinates 1JV2,[41] the open headpiece structure uses the αIIbβ3 coordinates 1TYE[84] and the I domain structures in closed and open conformations were generated using the αM I domain coordinates 1JLM[54] and 1IDO,[55] respectively. The additional sequence after the last helix of the I domain was modelled to highlight the conserved residues Ile (pink stick and lollipop) and Glu (white stick and triangle). The Ile is reported to be important for the SILEN,[140] and the Glu is reported to serve as an intrinsic ligand for the β I-like domain.[138] The major shape changes are indicated by arrows.

different conditions, which are required for binding to distinct ligands.[96,144–148] This may provide fine regulation of cell adhesion under different physiological conditions. We have shown that different αLβ2 affinity states are required for binding to its intercellular adhesion molecule (ICAM) ligands.[149] It is conceivable that different integrin conformers present different affinity states. Indeed, we were able to distinguish between a low, intermediate and high-affinity αLβ2 based on analyses that made use of conformational sensitive reporter antibodies and ligand-binding selectivity (and unpublished observations).[110,149] The physiological relevance of the intermediate- and high-affinity αLβ2 is under investigation. An intermediate-affinity αLβ2 with an extended conformation that promoted leukocyte rolling adhesion in shear flow was reported,[150] and the cytosolic protein talin could be involved in this process.[151]

It was proposed that the low-affinity integrin has an obtuse bent conformation, the intermediate-affinity integrin is extended but has a closed headpiece (no hybrid domain swing-out), and the high-affinity integrin is extended with an open headpiece (with hybrid domain swing-out; Figure 13.4).[40,84] At present, no evidence demonstrates an extended integrin with an open headpiece promoted by cytosolic molecule(s) interacting with integrin cytoplasmic tails. This integrin

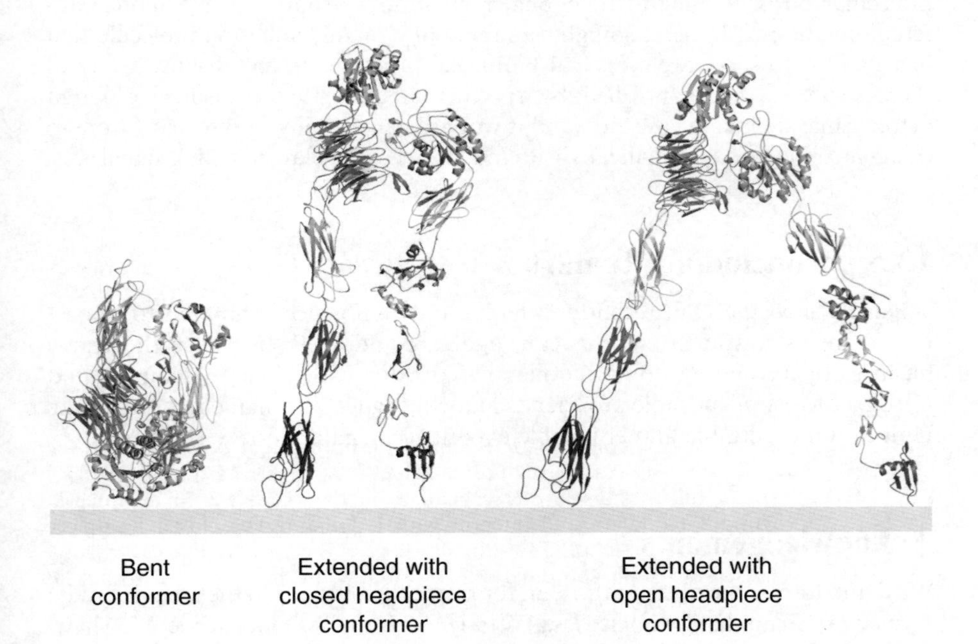

Figure 13.4 The conversion of integrin from a bent conformation to an extended conformation when activated. For illustration the αLβ2 conformers were generated by Modeller8v1 and PyMOL using the coordinates from crystal structures of αVβ3 1JV2 as a template.[41] The coordinates of β2 PSI-hybrid-IEGFs 2P26 and 2P28,[100] and the αL I domain 1LFA[56] were used to generate the structures of the PSI, hybrid domain, I-EGF1-3, and the αL domain.

conformation may depict that of a ligand-bound integrin instead. In this context, the change in conformation of an integrin during its activation followed by ligand binding could be summarized as follows. The binding of relevant cytosolic protein(s) to the integrin cytoplasmic tails results in tail separation, which triggers integrin extension. The extended integrin presents a closed headpiece that can bind to its ligand via the I domain. This leads to further downward motion of the last helix in the I domain and that of the I-like domain, and promotes hybrid domain swing-out that will subsequently displace the two cytoplasmic tails farther from each other. The increased separation of the cytoplasmic tails may allow docking of additional cytosolic molecules to the tails. The transmembrane domains of the integrins are reported to play a role in integrin homo-oligomer formation,[152] although it was not detected by others.[153] Notwithstanding, in the presence of a multivalent ligand, there is a propensity for the integrins to cluster when they bind the ligand. Thus, the conformational change of an integrin when it binds ligand and the reinforcement of the binding by the clustering of integrins serve as post ligand-binding events that are required to stabilize the adhesion contacts.

The knowledge gained from the numerous studies on integrins unravels the molecular basis of integrin functional regulation by shape changes under different conditions. It also highlights a family of dynamic adhesion molecule that has evolved to serve a panoply of biological processes in metazoans. Many of these processes are controlled by specific integrins that transduce different cytoplasmic signalling cascades, and it would be interesting in the near future to designate integrins systematically with individual molecular signalling signatures.

13.5 Concluding Remarks

When I visited the Unit recently, it had a quiet ambience compared to when I was there as a student seven years ago. I could still remember the tricky business of navigating down the narrow corridor with trolley fetching liquid nitrogen! Most memorable are friends and colleagues who made my stay at the Unit an unforgettable and enjoyable experience. Thank you.

Acknowledgements

We thank former members of the Unit for their contributions to the integrin study: Aymen Al-Shamkhani, Emilia Tng, Wendy A. Douglass, Jacqueline M. Shaw, Robert H. Hyland, Sarah L. Scarth, Sheila M. Nolan and Susannah E. Walters.

References

1. A. N. Barclay, M. H. Brown, S. K. A. Law, A. J. McKnight, M. G. Tomlinson and P. A. van der Merwe, The leucocyte antigen facts book, Academic Press, UK, 1997.

2. S. K. Law and R. P. Levine, *Proc. Natl. Acad. Sci. U.S.A.*, 1977, **74**, 2701.
3. S. K. Law, D. T. Fearon and R. P. Levine, *J. Immunol.*, 1979, **122**, 759.
4. T. Springer, G. Galfre, D. S. Secher and C. Milstein, *Eur. J. Immunol.*, 1979, **9**, 301.
5. M. A. Arnaout, R. F. Todd, 3rd, N. Dana, J. Melamed, S. F. Schlossman and H. R. Colten, *J. Clin. Invest.*, 1983, **72**, 171.
6. D. I. Beller, T. A. Springer and R. D. Schreiber, *J. Exp. Med.*, 1982, **156**, 1000.
7. F. Sanchez-Madrid, J. A. Nagy, E. Robbins, P. Simon and T. A. Springer, *J. Exp. Med.*, 1983, **158**, 1785.
8. V. Malhotra, N. Hogg and R. B. Sim, *Eur. J. Immunol.*, 1986, **16**, 1117–23.
9. N. Dana, R. F. Todd, 3rd, J. Pitt, T. A. Springer and M. A. Arnaout, *J. Clin. Invest.*, 1984, **73**, 153.
10. M. A. Arnaout, H. Spits, C. Terhorst, J. Pitt and R. F. Todd, 3rd, *J. Clin. Invest.*, 1984, **74**, 1291.
11. M. A. Arnaout, J. Pitt, H. J. Cohen, J. Melamed, F. S. Rosen and H. R. Colten, *N. Engl. J. Med.*, 1982, **306**, 693.
12. R. O. Hynes, *Biochim. Biophys. Acta.*, 1976, **458**, 73.
13. R. Pytela, M. D. Pierschbacher, M. H. Ginsberg, E. F. Plow and E. Ruoslahti, *Science*, 1986, **231**, 1559.
14. E. F. Plow, J. C. Loftus, E. G. Levin, D. S. Fair, D. Dixon, J. Forsyth and M. H. Ginsberg, *Proc. Natl. Acad. Sci. U.S.A.*, 1986, **83**, 6002.
15. J. W. Tamkun, D. W. DeSimone, D. Fonda, R. S. Patel, C. Buck, A. F. Horwitz and R. O. Hynes, *Cell*, 1986, **46**, 271.
16. S. Suzuki, W. S. Argraves, R. Pytela, H. Arai, T. Krusius, M. D. Pierschbacher and E. Ruoslahti, *Proc. Natl. Acad. Sci. U.S.A.*, 1986, **83**, 8614.
17. T. A. Springer, D. B. Teplow and W. J. Dreyer, *Nature*, 1985, **314**, 540.
18. T. K. Kishimoto, K. O'Connor, A. Lee, T. M. Roberts and T. A. Springer, *Cell*, 1987, **48**, 681.
19. S. K. Law, J. Gagnon, J. E. Hildreth, C. E. Wells, A. C. Willis and A. J. Wong, *EMBO J.*, 1987, **6**, 915.
20. S. K. A. Law, J. Gagnon, J. E. K. Hildreth, A. C. Willis and A. J. Wong, In: Leucocyte typing III: white cell differentiation antigens, ed. A. J. McMichael, Oxford University Press, UK, 1987.
21. Y. Takada, C. Huang and M. E. Hemler, *Nature*, 1987, **326**, 607.
22. Y. Takada, J. L. Strominger and M. E. Hemler, *Proc. Natl. Acad. Sci. U.S.A.*, 1987, **84**, 3239.
23. I. F. Charo, L. A. Fitzgerald, B. Steiner, S. C. Rall, Jr., L. S. Bekeart and D. R. Phillips, *Proc. Natl. Acad. Sci. U.S.A.*, 1986, **83**, 8351.
24. M. H. Ginsberg, J. Loftus, J. J. Ryckwaert, M. Pierschbacher, R. Pytela, E. Ruoslahti and E. F. Plow, *J. Biol. Chem.*, 1987, **262**, 5437.
25. S. Suzuki and Y. Naitoh, *EMBO J.*, 1990, **9**, 757.
26. H. Ramaswamy and M. E. Hemler, *EMBO J.*, 1990, **9**, 1561.
27. D. J. Erle, D. Sheppard, J. Breuss, C. Ruegg and R. Pytela, *Am. J. Respir. Cell Mol. Biol.*, 1991, **5**, 170.

28. D. J. Erle, C. Ruegg, D. Sheppard and R. Pytela, *J. Biol. Chem.*, 1991, **266**, 11009.
29. M. Moyle, M. A. Napier and J. W. McLean, *J. Biol. Chem.*, 1991, **266**, 19650.
30. C. M. Isacke and M.A. Horton, The Adhesion Molecules, Academic Press, UK, 2000.
31. T. Velling, M. Kusche-Gullberg, T. Sejersen and D. Gullberg, *J. Biol. Chem.*, 1999, **274**, 25735.
32. A. W. Dodds, X. D. Ren, A. C. Willis and S. K. Law, *Nature*, 1996, **379**, 177.
33. S. K. Law and G. M. Taylor, *Immunogenetics*, 1991, **34**, 341.
34. J. B. Weitzman, C. E. Wells, A. H. Wright, P. A. Clark and S. K. Law, *FEBS Lett.*, 1991, **294**, 97.
35. K. A. Davies, V. J. Toothill, J. Savill, N. Hotchin, A. M. Peters, J. D. Pearson, C. Haslett, M. Burke, S. K. Law and N. F. Mercer, *et al.*, *Clin. Exp. Immunol.*, 1991, **84**, 223.
36. A. H. Wright, W. A. Douglass, G. M. Taylor, Y. L. Lau, D. Higgins, K. A. Davies and S. K. Law, *Eur. J. Immunol.*, 1995, **25**, 717.
37. M. Neira, J. Rincon, H. Arias, S. K. Law and M. Patarroyo, *Eur. J. Haematol.*, 1997, **58**, 32.
38. S. M. Tan, M. C. Chung, O. L. Kon, S. Thiel, S. H. Lee and J. Lu, *Biochem. J.*, 1996, **319**, 329.
39. A. Al-Shamkhani and S. K. Law, *Eur. J. Immunol.*, 1998, **28**, 3291.
40. N. Nishida, C. Xie, M. Shimaoka, Y. Cheng, T. Walz and T. A. Springer, *Immunity*, 2006, **25**, 583.
41. J. P. Xiong, T. Stehle, B. Diefenbach, R. Zhang, R. Dunker, D. L. Scott, A. Joachimiak, S. L. Goodman and M. A. Arnaout, *Science*, 2001, **294**, 339.
42. B. H. Luo, C. V. Carman and T. A. Springer, *Annu. Rev. Immunol.*, 2007, **25**, 619.
43. W. S. Argraves, R. Pytela, S. Suzuki, J. L. Millan, M. D. Pierschbacher and E. Ruoslahti, *J. Biol. Chem.*, 1986, **261**, 12922.
44. T. Bogaert, N. Brown and M. Wilcox, *Cell*, 1987, **51**, 929.
45. M. A. Arnaout, B. Mahalingam and J. P. Xiong, *Annu. Rev. Cell Dev. Biol.*, 2005, **21**, 381.
46. A. L. Corbi, L. J. Miller, K. O'Connor, R. S. Larson and T. A. Springer, *EMBO J.*, 1987, **6**, 4023.
47. R. Pytela, *EMBO J.*, 1988, **7**, 1371.
48. A. L. Corbi, T. K. Kishimoto, L. J. Miller and T. A. Springer, *J. Biol. Chem.*, 1988, **263**, 12403.
49. M. A. Arnaout, S. K. Gupta, M. W. Pierce and D. G. Tenen, *J. Cell Biol.*, 1988, **106**, 2153.
50. M. Poncz, R. Eisman, R. Heidenreich, S. M. Silver, G. Vilaire, S. Surrey, E. Schwartz and J. S. Bennett, *J. Biol. Chem.*, 1987, **262**, 8476.
51. J. P. Girma, D. Meyer, C. L. Verweij, H. Pannekoek and J. J. Sixma, *Blood*, 1987, **70**, 605.
52. C. L. Verweij, P. J. Diergaarde, M. Hart and H. Pannekoek, *EMBO J.*, 1986, **5**, 1839.

53. A. Colombatti and P. Bonaldo, *Blood*, 1991, **77**, 2305.
54. J. O. Lee, L. A. Bankston, M. A. Arnaout and R. C. Liddington, *Structure*, 1995, **3**, 1333.
55. J. O. Lee, P. Rieu, M. A. Arnaout and R. Liddington, *Cell*, 1995, **80**, 631.
56. A. Qu and D. J. Leahy, *Proc. Natl. Acad. Sci. U.S.A.*, 1995, **92**, 10277.
57. M. Nolte, R. B. Pepinsky, S. Venyaminov, V. Koteliansky, P. J. Gotwals and M. Karpusas, *FEBS Lett.*, 1999, **452**, 379.
58. J. Emsley, C. G. Knight, R. W. Farndale, M. J. Barnes and R. C. Liddington, *Cell*, 2000, **101**, 47.
59. M. Michishita, V. Videm and M. A. Arnaout, *Cell*, 1993, **72**, 857.
60. A. M. Randi and N. Hogg, *J. Biol. Chem.*, 1994, **269**, 12395.
61. L. Zhou, D. H. Lee, J. Plescia, C. Y. Lau and D. C. Altieri, *J. Biol. Chem.*, 1994, **269**, 17075.
62. T. Q. Cai and S. D. Wright, *J. Biol. Chem.*, 1995, **270**, 14358.
63. D. S. Tuckwell, K. B. Reid, M. J. Barnes and M. J. Humphries, *Eur. J. Biochem.*, 1996, **241**, 732.
64. B. Leitinger and N. Hogg, *Mol. Biol. Cell*, 2000, **11**, 677.
65. P. Yalamanchili, C. Lu, C. Oxvig and T. A. Springer, *J. Biol. Chem.*, 2000, **275**, 21877.
66. T. Kamata, R. Wright and Y. Takada, *J. Biol. Chem.*, 1995, **270**, 12531.
67. P. Rieu, T. Sugimori, D. L. Griffith and M. A. Arnaout, *J. Biol. Chem.*, 1996, **271**, 15858.
68. T. A. Springer, *Proc. Natl. Acad. Sci. U.S.A.*, 1997, **94**, 65.
69. C. Lu, C. Oxvig and T. A. Springer, *J. Biol. Chem.*, 1998, **273**, 15138.
70. C. Huang, C. Lu and T. A. Springer, *Proc. Natl. Acad. Sci. U.S.A.*, 1997, **94**, 3156.
71. J. J. Calvete, A. Henschen and J. Gonzalez-Rodriguez, *Biochem. J.*, 1991, **274**, 63.
72. W. A. Douglass, R. H. Hyland, C. D. Buckley, A. Al-Shamkhani, J. M. Shaw, S. L. Scarth, D. L. Simmons and S. K. Law, *FEBS Lett.*, 1998, **440**, 414.
73. P. Stephens, J. T. Romer, M. Spitali, A. Shock, S. Ortlepp, C. G. Figdor and M. K. Robinson, *Cell Adhes. Commun.*, 1995, **3**, 375.
74. V. Bazil, I. Stefanova, I. Hilgert, H. Kristofova, S. Vanek and V. Horejsi, *Folia Biol. (Praha)*, 1990, **36**, 41.
75. L. Petruzzelli, L. Maduzia and T. A. Springer, *J. Immunol.*, 1995, **155**, 854.
76. S. M. Tan, R. H. Hyland, A. Al-Shamkhani, W. A. Douglass, J. M. Shaw and S. K. Law, *J. Immunol.*, 2000, **165**, 2574.
77. D. S. Tuckwell and M. J. Humphries, *FEBS Lett.*, 1997, **400**, 297.
78. C. Huang, Q. Zang, J. Takagi and T. A. Springer, *J. Biol. Chem.*, 2000, **275**, 21514.
79. J. Takagi, T. Kamata, J. Meredith, W. Puzon-McLaughlin and Y. Takada, *J. Biol. Chem.*, 1997, **272**, 19794.
80. S. M. Tan, S. E. Walters, E. C. Mathew, M. K. Robinson, K. Drbal, J. M. Shaw and S. K. Law, *FEBS Lett.*, 2001, **505**, 27.

81. S. M. Tan, M. K. Robinson, K. Drbal, Y. van Kooyk, J. M. Shaw and S. K. Law, *J. Biol. Chem.*, 2001, **276**, 36370.
82. E. Tng, S. M. Tan, S. Ranganathan, M. Cheng and S. K. Law, *J. Biol. Chem.*, 2004, **279**, 54334.
83. A. P. Mould, S. J. Barton, J. A. Askari, P. A. McEwan, P. A. Buckley, S. E. Craig and M. J. Humphries, *J. Biol. Chem.*, 2003, **278**, 17028.
84. T. Xiao, J. Takagi, B. S. Coller, J. H. Wang and T. A. Springer, *Nature*, 2004, **432**, 59.
85. N. A. Carrell, L. A. Fitzgerald, B. Steiner, H. P. Erickson and D. R. Phillips, *J. Biol. Chem.*, 1985, **260**, 1743.
86. L. V. Parise and D. R. Phillips, *J. Biol. Chem.*, 1985, **260**, 1750.
87. M. V. Nermut, N. M. Green, P. Eason, S. S. Yamada and K. M. Yamada, *EMBO J.*, 1988, **7**, 4093.
88. T. G. Goodman, M. E. DeGraaf, H. D. Fischer and M. L. Bajt, *J. Leukoc. Biol.*, 1998, **64**, 767.
89. N. C. Kaneider, A. J. Leger and A. Kuliopulos, *FEBS J.*, 2006, **273**, 4416.
90. J. P. Xiong, T. Stehle, R. Zhang, A. Joachimiak, M. Frech, S. L. Goodman and M. A. Arnaout, *Science*, 2002, **296**, 151.
91. B. D. Adair, J. P. Xiong, C. Maddock, S. L. Goodman, M. A. Arnaout and M. Yeager, *J. Cell Biol.*, 2005, **168**, 1109.
92. J. P. Xiong, T. Stehle, S. L. Goodman and M. A. Arnaout, *Blood*, 2003, **102**, 1155.
93. R. O. Hynes, *Cell*, 2002, **110**, 673.
94. N. Beglova, S. C. Blacklow, J. Takagi and T. A. Springer, *Nat. Struct. Biol.*, 2002, **9**, 282.
95. J. Takagi and T. A. Springer, *Immunol. Rev.*, 2002, **186**, 141.
96. A. Chigaev, A. M. Blenc, J. V. Braaten, N. Kumaraswamy, C. L. Kepley, R. P. Andrews, J. M. Oliver, B. S. Edwards, E. R. Prossnitz, R. S. Larson and L. A. Sklar, *J. Biol. Chem.*, 2001, **276**, 48670.
97. A. Chigaev, T. Buranda, D. C. Dwyer, E. R. Prossnitz and L. A. Sklar, *Biophys. J.*, 2003, **85**, 3951.
98. A. Chigaev, A. Waller, G. J. Zwartz, T. Buranda and L. A. Sklar, *J. Immunol.*, 2007, **178**, 6828.
99. R. S. Larson, T. Davis, C. Bologa, G. Semenuk, S. Vijayan, Y. Li, T. Oprea, A. Chigaev, T. Buranda, C. R. Wagner and L. A. Sklar, *Biochemistry*, 2005, **44**, 4322.
100. M. Shi, S. Y. Foo, S. M. Tan, E. P. Mitchell, S. K. Law and J. Lescar, *J. Biol. Chem.*, 2007, **282**, 30198.
101. P. E. Hughes, F. Diaz-Gonzalez, L. Leong, C. Wu, J. A. McDonald, S. J. Shattil and M. H. Ginsberg, *J. Biol. Chem.*, 1996, **271**, 6571.
102. S. Fagerholm, N. Morrice, C. G. Gahmberg and P. Cohen, *J. Biol. Chem.*, 2002, **277**, 1728.
103. S. C. Fagerholm, T. J. Hilden and C. G. Gahmberg, *Trends Biochem. Sci.*, 2004, **29**, 504.

104. S. C. Fagerholm, T. J. Hilden, S. M. Nurmi and C. G. Gahmberg, *J. Cell Biol.*, 2005, **171**, 705.
105. S. M. Nurmi, M. Autero, A. K. Raunio, C. G. Gahmberg and S. C. Fagerholm, *J. Biol. Chem.*, 2007, **282**, 968.
106. S. Tadokoro, S. J. Shattil, K. Eto, V. Tai, R. C. Liddington, J. M. de Pereda, M. H. Ginsberg and D. A. Calderwood, *Science*, 2003, **302**, 103.
107. K. L. Wegener, A. W. Partridge, J. Han, A. R. Pickford, R. C. Liddington, M. H. Ginsberg and I. D. Campbell, *Cell*, 2007, **128**, 171.
108. B. Garcia-Alvarez, J. M. de Pereda, D. A. Calderwood, T. S. Ulmer, D. Critchley, I. D. Campbell, M. H. Ginsberg and R. C. Liddington, *Mol. Cell*, 2003, **11**, 49.
109. M. Kim, C. V. Carman and T. A. Springer, *Science*, 2003, **301**, 1720.
110. Y. F. Li, R. H. Tang, K. J. Puan, S. K. Law and S. M. Tan, *J. Biol. Chem.*, 2007, **282**, 24310.
111. C. Geiger, W. Nagel, T. Boehm, Y. van Kooyk, C. G. Figdor, E. Kremmer, N. Hogg, L. Zeitlmann, H. Dierks, K. S. Weber and W. Kolanus, *EMBO J.*, 2000, **19**, 2525.
112. K. Katagiri, A. Maeda, M. Shimonaka and T. Kinashi, *Nat. Immunol.*, 2003, **4**, 741.
113. K. Katagiri, M. Hattori, N. Minato, S. Irie, K. Takatsu and T. Kinashi, *Mol. Cell Biol.*, 2000, **20**, 1956.
114. E. M. Lafuente, A. A. van Puijenbroek, M. Krause, C. V. Carman, G. J. Freeman, A. Berezovskaya, E. Constantine, T. A. Springer, F. B. Gertler and V. A. Boussiotis, *Dev. Cell*, 2004, **7**, 585.
115. J. Han, C. J. Lim, N. Watanabe, A. Soriani, B. Ratnikov, D. A. Calderwood, W. Puzon-McLaughlin, E. M. Lafuente, V. A. Boussiotis, S. J. Shattil and M. H. Ginsberg, *Curr. Biol.*, 2006, **16**, 1796.
116. O. D. Perez, D. Mitchell, G. C. Jager, S. South, C. Murriel, J. McBride, L. A. Herzenberg, S. Kinoshita and G. P. Nolan, *Nat. Immunol.*, 2003, **4**, 1083.
117. E. J. Peterson, M. L. Woods, S. A. Dmowski, G. Derimanov, M. S. Jordan, J. N. Wu, P. S. Myung, Q. H. Liu, J. T. Pribila, B. D. Freedman, Y. Shimizu and G. A. Koretzky, *Science*, 2001, **293**, 2263.
118. C. Krawczyk, A. Oliveira-dos-Santos, T. Sasaki, E. Griffiths, P. S. Ohashi, S. Snapper, F. Alt and J. M. Penninger, *Immunity*, 2002, **16**, 331.
119. W. T. Simonson, S. J. Franco and A. Huttenlocher, *J. Immunol.*, 2006, **177**, 7707.
120. Y. van Kooyk and C. G. Figdor, *Curr. Opin. Cell Biol.*, 2000, **12**, 542.
121. G. Bazzoni and M. E. Hemler, *Trends Biochem. Sci.*, 1998, **23**, 30.
122. C. V. Carman and T. A. Springer, *Curr. Opin. Cell Biol.*, 2003, **15**, 547.
123. M. Kim, C. V. Carman, W. Yang, A. Salas and T. A. Springer, *J. Cell Biol.*, 2004, **167**, 1241.
124. C. Xie, M. Shimaoka, T. Xiao, P. Schwab, L. B. Klickstein and T. A. Springer, *Proc. Natl. Acad. Sci. U.S.A.*, 2004, **101**, 15422.

125. M. Shi, K. Sundramurthy, B. Liu, S. M. Tan, S. K. Law and J. Lescar, *J. Biol. Chem.*, 2005, **280**, 30586.
126. J. Takagi, K. Strokovich, T. A. Springer and T. Walz, *EMBO J.*, 2003, **22**, 4607.
127. N. Hogg and P. A. Bates, *Matrix Biol.*, 2000, **19**, 211.
128. J. Chen, A. Salas and T. A. Springer, *Nat. Struct. Biol.*, 2003, **10**, 995.
129. J. Chen, J. Takagi, C. Xie, T. Xiao, B. H. Lu and T. A. Springer, *J. Biol. Chem.*, 2004, **279**, 55556.
130. A. P. Mould, S. J. Barton, J. A. Askari, S. E. Craig and M. J. Humphries, *J. Biol. Chem.*, 2003, **278**, 51622.
131. N. Dana, L. K. Clayton, D. G. Tennen, M. W. Pierce, P. J. Lachmann, S. A. Law and M. A. Arnaout, *J. Clin. Invest.*, 1987, **79**, 1010.
132. N. Hogg, M. P. Stewart, S. L. Scarth, R. Newton, J. M. Shaw, S. K. Law and N. Klein, *J. Clin. Invest.*, 1999, **103**, 97.
133. E. C. Mathew, J. M. Shaw, F. A. Bonilla, S. K. Law and D. A. Wright, *Clin. Exp. Immunol.*, 2000, **121**, 133.
134. J. M. Shaw, A. Al-Shamkhani, L. A. Boxer, C. D. Buckley, A. W. Dodds, N. Klein, S. M. Nolan, I. Roberts, D. Roos, S. L. Scarth, D. L. Simmons, S. M. Tan and S. K. Law, *Clin. Exp. Immunol.*, 2001, **126**, 311.
135. D. Roos, C. Meischl, M. de Boer, S. Simsek, R. S. Weening, O. Sanal, I. Tezcan, T. Gungor and S. K. Law, *Exp. Hematol.*, 2002, **30**, 252.
136. G. Uzel, E. Tng, S. D. Rosenzweig, A. P. Hsu, J. M. Shaw, M. E. Horwitz, G. F. Linton, S. M. Anderson, M. R. Kirby, J. B. Oliveira, M. R. Brown, T. A. Fleisher, S. K. Law and S. M. Holland, *Blood*, 2007.
137. M. Cheng, S. Y. Foo, M. L. Shi, R. H. Tang, L. S. Kong, S. K. Law and S. M. Tan, *J. Biol. Chem.*, 2007, **282**, 18225.
138. W. Yang, M. Shimaoka, A. Salas, J. Takagi and T. A. Springer, *Proc. Natl. Acad. Sci. U.S.A.*, 2004, **101**, 2906.
139. M. Shimaoka, T. Xiao, J. H. Liu, Y. Yang, Y. Dong, C. D. Jun, A. McCormack, R. Zhang, A. Joachimiak, J. Takagi, J. H. Wang and T. A. Springer, *Cell*, 2003, **112**, 99.
140. J. P. Xiong, R. Li, M. Essafi, T. Stehle and M. A. Arnaout, *J. Biol. Chem.*, 2000, **275**, 38762.
141. J. Emsley, S. L. King, J. M. Bergelson and R. C. Liddington, *J. Biol. Chem.*, 1997, **272**, 28512.
142. T. Vorup-Jensen, C. Ostermeier, M. Shimaoka, U. Hommel and T. A. Springer, *Proc. Natl. Acad. Sci. U.S.A.*, 2003, **100**, 1873.
143. S. E. Walters, R. H. Tang, M. Cheng, S. M. Tan and S. K. Law, *Biochem. Biophys. Res. Commun.*, 2005, **337**, 142.
144. B. Savage and Z. M. Ruggeri, *J. Biol. Chem.*, 1991, **266**, 11227.
145. D. R. Phillips, I. F. Charo and R. M. Scarborough, *Cell*, 1991, **65**, 359.
146. B. Savage, S. J. Shattil and Z. M. Ruggeri, *J. Biol. Chem.*, 1992, **267**, 11300.
147. A. Masumoto and M. E. Hemler, *J. Biol. Chem.*, 1993, **268**, 228.
148. A. R. de Fougerolles and T. A. Springer, *J. Exp. Med.*, 1992, **175**, 185.

149. R. H. Tang, E. Tng, S. K. Law and S. M. Tan, *J. Biol. Chem.*, 2005, **280**, 29208.
150. A. Salas, M. Shimaoka, A. N. Kogan, C. Harwood, U. H. von Andrian and T. A. Springer, *Immunity*, 2004, **20**, 393.
151. R. Shamri, V. Grabovsky, J. M. Gauguet, S. Feigelson, E. Manevich, W. Kolanus, M. K. Robinson, D. E. Staunton, U. H. von Andrian and R. Alon, *Nat. Immunol.*, 2005, **6**, 497.
152. R. Li, N. Mitra, H. Gratkowski, G. Vilaire, R. Litvinov, C. Nagasami, J. W. Weisel, J. D. Lear, W. F. DeGrado and J. S. Bennett, *Science*, 2003, **300**, 795.
153. B. H. Luo, C. V. Carman, J. Takagi and T. A. Springer, *Proc. Natl. Acad. Sci. U.S.A.*, 2005, **102**, 3679.
154. O. Vinogradova, A. Velyvis, A. Velyviene, B. Hu, T. Haas, E. Plow and J. Qin, *Cell*, 2002, **110**, 587.

Section 5
Immunogenetics and Major Histocompatibility Complex Class III Analysis

CHAPTER 14

Molecular Genetics of the Major Histocompatibility Complex Class III Region

R. DUNCAN CAMPBELL,[a] WENDY THOMSON[b] AND BERNARD MORLEY[c]

[a] Department of Physiology, Anatomy and Genetics, University of Oxford, Oxford, OX1 3QX, UK; [b] Translational Medicine, University of Manchester, Manchester, M13 9PT,UK; [c] Division of Medicine, Imperial College, London, W12 0NN, UK

14.1 Foreword by Duncan Campbell

When I arrived in Oxford in 1978 to carry out a post-doctorate with Rod Porter in the MRC Immunochemistry Unit I was excited about the prospect of working on the early activation events of the human complement system and, in particular, on the role of C4. John Goers with Porter had been looking at the assembly of early components of complement on antibody–antigen aggregates and on antibody-coated erythrocytes. Their data suggested that, although the binding of C4 is about 12 times greater to cells than in aggregates, the effective number of C4 molecules as judged by their ability to form a C3 convertase was much more similar.[1] They hypothesized that activated C4 could bind to the antibody molecule, possibly to the Fab portion, which then acted as a site for activated C2 to bind to form the C3 convertase. With Alister Dodds, not only was I able to show that C4 was able to bind to the Fab portion of IgG in antibody–antigen aggregates, we were also able to determine that C4 bound covalently *via* its alpha chain to the Fd region of the heavy chain of IgG.[2] Also,

Molecular Aspects of Innate and Adaptive Immunity
Edited By Kenneth BM Reid and Robert B Sim

Published by the Royal Society of Chemistry, www.rsc.org

we were able to provide evidence that this interaction was mediated by a reactive acyl group released from a thiolester bond in the alpha chain of C4 when C4 was activated by C1.[2] This work involved isolating C4 from plasma, converting it into C4b and then into C4c and C4d, and purifying the C4d fragment followed by cleavage with cyanogen bromide to isolate peptides for sequencing by the Edman degradation technique. Given the amount of C4 in plasma, and its lability, this was no easy task and it took us about 12 months to generate about 100 amino acids of sequence.[3,4] With the predicted size of C4 being about 1700 amino acids it was obvious that it would take years to sequence the protein completely using classical protein-sequencing methodology.

At that time the first papers to describe the use of degenerate oligonucleotides to screen cDNA libraries for cDNA clones were being published.[5] That, together with advances in DNA-sequencing technology using either the Maxam and Gilbert[6] or Sanger[7] sequencing approaches indicated that there were major advantages to obtaining cDNA clones to proteins if we wanted to determine their amino acid sequence. Also, having cDNA clones would allow us to isolate and characterize the corresponding genes. Porter had become very interested in the molecular genetics of the early complement components C4, C2 and factor B as the proteins were all polymorphic and the corresponding genes mapped to the human major histocompatibility complex (MHC).[8–11] This region of the human genome had been shown to be intimately involved in the immune response to pathogens and was where susceptibility to a number of human autoimmune diseases mapped.[12,13] In 1980 Porter suggested that I should become involved in setting up molecular biology techniques in the Unit. I was able to spend six months in George Brownlees' laboratory (Sir William Dunn School of Pathology, Oxford) and isolated a 515 base pair (bp) cDNA clone to factor B using a 14 bp oligonucleotide probe synthesized by Porter when he took a mini-sabbatical during the long vacation.[14] This cDNA clone, together with those isolated for C4[15] and C2,[16] allowed the identification of cosmid clones that contained the corresponding genes. These were eventually linked up to produce a 100 kb contig of genomic DNA,[17] which established the order and orientation of the complement genes in the Class III region. Subsequently, the isolation of full-length cDNA clones for factor B,[18] C4[19] and C2[20] paved the way for much work in the Unit detailing the exon structures as well as the nature of the polymorphisms in these three complement genes.

14.2 Physical Mapping of the Major Histocompatibility Complex

The human MHC is located on the short arm of chromosome 6 in the distal portion of the 6p21.3 band. Classically, it is split into three major linked gene clusters. The Class I loci [human leukocyte antigen (HLA)-A, -B, -C] and the Class II loci (HLA-DR, -DQ, -DP) encode cell-surface glycoproteins that are highly polymorphic and act as restriction elements in the recognition and interaction of regulatory and effector T lymphocytes with their target cells.[21] The Class I and Class II loci are separated by the Class III region. Having been able to align the complement genes with respect to one another, it became of major interest to establish their location in relation to the flanking Class I and Class II regions and also whether there were other genes in the Class III region, what they encoded and whether they could be involved in autoimmune disease susceptibility. Genetic studies had suggested that the Class III region could extend over 1000 kb.[22] Thus, trying to clone this amount of DNA in 1985 was a mammoth task. At that time a number of papers appeared in the literature,[23,24] including one from Ed Southern,[25] which described the use of pulsed field gel electrophoresis (PFGE) to separate very large fragments of DNA of the order of hundreds of kilobases in size. During the summer of 1985 we set about using the technique to try and detect large DNA fragments containing the C4 genes. Porter was very excited about this as we realized if this worked it would allow us to embark on a much more ambitious project to clone the Class III region. The first Southern blot from a PFGE gel was generated in early September 1985, hybridized with a C4 probe and put down to autoradiograph just as Porter left to go on holiday after his retirement party. Had it not been for his untimely death on his way to France he would have seen the results of that first blot which showed that we could detect the C4 genes in single DNA fragments of 130 kb. That initial result gave us the impetus to establish PFGE in the Unit, which was used extensively over the next few years.

At that time there were very few genomic DNA libraries available to clone large segments of the human genome, so we constructed our own cosmid library using genomic DNA from a HLA homozygous consanguineous cell line.[26] This proved an excellent resource to isolate overlapping cosmid clones that extended from the complement genes toward the flanking Class I and II regions.[26] The cosmid clones also provided a rich source of probes for PFGE to build up a physical map not just of the Class III region but also of the whole MHC.[27] This map oriented not only the complement gene cluster with respect to the Class II DR region, with the C2 gene being on the telomeric side of the 21-hydroxylase B (CYP21B) gene, but it also defined the positions of the genes for tumour necrosis factor (TNF) and lymphotoxin in the human MHC.

Extensive use of PFGE was made to observe directly the organization and arrangement of the C4 and CYP21A/B genes in the Class III region,[28] and to characterize the Class III and Class II regions in different HLA haplotypes.[29–31] During this analysis it became apparent that many of the restriction enzyme sites for the rarely cutting restriction enzymes being used, instead of being randomly distributed, appeared to cluster in the genome. These clustered sites marked the location of CpG islands, CpG-rich sequences found at the 5′ end of ubiquitously expressed genes.[32] This, together with the use of genomic probes in Northern blot analysis, resulted in the discovery of 14 new genes between the C2 and HLA-B genes,[26,33] and seven new genes between the CYP21B and DRA gene.[34] However, although the cloning and mapping of novel genes in the Class III region proved very successful it was not until cosmid and bacterial artificial chromosome (BAC) genomic clones that covered the region were sequenced and fully annotated that all the genes were uncovered.[35] This is partly because some of the genes, such as G6B,[36] were very restricted in their pattern of expression and were not revealed in some of the more general gene-identification approaches adopted.

14.3 Sequence Analysis of the MHC and Annotation of the Genes

With the improvements in DNA sequencing technology in the 1990s we set up high-throughput DNA sequencing in the Unit and embarked on sequencing a number of our cosmid clones. Careful annotation of the DNA sequence led to the discovery of yet more novel genes in the genomic inserts of these cosmids and indicated that the DNA sequencing approach would prove very rewarding.[37–40] The DNA sequence we generated in the Class III region formed part of a much larger international effort to sequence the whole human MHC and in 1999 The MHC Sequencing Consortium reported the first complete sequence and gene map of the MHC.[35] This revealed that the human MHC spans 3.6 Mbp of DNA and contains 224 loci of which 129 are expressed. The Class III region contains 64 genes in 900 kb (Figure 14.1), of which 61 are expressed; with an average gene size of about 9 kb and an average intergenic distance of just under 3 kb, it is the most gene-dense region in the human genome.[41]

The MHC is noteworthy for containing the most polymorphic genes in the genome,[42,43] for displaying marked linkage disequilibrium[44] and for the large number of human diseases with which it is associated. All of these features make the MHC an excellent region in which to study key aspects of genome biology. As a consequence the MHC has been sequenced in a number of different HLA haplotypes to catalogue all the single nucleotide polymorphisms (SNPs) that might be the basis of MHC-associated diseases.[45–47] There has also been a major initiative to identify, catalogue and interpret genome-wide DNA methylation phenomena across the MHC.[48] This has revealed that the methylation profile of the MHC is bimodal, with over 90% of the amplicons being either relatively hypomethylated or relatively hypermethylated. There were also distinct tissue-specific methylation profiles.

The MHC has also provided a model for studies of gene paralogy as there are striking examples of clusters of MHC-paralogous genes on human chromosomes 1, 9 and 19.[49–51] For example, the neurogenic locus notch homologue protein 4 (NOTCH4) gene in the Class III region is homologous to the NOTCH1 gene on chromosome 9q34, the NOTCH2 gene on chromosome 1p13 and the NOTCH3 gene on 19p13.2. However, recently Horton *et al.*[52] identified 88 genes in the extended MHC that have a combined total of 791 putative paralogues elsewhere in the human genome. Their data demonstrate that MHC paralogues cluster in, but are not restricted to, these three previously known MHC-paralogous regions.

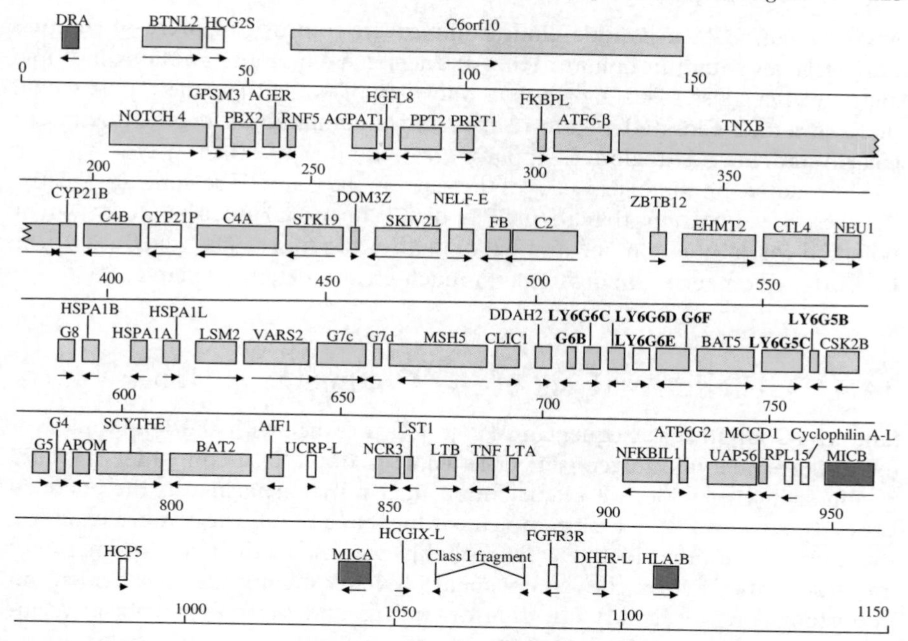

Figure 14.1 Gene map of the human MHC Class III region. Genes in yellow encode secreted or cell surface proteins, while those in blue encode intracellular proteins. The flanking Class I and Class II genes are in green, while pseudogenes are white. Arrows indicate orientation of the gene.

Comparative analysis between the human and other genomes can yield valuable information regarding gene structure and function. Alignment of the human and mouse Class III regions has revealed a high level of similarity across the whole region.[41] The comparative analyses have identified conserved sequence blocks that provide insights into gene structure, alternative RNA splicing exons and putative regulatory regions. Mostly, the gene order, the gene structures and the exon sequences are well conserved between human and mouse. Much less sequence similarity is observed in intragenic and intergenic sequences, which cannot be aligned. However, some non-coding sequences, mainly in the 5′ region or in the first intron of individual genes, are conserved. These conserved non-coding sequences will be of great biological importance as they most likely contain regulatory elements involved in the control of gene expression.

Since divergence of human and mouse from the last common ancestor, the Class III region has remained a relatively stable region of the genome. On the centromeric side of NOTCH4, mouse has expanded the copy number of the butyrophilin gene family in comparison to human, which has a single butyrophilin gene.[53] However, human has extensively duplicated the Class II DRB genes. At the telomeric end of the Class III region, between the UAP56 and HLA-B genes, the human genome has two copies of the MIC (MIC-A and

MIC-B) gene, which is not found in mouse. In contrast, in place of the two Class I genes found in human, HLA-B and HLA-C, mouse contains multiple copies of the Class I H2-Q genes and one copy of the H2-D gene.[54] The theme of a conserved Class III region relative to the dynamically evolving adjacent Class I and Class II regions is apparent in a number of other species.[55]

The putative physiological roles of the genes in the Class III region vary greatly, but there are indications that about 30 of the 61 functional genes have a known or potential role in the immune and inflammatory responses. The functions of the Class III region genes are described in much greater detail in Chapter 15.

14.4 Genetic Analysis of the Complement Genes

The MHC Haplotype Sequencing Project,[46] together with the HapMap Project,[56] have identified a considerable number of SNPs within the Class III region, many of which result in amino acid polymorphisms in the encoded proteins. However, the three complement loci have significantly more variation than any of the other loci. This is probably because of their role in immunity and the antiquity of the complement system. Complement provided an important defence against invading organisms and the polymorphism facilitated a response to multiple variable insults.

14.4.1 Polymorphism of C4

The C4 locus is highly polymorphic, not only with respect to the number of variants, but also to the type of variant, ranging from single base changes through to copy number variation (CNV).[57] In fact, the recent explosion in interest of CNV, particularly with a link to disease, was preceded by early studies on C4, although at the time this was thought to be unique to the C4 genes, rather than a general phenomenon.

The C4 genes form part of the RCCX module that includes the flanking serine/threonine protein kinase 19 (STK19; RP), CYP21 and TNX genes.[58,59] Although the modules themselves show variation in number, the wealth of polymorphism present in the C4 genes is not repeated in the other genes. In the gene arrangement that was originally described,[17,60] two C4 genes exist in tandem, with one C4A gene and one C4B gene. Thus, the first level of polymorphism is both structurally and functionally important; the allotypes, C4A and C4B. The difference between C4A and C4B is due to four amino acids (C4A – ProCysProValLeuAsp, C4B – LeuSerProValIleHis) in the C4d region of the protein.[61,62] These residues not only confer a propensity to react with specific nucleophilic groups, but also affect the mechanism through which the transacetylation takes place.[63,64] Thus, C4A is thought to play a greater role in opsonization and solubilization of immune aggregates, and hence in immunoclearance, while C4B plays more of a role in the continued activation of complement pathways.[65–67]

C4A and C4B are the most obvious manifestations of polymorphism at this locus with the two most common variants being C4A3 and C4B1.[68] There are, however, many alleles of both genes as a result of non-synonymous base changes, the most obvious of which underlie the Chido-Rogers blood group antigens.[69,70] Numerous other variants have been described, most of which do not affect function.[71] The C4A6 variant is of additional interest, however, since this variant, arising as a result of Arg477 being replaced by Trp, has lost the ability to bind to C5.[72,73]

The C4 genes also show length variation, there being a long (20.6 kb) and short (14.2 kb) variant of both C4A and C4B. These differences result from the insertion in intron 9 of an endogenous retrovirus, HERV-K of 6.36 kb.[74–76] Three-quarters of C4 genes are of the long variant and the fact that an identical variation occurs in both C4A and C4B emphasizes the common ancestry of these two genes.

Although the structure of the C4 locus as originally described[17] suggested that both C4A and C4B were invariably found in single copy, in fact only 55% of Caucasians have this arrangement, with up to a third showing a single copy or three copies on a chromosome and up to four having been identified in a very small minority of individuals.[77–80] The variation is loosely related to haplotype and is centred around the RCCX module. These modules take the form of a long (32.7 kb) module or a short (26.3 kb) module, depending on the C4 gene.[80] Often, the higher copy number is linked to disease, specifically the development of systemic lupus erythematosus (SLE), an archetypal immune complex clearance disease, although it is still highly contentious whether this is just a bystander effect of the MHC haplotype. The mechanism of this variation is presumably through mispairing at meiosis resulting in chromosomes carrying one or three copies and rarely four or none.

14.4.2 Deficiency of C4

Partial deficiency of C4 is extremely common, with 35% of individuals having one non-expressing locus, 8–10% with two and 1% with three.[81–85] This effect has a strong ethnic bias with 20% of Caucasians heterozygous for C4A deficiency, and 10% of Caucasians having a short C4B without C4A, while only 1% of Asians show this arrangement.[86] The link between partial deficiency and disease, specifically immune complex disease, has not been fully established and is still a matter of some debate.[87–89] However, with total deficiency of C4, of which 24 cases have been described, three-quarters are associated with a severe form of SLE.[89,90]

Null alleles can arise in three ways. In approximately 40% of cases the entire gene is deleted. This is part of the process of CNV, as described above. The second mechanism is single base changes. Thus, Rupert *et al.*[91] described a C4A null allele caused by a 2 bp insertion at exon 29 and a C4B null allele caused by a single nucleotide deletion at codon 522, giving a frame shift and premature termination. The third mechanism is the loss of C4B through gene conversion to

C4A.[92] However, it is also possible for a functional C4 deficiency to arise because of mutations in other genes. For example, hereditary angioedema, which is a functional deficiency of C1-inhibitor leading to uncontrolled activation of the classical pathway, results in chronically low levels of C4 and C2.[93]

Mechanistic information on C4 deficiency and its role in SLE has been provided by the development of mouse models of C4 deficiency in which the C4 locus has been 'knocked-out' by targeted deletion. These animals have a profound defect in the antibody response to T cell dependent antigen; they fail to show isotype switching, while B cell signalling is normal.[94] The mice also develop autoantibodies and a concomitant glomerulonephritis.[95] However, there is an element of doubt to this claim.[96] It may be that a necessary prerequisite is a mixed mouse background of 129/C57BL/6,[86] which suggests that absence of C4 contributes to the development of autoantibodies and nephritis, but is insufficient alone to cause severe disease. Interestingly, the C3–C4 double knockout has the same phenotype as the C4 knockout, which implies that C4 is crucial and protective from disease while C3 is not important in the development of lupus.[94]

14.4.3 Polymorphism and Deficiency of C2

C2 has three polymorphic variants (C2C, C2A, C2B) of which the common form is C2C (frequency of 0.97 in Caucasians), and a number of other very rare variants.[97–99] None is thought to affect function. The null allele frequency for C2 is 1% of the population and it is the most common complement deficiency.[100] There is an increase in the development of SLE in totally deficient individuals, but this affects less than 10% of the total deficients and it is a much milder form of lupus than that in the C4 deficient individuals.[101,102] Thus the role of C2 in immune complex clearance is less obvious. There are two routes to C2 deficiency. When there is no detectable C2, either in the serum or intracellularly, this is referred to as Type I. It results from a 28bp deletion at the 3′ end of exon 6. Consequently, exon 6 is spliced out of the mRNA to give a frame shift and hence premature termination.[103] Type II deficiency is caused by a number of different non-synonymous mutations.[104,105] A mutant polypeptide is produced, but is usually retained intracellularly.

14.4.4 Polymorphism and Deficiency of Factor B

There are four main alleles of factor B (S, F, S0.7, F1) with the S allele being the most common in Caucasians (frequency of 0.73).[106] The F/S variants result from a Gln/Arg polymorphism at residue 32,[107] which causes the F variant to have about 30–70% of the haemolytic activity of the S variant.[108] Interestingly, they also show polymorphism within the variant caused by differences in sialic acid content.[109] Recently, haemolytic uraemic syndrome (HUS), which is an important cause of acute renal failure in children, has been shown to be caused by gain-of-function mutations in factor B at amino acids 286 (Phe286Leu) and

323 (Lys323Glu).[110] Functional analyses demonstrated that these mutations result in the enhanced formation of the C3bBb convertase or increased resistance to inactivation by complement regulatory proteins. In addition, two variants of factor B (Leu9His and Arg32Gln) and one variant of C2 (Glu318Asp) have been reported to confer a significantly reduced risk in individuals to develop age-related macular degeneration.[111,112]

No true homozygous deficient individual has ever been identified[113,114] and it was proposed that this might be because of the lethality of the homozygous situation. However, targeted deletion of the factor B locus in mice has proved this hypothesis to be false since the factor B-deficient mice do not show any phenotype,[115,116] although a suppression of expression of the surrounding genes, C2 and NELF-E, was observed.

14.5 Disease Association Studies

The high density of genes, clustering of immune-related genes and extreme degree of polymorphism across the MHC has ensured that this region of the genome has remained of key focus in terms of the identification of disease susceptibility loci. However, these same factors, combined with the strong linkage disequilibrium that exists across the MHC, have also been instrumental in slowing down progress in this area. Most, if not all, of the autoimmune and inflammatory conditions now have well-established HLA associations.[117] However, even though some of these associations were described over 30 years ago, it is still true to say that the precise molecular mechanism by which they exert their effect remains unclear. Indeed, the tight linkage disequilibrium across the MHC means that we cannot even be certain that the association seen between a given disease and any particular HLA allele does not simply reflect linkage disequilibrium with the true causative variant, located elsewhere in the MHC. The majority of the autoimmune diseases are complex genetic diseases in which multiple genetic factors are each thought to contribute a relatively small amount to the overall familial risk. The HLA associations seen in different autoimmune diseases vary in the locus involved, the alleles involved and the relative contribution HLA is thought to have in terms of overall familial inheritance. What is also clear is that, in both early linkage studies and more recent genome wide association (GWA) studies of autoimmune diseases, the peak of linkage and/or association across the MHC is broad and it is widely believed that other genes within the MHC, particularly in the Class III region, may play a role in autoimmune disease susceptibility. These disease susceptibility loci in the Class III region may, in some cases, actually represent the true causative variant, whereas in others they may represent a second or even third independent effect.

14.5.1 Candidate Gene Studies

As with other areas of the genome the primary approach taken to identify disease susceptibility loci has been a candidate gene approach. The initial strategy was to select candidate genes within the Class III region based largely on function and/or availability of known polymorphisms, with the majority of studies focussing on TNF, C4 and factor B (discussed above), allograft inflammatory factor 1 (AIF1), nuclear factor of kappa light chain gene enhancer in B cells inhibitor-like 1 (NFKBIL1) and AGER. Unfortunately, many of the candidate gene association studies undertaken to date have been limited in relation to small sample sizes, across differing ethnic groups, and often not taking into account linkage disequilibrium with known HLA-associated alleles, which has made interpretation of the results very difficult.

A number of polymorphisms have been described within the TNF gene, but the most commonly studied ones have been the -238, -308 and -863 bp polymorphisms which reside within the promoter region of the gene. Numerous studies have considered the role of TNF polymorphisms in rheumatoid arthritis (RA). Some studies find no association,[118,119] some find a positive association[120–122] and others suggest that TNF is associated more with disease severity and/or outcome than with disease susceptibility.[123–125] A recent comprehensive review and meta-analysis of the role of the TNF -308 bp polymorphism in RA suggests that the -308A/G polymorphism represents a major risk factor for RA in Latin Americans, but not in Europeans.[126]

TNF polymorphisms have also been associated with SLE in multiple populations, including Caucasian,[127,128] Mexican[129] and Thai,[130] although the associated SNP or haplotype varies between studies. A recent review and meta-analysis of studies looking at the -308 bp polymorphism in SLE in different ethnic groups revealed that the association appears to be restricted to European populations, with no association in African or Asian populations.[131] A similar story has emerged in relations to ankylosing spondylitis (AS) for which some studies report a positive association[132–134] and others report no association.[135,136] Again, the interpretation of these findings is extremely difficult as many are underpowered, particularly when linkage disequilibrium is taken into account.

Association has been described for two SNPs within the promoter of NFKBIL1 in Japanese populations,[137] but these have not been confirmed in UK[138] or Spanish[139] cohorts. A more recent Japanese study has suggested that the association is with a promoter haplotype rather than with any single SNP.[140] The genetic data to support a role for NFKBIL1 are, therefore, once again conflicting. However, recent data that show expression of NFKBIL1 in activated cells in the rheumatoid synovium[141] suggest that a role for this gene in RA should not be ruled out.

Despite evidence for a role for the advanced glycosylation end-products receptor (AGER) in the amplification of pro-inflammatory responses[142] and that one specific polymorphism of AGER, Gly82Ser, has been shown to influence its function,[143] there have been limited studies in terms of disease association. A recent study of Swedish early onset RA patients found no association with the Gly28Ser polymorphism independent of HLA-DRB1.[144]

14.5.2 Multiple Marker Studies

An alternative strategy to the candidate gene approach has been to use multiple markers that span the whole MHC, the Class III region or the whole genome to pinpoint a disease susceptibility locus. Initially, the focus in relation to the MHC was on the use of microsatellite markers, which have been used to localize disease susceptibility loci in the Class III region in a number of studies, such as RA,[145–147] AS[148] and type 1 diabetes (T1D).[149] While the majority of microsatellite-based studies have implicated additional susceptibility loci within the Class III region, the recurring issues of small sample sizes and linkage disequilibrium mean that, although they have not been able to identify the causative variant(s), they have been very useful in narrowing down the region that needs to be analyzed further from a megabase to a few hundred kilobases. Greater success has been achieved by either combining microsatellite markers and specific SNP markers in genotyping studies or by using specific SNP markers on their own. For example, Newton *et al.*,[150] in addition to confirming the known Class II association in RA, were able to map a RA susceptibility locus to a 126 kb segment of the Class III region between the heat shock protein HSP70 and BAT2 loci that contains 13 genes, while Sims *et al.*[151] were able to map an AS susceptibility locus to a 270 kb segment that contains 23 genes, including the C4 genes independent of the HLA-B27 association. In SLE Fernando *et al.*[152] have found associations with two distinct and independent variants within the Class II (HLA-DRB1*0301) and Class III (SKIV2L) regions in UK SLE trios.

Perhaps the most exciting advance in recent times in terms of the identification of susceptibility loci for complex genetic diseases has been our ability to undertake GWA studies. The most comprehensive GWA study conducted to date is that undertaken by the Wellcome Trust Case Control Consortium.[153] This study involved examining ~2000 individuals from each of seven complex diseases and a shared set of ~3000 controls using the Affymetrix Gene Chip 500 k Mapping Array Set. Both RA and TID were among the diseases studied, and in both cases the most significant effect sizes were found with SNPs within the MHC. In fact, the GWAs undertaken to date in terms of TID,[154–156] RA,[157–159] SLE,[160] AS[161,162] and multiple sclerosis (MS)[163] all report a major effect with SNPs within the MHC. This will require further detailed analysis.

14.6 Conclusions

Since the work on cloning the Class III region in the Immunochemistry Unit was initiated in 1985 a huge amount of information has been assembled on the genes that are located in the Class III region, the functions of their protein products, the polymorphisms they display and how these polymorphisms might be implicated in susceptibility to the many autoimmune and inflammatory diseases associated with the MHC. The GWA studies have reported association with multiple SNPs across a wide region of the MHC and further analysis is required to disentangle how much of the effect seen is due to known effects of HLA alleles and how much can be accounted for by novel MHC susceptibility loci, particularly in the Class III region. Clearly, much remains to be done in relation to defining completely all the causative SNP variants and their roles in disease pathogenesis. This will be helped by the development of novel statistical approaches that take account of the known HLA associations, high gene density and strong linkage disequilibrium. Also, while the SNPs utilised within the GWA studies provide good coverage across most of the genome this is not the case across the MHC. The number of SNPs required needs to be greatly increased if the complexities of this region are to be unravelled. Over the past few years a number of projects directed towards addressing this issue[35,45–47] have facilitated the construction of high-resolution SNP maps across the MHC.[164,165] Use of these high-resolution SNP maps in large, well-powered disease association studies alongside the use of novel statistical approaches will be required if our goal to localize truly independent disease susceptibility loci in the MHC Class III region is to be met.

References

1. J. W. Goers and R. R. Porter, *Biochem. J.*, 1978, **175**, 675.
2. R. D. Campbell, A. W. Dodds and R. R. Porter, *Biochem. J.*, 1980, **189**, 67.
3. R. D. Campbell, J. Gagnon and R. R. Porter, *Biochem. J.*, 1981, **199**, 359.
4. D. N. Chakravarti, R. D. Campbell and J. Gagnon, *FEBS Lett.*, 1983, **154**, 387.
5. S. V. Suggs, R. B. Wallace, T. Hirose, E. H. Kawashima and K. Itakurs, *Proc. Natl. Acad. Sci. USA*, 1981, **78**, 6613.
6. A. M. Maxam and W. Gilbert, *Methods Enzymol.*, 1980, **65**, 499.
7. F. Sanger, S. Nicklen and A. R. Coulson, *Proc. Natl. Acad. Sci. USA*, 1977, **74**, 5463.
8. C. A. Alper, In: *The Role of the major histocompatibility complex in immunobiology*, ed. M. E. Dorf, Garland, New York, 1981, p. 173.
9. R. R. Porter, *Mol. Biol. Med.*, 1983, **1**, 161.
10. R. R. Porter, *CRC Crit. Rev. Biochem.*, 1984, **16**, 1.
11. R. D. Campbell, M. C. Carroll and R. R. Porter, *Adv. Immunol.*, 1985, **38**, 203.
12. J. L. Tiwari and P. I. Terasaki, In: *HLA and Disease Associations*, Springer-Verlag, New York, 1985, p. 1.
13. R. Batchelor and A. McMichael, *Brit. Med. Bull.*, 1987, **43**, 156.
14. R. D. Campbell and R. R. Porter, *Proc. Natl. Acad. Sci. USA*, 1983, **80**, 4464.
15. M. C. Carroll and R. R. Porter, *Proc. Natl. Acad. Sci. USA*, 1983, **80**, 264.
16. D. R. Bentley and R. R. Porter, *Proc. Natl. Acad. Sci. USA*, 1984, **81**, 1212.
17. M. C. Carroll, R. D. Campbell, D. R. Bentley and R. R. Porter, *Nature*, 1984, **307**, 237.
18. B. J. Morley and R. D. Campbell, *EMBO J.*, 1984, **3**, 153.

19. T. Belt, M. C. Carroll and R. R. Porter, *Cell*, 1984, **36**, 907.
20. D. R. Bentley, *Biochem. J.*, 1986, **239**, 339.
21. *Immunol. Rev.*, 1999, **172**, 5.
22. E. B. Robson and L. U. Lamm, *Cytogenet. Cell Genet.*, 1984, **37**, 47.
23. D. C. Schwartz and C. R. Cantor, *Cell*, 1984, **37**, 67.
24. G. F. Carle and M. V. Olson, *Nucleic Acids Res.*, 1984, **12**, 5647.
25. E. M. Southern, R. Anand, W. R. Brown and D. S. Fletcher, *Nucleic Acids Res.*, 1987, **15**, 5925.
26. C. A. Sargent, I. Dunham and R. D. Campbell, *EMBO J.*, 1989, **8**, 2305.
27. I. Dunham, C. A. Sargent, J. Trowsdale and R. D. Campbell, *Proc. Natl. Acad. Sci. USA*, 1987, **84**, 7237.
28. I. Dunham, C. A. Sargent, R. L. Dawkins and R. D. Campbell, *J. Exp. Med.*, 1989, **169**, 1803.
29. I. Dunham, C. A. Sargent, R. L. Dawkins and R. D. Campbell, *Genomics*, 1989, **5**, 787.
30. I. Dunham, C. A. Sargent, E. Kendall and R. D. Campbell, *Immunogenetics*, 1990, **32**, 175.
31. E. Kendall, J. A. Todd and R. D. Campbell, *Immunogenetics*, 1991, **34**, 349.
32. W. R. Brown and A. P. Bird, *Nature*, 1986, **322**, 477.
33. C. A. Sargent, I. Dunham, J. Trowsdale and R. D. Campbell, *Proc. Natl. Acad. Sci. USA*, 1989, **86**, 1968.
34. E. Kendall, C. A. Sargent and R. D. Campbell, *Nucl. Acids Res.*, 1990, **18**, 7251.
35. The MHC Sequencing Consortium, *Nature*, 1999, **401**, 921.
36. S. A. Newland, I. C. Macaulay, A. R. Floto, E. C. de Vet, W. H. Ouwehand, N. A. Watkins, P. A. Lyons and R. D. Campbell, *Blood*, 2007, **109**, 4806.
37. M. A. Albertella, H. Jones, W. Thomson, M. G. Olavesen and R. D. Campbell, *Genomics*, 1996, **36**, 240.
38. M. A. Albertella, H. Jones, W. Thomson, M. G. Olavesen, M. Neville and R. D. Campbell, *DNA Seq.*, 1996, **7**, 9–12.
39. M. J. Neville and R. D. Campbell, *J. Immunol.*, 1999, **162**, 4745.
40. G. Ribas, M. Neville, J. L. Wixon, J. Cheng and R. D. Campbell, *J. Immunol.*, 1999, **163**, 278.
41. T. Xie, L. Rowen, B. Aguado, M. E. Ahearn, A. Madan, S. Qin, R. D. Campbell and L. Hood, *Genome Research*, 2003, **13**, 2621.
42. D. E. Geraghty, R. Daza, L. M. Williams, Q. Vu and A. Ishitani, *Immunol. Rev.*, 2002, **190**, 69.
43. J. Trowsdale and P. Parham, *Eur. J. Immunol.*, 2004, **34**, 7.
44. A. Stenzel, T. Lu, W. A. Koch, J. Hampe, S. M. Guenther, F. de la Vega, M. Krawczak and S. Schreiber, *Hum. Genet.*, 2004, **114**, 377.
45. J. Traherne, R. Horton, A. N. Roberts, M. Miretti, M. Hurles and C. A. Stewart, *et al.*, *PLoS Genetics*, 2006, **2**, 81.
46. C. A. Stewart, R. Horton, R. Allcock, J. Ashurst, A. Atrazhev, P. Coggill and I. Dunham, *et al.*, *Genome Res.*, 2004, **14**, 1176.

47. W. P. Smith, Q. Vu, S. S. Li, J. A. Hansen, L. P. Zhao and D. E. Geraghty, *Genomics*, 2006, **87**, 561.
48. V. Rakyan, T. Hildmann, K. L. Novik, J. Lewin, J. Tost and A. V. Cox, *et al.*, *PLoS Biol.*, 2004, **2**, 2170.
49. A. L. Hughes, *Mol. Biol. Evol.*, 1998, **15**, 854.
50. M. F. Flajnik and M. Kasahara, *Immunity*, 2001, **15**, 351.
51. L. Abi-Rached, A. Gilles, T. Shiina, P. Pontarotti and H. Inoko, *Nature Genet.*, 2002, **31**, 100.
52. R. Horton, L. Wilming, V. Rand, R. Lovering, E. Bruford, V. Khodiyar, M. J. Lush, S. Povey, C. Talbot, M. W. Wright, H. Wain, J. Trowsdale, A. Ziegler and S. Beck, *Nature Rev. Genet.*, 2004, **5**, 889.
53. M. Stammers, L. Rowen, D. Rhodes, J. Trowsdale and S. Beck, *Immunogenetics*, 2000, **51**, 373.
54. A. Kumanovics, A. Madan, S. Qin, L. Rowen, L. Hood and K. Fischer-Lindahl, *Immunogenetics*, 2002, **54**, 479.
55. J. Kelley, L. Walter and J. Trowsdale, *Immunogenetics*, 2005, **56**, 683.
56. The International HapMap Consortium, *Nature*, 2005, **437**, 1299.
57. D. E. Isenman, In: *The Complement Factsbook*, eds. B. J. Morley and M. J. Walport, Academic Press, London, 2000, p. 95.
58. L. M. Shen, L. C. Wu, S. Sanlioglu, R. Chen, A. R. Mendoza, A. W. Dangel, M. C. Carroll, W. B. Zipf and C. Y. Yu, *J. Biol. Chem.*, 1994, **269**, 8466.
59. Z. Yang, A. R. Mendoza, T. R. Welch, W. B. Zipf and C. Y. Yu, *J. Biol. Chem.*, 1999, **274**, 12147.
60. G. J. O'Neill, S. Y. Yang and B. Dupont, *Proc. Natl. Acad. Sci. USA*, 1978, **75**, 5165.
61. K. T. Belt, C. Y. Yu, M. C. Carroll and R. R. Porter, *Immunogenetics*, 1985, **21**, 173.
62. C. Y. Yu, K. T. Belt, C. M. Giles, R. D. Campbell and R. R. Porter, *EMBO J.*, 1986, **5**, 2873.
63. A. W. Dodds, X. D. Ren, A. C. Willis and S. K. Law, *Nature*, 1996, **379**, 177.
64. S. K. A. Law and A. W. Dodds, *Protein Sci.*, 1997, **6**, 263.
65. J. A. Schifferli, G. Steiger, J. P. Paccaud, A. G. Sjoholm and G. Hauptmann, *Clin. Exp. Immunol.*, 1986, **63**, 473.
66. B. D. Reilly and C. Mold, *Clin. Exp. Immunol.*, 1997, **110**, 310.
67. M. J. Walport, *N. Eng. J. Med.*, 2001, **344**, 1058.
68. G. Mauff, C. A. Alper, R. Dawkins, G. Doxiadis, C. M. Giles, G. Hauptmann, C. Rittner and P. M. Schneider, *Complement Inflamm.*, 1990, **7**, 261.
69. D. E. Isenman and J. R. Young, *J. Immunol.*, 1984, **132**, 3019.
70. C. Y. Yu, R. D. Campbell and R. R. Porter, *Immunogenetics*, 1988, **27**, 399.
71. C. Y. Yu, *J. Immunol.*, 1991, **146**, 1057.
72. M. J. Anderson, C. M. Milner, R. G. Cotton and R. D. Campbell, *J. Immunol.*, 1992, **148**, 2795.

73. R. O. Ebanks, A. S. Jaikaran, M. C. Carroll, M. J. Anderson, R. D. Campbell and D. E. Isenman, *J. Immunol.*, 1992, **148**, 2803.
74. M. Tassabehji, T. Strachan, M. Anderson, R. D. Campbell, S. Collier and M. Lako, *Nucleic Acids Res.*, 1994, **22**, 5211.
75. A. W. Dangel, A. R. Mendoza, B. J. Baker, C. M. Daniel, M. C. Carroll, L. C. Wu and C. Y. Yu, *Immunogenetics*, 1994, **40**, 425.
76. P. M. Schneider, K. Witzel-Schlomp, K. Rittner and L. Zhang, *Immunogenetics*, 2001, **53**, 1.
77. B. Olaisen, P. Teisberg and R. Jonassen, *Immunobiology*, 1980, **158**, 82.
78. P. Teisberg, R. Jonassen, B. Mevag, T. Gedde-Dahl and B. Olaisen, *Ann. Hum. Genet.*, 1988, **52**, 77.
79. C. A. Blanchong, B. Zhou, K. L. Rupert, E. K. Chung, K. N. Jones, J. F. Sotos, W. B. Zipf, R. M. Rennebohm and C. Y. Yu, *J. Exp. Med.*, 2000, **191**, 2183.
80. E. K. Chung, Y. Yang, R. M. Rennebohm, M. L. Lokki, G. C. Higgins, K. N. Jones, B. Zhou, C. A. Blachong and C. Y. Yu, *Am. J. Hum. Genet.*, 2002, **71**, 823.
81. G. Hauptmann, G. Tappeiner and J. A. Schifferli, *Immunodefic. Rev.*, 1988, **1**, 3.
82. G. Barba, C. Rittner and P. M. Schneider, *J. Clin. Invest.*, 1993, **91**, 1681.
83. K. E. Sullivan, N. A. Kim, D. Goldman and M. A. Petri, *J. Rheumatol.*, 1999, **26**, 2144.
84. G. N. Fredrikson, B. Gullstrand, P. M. Schneider, K. Witzel-Schlomp, A. G. Sjoholm, C. A. Alper, Z. Awdeh and L. Truedsson, *Hum. Immunol.*, 1998, **59**, 713.
85. M. L. Lokki, A. Circolo, P. Ahokas, K. L. Rupert, C. Y. Yu and H. R. Colten, *J. Immunol.*, 1999, **162**, 3687.
86. M. J. Lewis and M. Botto, *Autommunity*, 2006, **39**, 367.
87. J. P. Atkinson, In: *Systemic Lupus Erythematosus*, ed. R.G. Lahita, Academic Press, San Diego, 1999, p. 91.
88. J. S. Navratil, L. C. Korb and J. M. Ahearn, *Immunopharmacol.*, 1999, **42**, 47.
89. Y. Yang, E. Chung, B. Zhou, K. Lhotta, L. A. Hebert, D. Birmingham, B. Rovin and C. Y. Yu, *Curr. Dir. Autoimmun.*, 2004, **7**, 98.
90. M. C. Pickering, M. Botto, P. R. Taylor, P. J. Lachmann and M. J. Walport, *Adv. Immunol.*, 2000, **76**, 227.
91. K. L. Rupert, J. M. Moulds, Y. Yang, F. C. Arnett, R. W. Warren, J. D. Reveille, B. L. Myones, C. Blachong and C. Y. Yu, *J. Immunol.*, 2002, **169**, 1570.
92. L. Braun, P. M. Schneider, C. M. Giles, J. Bertrams and C. Rittner, *J. Exp. Med.*, 1990, **171**, 129.
93. V. Agnello, *Arthritis Rheum.*, 1978, **21**, S146.
94. M. B. Fischer, M. Ma, S. Georg, X. Zhou, J. Xia, O. Finco, S. Han, G. Kelsoe, R. G. Howard, T. L. Rothstein, E. Kremmer, F. S. Rosen and M. C. Carroll, *J. Immunol.*, 1996, **157**, 549.
95. Z. Chen, S. B. Koralov and G. Kelsoe, *J. Exp. Med.*, 2000, **192**, 1339.

96. E. Paul, O. O. Pozdnyakova, E. Mitchell and M. C. Carroll, *Eur. J. Immunol.*, 2002, **32**, 2672.
97. C. A. Alper, *J. Exp. Med.*, 1976, **144**, 1111.
98. T. Meo, J. Atkinson, M. Bernoco, D. Bernoco and R. Ceppellini, *Proc. Natl. Acad. Sci. USA*, 1977, **74**, 1672.
99. D. R Bentley, R. D. Campbell and S. J. Cross, *Immunogenetics*, 1985, **22**, 377.
100. V. Agnello, *Medicine*, 1978, **57**, 1.
101. D. Glass, D. Raum, D. Gibson, J. S. Stillman and P. H. Schur, *J. Clin. Invest.*, 1976, **58**, 853.
102. C. A. Johnson, P. Densen, R. A. Wetsel, F. S. Cole, N. E. Goeken and H. R. Colten, *New Eng. J. Med.*, 1992, **326**, 871.
103. C. A. Johnson, P. Densen, R. K. Hurford, H. R. Colten and R. A. Wetsel, *J. Biol. Chem.*, 1992, **267**, 9347.
104. R. A. Wetsel, J. Kulics, M. L. Lokki, P. Kiepiela, H. Akama, C. A. Johnson, P. Densen and H. R. Colten, *J. Biol. Chem.*, 1996, **271**, 5824.
105. Z. B. Zhu, T. P. Atkinson and J. E. Volanakis, *J. Immunol.*, 1998, **161**, 578.
106. C. A. Alper, T. Boenisch and L. Watson, *J. Exp. Med.*, 1972, **135**, 68.
107. D. R. Bentley and R. D. Campbell, *Biochem. Soc. Symp.*, 1986, **51**, 7.
108. M. L. Lokki and S. A. Koskimies, *Immunogenetics*, 1991, **34**, 242.
109. G. Garnier, C. Davrinche, R. Charlionet and M. Fontaine, *Complement*, 1988, **5**, 77.
110. E. G. de Jorge, C. Harris, J. Esparza-Gordillo, L. Carreras, E. Arranz, C. Garrido, M. Lopez-Trascasa, P. Sanchez-Corral, B. P. Morgan and S. Rodríguez de Cordoba, *Proc. Natl. Acad. Sci. USA*, 2007, **104**, 240.
111. B. Gold, J. Merriman, J. Zernant, L. Hancox, A. J. Taiber and K. Gehrs, et al., *Nature Genet.*, 2006, **38**, 458.
112. K. Spencer, M. Hauser, L. Olson, S. Schmidt, W. Scott, P. Gallins, A. Agarwal, E. Postel, M. Pericak-Vance and J. Haines, *Hum. Mol. Genet.*, 2007, **16**, 1986.
113. J. M. Thurman, D. Ljubanovic, C. Edelstein, G. Gilkeson and V. M. Holers, *J. Immunol.*, 2003, **170**, 1517.
114. M. A. Hietala, K. Nandakumar, L. Persson, S. Fahlen, R. Holmdahl and M. Pekna, *Eur. J. Immunol.*, 2004, **34**, 1208.
115. M. Matsumoto, W. Fukuda, A. Circolo, J. Goellner, J. Strauss-Scoenberger, X. Wang, S. Fujita, T. Hidvegi, D. Chaplin and H. R. Colten, *Proc. Natl. Acad. Sci. USA*, 1997, **94**, 8720.
116. P. R. Taylor, J. T. Nash, E. Theodoridis, A. Bygrave, M. J. Walport and M. Botto, *J. Biol. Chem.*, 1998, **273**, 1699.
117. *HLA in Health and Disease*, eds R. Lechler and A. Warren, Academic Press, London, 2000, p. 3.
118. S. F. Lo, C. M. Huang, M. C. Wu, J. Y. Wu and F. J. Tsai, *Rheumatol. Int.*, 2003, **23**, 151.
119. Z. Rezaieyazdi, J. T. Afshari, M. Sandooghi and F. Mohajer, *Rheumatol. Int.*, 2007, **28**, 189.

120. R. Chen, M. Fang, Q. Cai, S. Duan, K. Lv, N. Cheng, D. Ren, J. Shen, D. He, L. He and S. Sun, *Rheumatol. Int.*, 2007, **28**, 121.
121. L. M. Gomez, E. A. Ruiz-Narváez, R. Pineda-Tamayo, A. Rojas-Villarraga and J. M. Anaya, *Clin. Exp. Rheumatol.*, 2007, **25**, 443.
122. N. Hirankarn, Y. Avihingsanon and J. Wongpiyabovorn, *Int. J. Immunogenet.*, 2007, **34**, 425.
123. B. M. Brinkman, T. W. Huizinga, S. S. Kurban, E. A. van der Velde, G. M. Schreuder, J. M. Hazes, F. C. Breedveld and C. L. Verweij, *Br. J. Rheumatol.*, 1997, **36**, 516.
124. D. Khanna, H. Wu, G. Park, V. Gersuk, R. H. Gold, G. T. Nepom, W. K. Wong, J. T. Sharp, E. F. Reed, H. E. Paulus and B. P. Tsao, *Arthritis Rheum.*, 2006, **54**, 1105.
125. P. Nemec, M. Pavkova-Goldbergova, M. Stouracova, A. Vasku, M. Soucek and J. Gatterova, *Clin. Rheumatol.*, 2007, **27**, 59.
126. Y. H. Lee, J. D. Ji and G. G. Song, *J. Rheumatol.*, 2007, **34**, 43.
127. S. D'Alfonso, G. Colombo, S. Della Bella, R. Scorza and P. Momigliano-Richiardi, *Tissue Antigens*, 1996, **47**, 551.
128. C. G. Parks, J. P. Pandey, M. A. Dooley, E. L. Treadwell, E. W. St Clair, G. S. Gilkeson, C. A. Feghali-Bostwick and G. S. Cooper, *Hum. Immunol.*, 2004, **65**, 622.
129. J. Zúñiga, G. Vargas-Alarcón, G. Hernández-Pacheco, C. Portal-Celhay, J. K. Yamamoto-Furusho and J. Granados, *Genes Immun.*, 2001, **2**, 363.
130. N. Hirankarn, Y. Avihingsanon and J. Wongpiyabovorn, *Int. J. Immunogenet.*, 2007, **34**, 425.
131. Y. H. Lee, J. B. Harley and S. K. Nath, *Eur. J. Hum. Genet.*, 2006, **14**, 364.
132. A. Milicic, F. Lindheimer, S. Laval, M. Rudwaleit, H. Ackerman, P. Wordsworth, T. Hohler and M. A. Brown, *Genes Immun.*, 2000, **1**, 418.
133. T. Höhler, T. Schäper, P. M. Schneider, K. H. Büschenfelde and E. Märker-Hermann, *Arthritis Rheum.*, 1998, **41**, 1489.
134. X. Zhu, Y. Wang, L. Sun, Y. Song, F. Sun, L. Tang, Z. Huo, J. Li and Z. Yang, *Ann. Rheum. Dis.*, 2007, **66**, 1419.
135. G. M. Verjans, B. M. Brinkman, C. E. Van Doornik, A. Kijlstra and C. L. Verweij, *Clin. Exp. Immunol.*, 1994, **97**, 45.
136. S. González, J. C. Torre-Alonso, J. Martínez-Borra, J. A. Fernández Sánchez, A. López-Vazquez, A. Rodríguez Pérez and C. López-Larrea, *J. Rheumatol.*, 2001, **28**, 1288.
137. K. Okamoto, S. Makino, Y. Yoshikawa, A. Takaki, Y. Nagatsuka, M. Ota, G. Tamiya, A. Kimura, S. Bahram and H. Inoko, *Am. J. Hum. Genet.*, 2003, **72**, 303.
138. R. Kilding, M. M. Iles, J. M. Timms, J. Worthington and A. G. Wilson, *Arthritis Rheum.*, 2004, **50**, 763.
139. L. Collado, B. Rueda, R. Cáliz, B. Torres, A. García, A. Nuñez-Roldan, M. F. González-Escribano and J. Martin, *Arthritis Rheum.*, 2004, **50**, 2032.

140. H. Shibata, M. Yasunami, N. Obuchi, M. Takahashi, Y. Kobayashi, F. Numano and A. Kimura, *Hum. Immunol.*, 2006, **67**, 363.
141. D. Greetham, C. D. Ellis, D. Mewar, U. Fearon, S. N. Ultaigh, D. J. Veale, F. Guesdon and A. G. Wilson, *Hum. Mol. Genet.*, 2007, **16**, 3027.
142. M. A. Hofmann, S. Drury, C. Fu, W. Qu, A. Taguchi, Y. Lu, C. Avila, N. Kambham, A. Bierhaus, P. Nawroth, M. F. Neurath, T. Slattery, D. Beach, J. McClary, M. Nagashima, J. Morser, D. Stern and A. M. Schmidt, *Cell*, 1999, **97**, 889.
143. M. A. Hofmann, S. Drury, B. I. Hudson, M. R. Gleason, W. Qu, Y. Lu, E. Lalla, S. Chitnis, J. Monteiro, M. H. Stickland, L. G. Bucciarelli, B. Moser, G. Moxley, S. Itescu, P. J. Grant, P. K. Gregersen, D. M. Stern and A. M. Schmidt, *Genes Immun.*, 2002, **3**, 123.
144. M. M. Steenvoorden, A. H. van Mil, G. Stoeken, R. A. Bank, R. R. Devries, T. W. Huizinga, J. Degroot and R. E. Toes, *Rheumatology*, 2006, **45**, 488.
145. D. P. Singal, J. Li and Y. Zhu, *Clin. Exp. Rheumatol.*, 2000, **18**, 485.
146. T. Barnetche, A. Constantin, P. A. Gourraud, M. Abbal, J. G. Garnier, S. Cantagrel and A. Cambon-Thomsen, *Tissue Antigens*, 2006, **68**, 390.
147. G. Tamiya, M. Shinya, T. Imanishi, T. Ikuta, S. Makino and K. Okamoto, *et al.*, *Hum. Mol. Genet.*, 2005, **14**, 2305.
148. M. A. Brown, K. D. Pile, L. G. Kennedy, R. D. Campbell, L. Andrew, R. March, J. L. Shatford, D. E. Weeks, A. Calin and B. P. Wordsworth, *Arthritis Rheum.*, 1998, **41**, 588.
149. M. Herr, F. Dudbridge, P. Zavattari, F. Cucca, C. Guja, R. March, R. D. Campbell, A. H. Barnett, S. C. Bain, J. A. Todd and B. P. Koeleman, *Hum. Mol. Genet.*, 2000, **9**, 1291.
150. J. Newton, S. Harney, A. Timms, A.-M. Sims, K. Rockett, C. Darke, B. P. Wordsworth, D. Kwiatkowski and M. A. Brown, *Arthritis Rheum.*, 2004, **50**, 2122.
151. A.-M. Sims, M. Bernardo, I. Herzberg, L. Bradbury, A. Calin, B. P. Wordsworth, C. Darke and M. A. Brown, *Genes Immun.*, 2007, **8**, 115.
152. M. Fernando, C. Stevens, P. Sabeti, E. Walsh, A. McWhinnie, A. Shah, T. Green, J. Rioux and T. Vyse, *PLoS Genet.*, 2007, **3**, e192.
153. Wellcome Trust Case Control Consortium, *Nature*, 2007, **447**, 661.
154. D. J. Smyth, J. D. Cooper, R. Bailey, S. Field, O. Burren, L. J. Smink, C. Guja, C. Ionescu-Tirgoviste, B. Widmer, D. B. Dunger, D. A. Savage, N. M. Walker, D. G. Clayton and J. A. Todd, *Nature Genet.*, 2006, **38**, 617.
155. H. Hakonarson, S. F. Grant, J. P. Bradfield, L. Marchand, C. E. Kim and J. T. Glessner, et al., *Nature*, 2007, **448**, 591.
156. S. Nejentsev, J. Howson, N. Walker, J. Szeszko, S. Fields and H. Stevens, *et al.*, *Nature*, 2007, **450**, 887.
157. S. John, N. Shephard, G. Liu, E. Zeggini, M. Cao, W. Chen, N. Vasavda, T. Mills, A. Barton, A. Hinks, S. Eyre, K. W. Jones, W. Ollier, A. Silman, N. Gibson, J. Worthington and G. C. Kennedy, *Am. J. Hum. Genet.*, 2004, **75**, 54.

158. S. Steer, V. Abkevich, A. Gutin, H. J. Cordell, K. L. Gendall and M. E. Merriman, *et al.*, *Genes Immun.*, 2007, **8**, 57.
159. R. M. Plenge, M. Seielstad, L. Padyukov, A. T. Lee, E. F. Remmers and B. Ding, *et al.*, *N. Engl. J. Med.*, 2007, **357**, 1199.
160. P. Forabosco, J. Gorman, C. Cleveland, J. Kelly, W. Ortmann and C. Johansson, *et al.*, *Genes Immun.*, 2006, **7**, 609.
161. Wellcome Trust Case Control Consortium, *Nature Genet.*, 2007, **39**, 1329.
162. K. W. Carter, A. Pluzhnikov, A. E. Timms, C. Miceli-Richard, C. Bourgain, B. P. Wordsworth, H. Jean-Pierre, N. J. Cox, L. J. Palmer, M. Breban, J. D. Reveille and M. A. Brown, *Rheumatology*, 2007, **46**, 763.
163. International Multiple Sclerosis Genetics Consortium, *N. Engl. J. Med.*, 2007, **357**, 851.
164. M. M. Miretti, E. C. Walsh, X. Ke, M. Delgado, M. Griffiths, S. Hunt, J. Morrison, P. Whittaker, E. S. Lander, L. R. Cardon, D. R. Bentley, J. D. Rioux, S. Beck and P. Deloukas, *Am. J. Hum. Genet.*, 2005, **76**, 634.
165. P. I. de Bakker, G. McVean, P. C. Sabeti, M. M. Miretti, T. Green and J. Marchini, *et al.*, *Nature Genet.*, 2006, **38**, 1166.

CHAPTER 15

Functional Characterization of Major Histocompatibility Complex Class III Region Genes

R. DUNCAN CAMPBELL,[a] CAROLINE M. MILNER[b] AND BEGOÑA AGUADO[c]

[a] Department of Physiology, Anatomy and Genetics, University of Oxford, Oxford, OX1 3QX, UK; [b] Faculty of Life Sciences, University of Manchester, Manchester, M13 9PT, UK; [c] Centro de Biología Molecular Severo Ochoa (CBMSO), CSIC, Madrid, Spain

15.1 Introduction

The major histocompatibility complex (MHC) is a key region of the human genome that extends over 3.6 Mbp on human chromosome 6p21.3. It was originally identified as containing the genes that encode the Class I and Class II cell surface molecules critical for our immune system to mount responses to a variety of different pathogens, including viruses and bacteria. A large body of work over the past 30 years has defined the location of these genes and their relative order, the functions of the encoded proteins in antigen processing and presentation, and that some of the genes display a high degree of polymorphism in the human population.[1–4] In total 224 loci have been located in the MHC, of which 129 are known to be expressed.[5] Continued interest in the MHC is driven by its genetic association with a wide range of human pathologies, which include major autoimmune diseases [such as type 1 diabetes (T1D) and rheumatoid arthritis (RA)], infectious diseases (such as malaria), and the most common primary immunodeficiency, IgA deficiency.[6–10] Allelic variants of the Class II and, to a much lesser

Molecular Aspects of Innate and Adaptive Immunity
Edited By Kenneth BM Reid and Robert B Sim

Published by the Royal Society of Chemistry, www.rsc.org

extent, the Class I molecules have been implicated as disease-susceptibility loci. However, even in the best and most comprehensively studied conditions, such as T1D[6,7] and RA,[8] it has still proved very difficult to define the full contribution of the MHC to disease susceptibility. The main reasons for this are the complexity of associations where several MHC region genes might be involved, and the strong linkage disequilibrium that exists between allelic variants of genes in the MHC.

In 1984, having mapped the human complement genes C2, C4 and factor B relative to one another in a 160 kb contig of overlapping cosmid clones in the Class III region of the MHC,[11] we became much more interested in trying to map the Class III region as a whole in order to define genes that might be involved in susceptibility to MHC-associated diseases. At that time the only other loci known to lie within the MHC Class III region were the gene that encoded the steroid enzyme 21-hydroxylase required for cortisol biosynthesis, mutations in which are associated with congenital adrenal hyperplasia, and a gene linked to a neuraminidase deficiency. Subsequently, in 1985, it was shown that genes that encoded steroid 21-hydroxylase lay immediately 3′ of each of the two C4 genes.[12,13] Over the next 15 years the Class III region was cloned in overlapping cosmid and bacterial artificial chromosome (BAC) clones, which culminated in completion of the DNA sequence in 1999.[5] During that time a number of different techniques were used to identify coding sequences, including CpG-island mapping, screening of cDNA libraries using genomic probes, Northern blot analysis and exon trapping, as well as annotation of the DNA sequence using exon-prediction programs (for a detailed discussion see Chapter 14). The MHC Class III region is now known to contain 64 genes in 900 kb of DNA, making it the most gene-dense region in the human genome, with an average gene size of <9 kb and an average intergenic distance of just under 3 kb. Figure 14.1 illustrates the location of these genes together with the flanking Class I and Class II genes, while Table 15.1 shows some of the features of the encoded proteins.

In this chapter we focus on those genes that encode proteins with a known or suspected role in the immune and inflammatory responses, in the response of cells to stress, in protein ubiquitination and in transcriptional and/or translational control. Recent reviews that describe the functions of those not dealt with here can be found for tenascin X,[14] steroid 21-hydroxylase,[15] valyl-tRNA synthetase,[16] MSH5,[17] chloride intracellular channel 1,[18] dimethyl-arginine dimethyl-aminohydrolase 2,[19] 1-acyl-sn-glycerol-3-phosphate acetyl transferase 1,[20] palmitoyl-protein thioesterase 2,[21] apolipoprotein M,[22] pre-B-cell leukaemia transcription factor 2,[23] and G-protein-signalling modulator 3,[24] while some aspects of the complement proteins C4, C2 and factor B are described in detail in Chapters 5, 7 and 14.

15.2 Genes that Encode Cell Surface Receptors

15.2.1 G6B

The human G6B gene encodes at least five different splice variants, including two transmembrane isoforms, G6B-A and -B, and three secreted isoforms,

Table 15.1 Features of the proteins encoded by MHC Class III region genes.

Gene	*Alternative Name*	*Length of ORF*	*Molecular Weight (kDa)*	*Localisation*	*Function*
MCCD1		119	13	Mitochondria	Unknown
UAP56	BAT1	426	47	Nucleus	mRNA splicing and export
ATP6G2	ATP6G, NG38	118	13	Vacuoles	G2 subunit of vacuolar ATPase H^+ pump
NFKBIL1	IKBL	358	39	Nuclear speckles	Binds mRNA/role in mRNA processing
LTA	TNF-B, TNFSF1	205	25	Membrane associated	Cytokine/role in lymphoid organ development and germinal centre formation
TNF	TNF-A, TNFSF2	233	17	Secreted and membrane associated	Cytokine/involved in inflammation and immunomodulation
LTB	TNF-C, TNFSF3	306	33	Cell membrane	Anchors LTA to cell membrane
LST1	B144	Multiple	Various	Cell membrane/ cytosol	Unknown
NCR3	NKp30, 1C7	Multiple	30–42	Cell membrane	Natural killer cell receptor
AIF1	G1, IBA1	147	17	Cytosol	Binds Ca^{2+}/involved in macrophage activation
BAT2	G2	2157	229	Nucleus	Involved in regulation of pre-mRNA splicing
SCYTHE	BAT3, G3, BAG6	1126	124	Nucleus	Negative regulator of apoptosis
APOM	G3A, NG20	223	25	Plasma	Lipid transport
G4	C6orf47, NG34	294	32	Unknown	Unknown
G5	BAT4, GPATCH10	356	39	Unknown	Unknown
CSK2B	G5a, Phosvitin	215	24	Cytosol	Casein kinase II beta subunit/ involved in various cellular processes

LY6G5B		146	24	Secreted	Unknown
LY6G5C		225	15	Secreted	Unknown
BAT5	NG26	558	61	Cell membrane	Interacts with proteins involved in RNA processing
G6F	C6 or F21, NG32	297	35	Cell membrane	Activatory receptor on platelets
LY6G6E		Pseudogene			
LY6G6D	NG25	133	10–19	Cell membrane	Unknown
LY6G6C	NG24	125	18–26	Cell membrane	Unknown
G6B	C6 or F25, NG31	241	30	Cell membrane	Inhibitory receptor on platelets
DDAH2	G6A, NG30	285	32	Cytosol	Dimethyl-arginine Dimethyl-AminoHydrolase 2
CLIC1	NCC27, G6	241	27	Nuclear membrane	Nuclear chloride ion channel
MSH5	G7, MutS	835	92	Nucleus	Chromosome pairing in meiosis
G7d	C6orf26, NG23	148	16	Unknown	Unknown
G7c	C6orf27, NG37	891	98	Unknown	Unknown
VARS2	G7A	1265	139	Cytosol	Valyl tRNA synthetase
LSM2	G7B	95	11	Cytosol	Associated with the U6 SnRNP and with mRNA decapping factors
HSPA1L	Hsp70-Hom	641	70	Cytosol	HSP70, chaperone in recovery of cells from stress
HSPA1A	Hsp70-1	641	70	Cytosol	HSP70, chaperone in recovery of cells from stress
HSPA1B	Hsp70-2	641	70	Cytosol	HSP70, chaperone in recovery of cells from stress
G8	C6orf48	Unknown	Unknown	Unknown	Unknown
Neu1	G9	415	46	Lysozomes	Sialidase/regulates sialic acid content of glycoproteins/glycolipids
CTL4	SLC44A4, NG22	709	78	Unknown	Choline transporter-like protein
EHMT2	G9A/NG36, BAT8	1190	131	Nucleus	Histone methyltransferase/involved in regulation of gene expression
ZBTB12	G10, NG35	427	47	Unknown	Zn finger containing protein
C2		752	102	Plasma	Serine protease/involved in Complement classical pathway

Table 15.1 (*Continued*).

Gene	*Alternative Name*	*Length of ORF*	*Molecular Weight (kDa)*	*Localisation*	*Function*
Factor B		764	90	Plasma	Serine protease/involved in Complement alternative pathway
NELF-E	RD	380	42	Nucleus	Part of NELF complex/inhibits transcriptional elongation
SKIV2L	G11a, SKI2W	1246	138	Nucleus	RNA helicase/repression of translation and normal 3′ degradation
DOM3Z	DOM3L, NG6	396	45	Nucleus	Unknown
STK19	G11	364	41	Nucleus	Unknown
C4A		1744	200	Plasma	Involved in Complement classical pathway activation
CYP21P	CYP21A	Pseudogene			
C4B		1744	200	Plasma	Involved in Complement classical pathway activation
CYP21B	21-OHase	495	55	Membrane	Cytochrome P450 21-hydroxylase/required for cortisol biosynthesis
TNXB	TN-X, XB	4289	470	Extracellular matrix	Involved in collagen deposition/stabilisation in extracellular matrix
ATF6β	G13, CREBL1	700	77	Endoplasmic reticulum	Regulates the expression of stress response genes
FKBPL	DIR1, WISp39, NG7	349	39	Nucleus	Upregulated in cells exposed to DNA damaging agents
PRRT1	C6orf31, NG5	306	34	Unknown	Unknown
PPT2	G14	302	34	Lysosomes	Palmitoyl-protein thioesterase-1
EGFL8	NG3	293	33	Unknown	Unknown
AGPAT1	G15, LPAATα	283	32	Endoplasmic reticulum	Lysophosphatidic acid acyltransferase alpha
RNF5	G16, RMA1, NG2	180	20	Endoplasmic reticulum	E3 ubiquitin ligase

AGER	RAGE	404	50	Cell membrane	Receptor for advanced glycosylation end products of proteins
PBX2	G17	430	48	Nucleus	Homeobox domain-containing transcription factor
GPSM3	G18, AGS4	160	18	Cytoplasm	GoLoco motif-containing protein involved in G protein signalling
NOTCH4	INT3	2003	210	Cell membrane	Involved in vascular development and remodelling
C6orf10		563	62	Unknown	Unknown
HCG2S		Pseudogene			
BTNL2		455	50	Cell membrane	Negative co-stimulatory molecule on T cells

G6B-C, -D and -E.[25,26] The two principal cell surface isoforms G6B-A and -B share a common extracellular N-terminal Ig-like domain and transmembrane region, but differ completely in the sequences of their C-terminal cytoplasmic tails.[26] G6B-B is classified as an inhibitory receptor as it contains two Tyr residues in its cytoplasmic tail (at positions 211 and 237) within immunoreceptor tyrosine-based inhibitory motifs (ITIMs) that possess an SH2-binding domain. When the Tyr residues in the ITIMs are phosphorylated the G6B-B isoform is able to recruit the SH2 domain-containing protein-tyrosine phosphatases SHP1 and SHP2,[26] which dephosphorylate phospho-proteins, leading to negative modulation of signalling cascades. The extracellular domain of G6B binds heparin tightly.[27] The G6B protein has been found to be expressed on the surface of platelets, where it attenuates collagen related peptide (CRP)- and adenosine diphosphate (ADP)-induced platelet activation and aggregation *via* association with SHP1.[28,29] It has also been suggested that the G6B-B receptor is involved in regulation of the immune response by $CD4^+$ T cell-mediated and interleukin 4 (IL-4)-induced regulatory mechanisms.[30]

15.2.2 G6F

The human G6F gene encodes a type-I transmembrane glycoprotein, which contains a putative leader peptide, an extracellular region that shares sequence similarity with V-type immunoglobulin (Ig)-like domains of other proteins, a transmembrane segment and a short cytoplasmic tail with a single Tyr (position 281) phosphorylation site.[31] Using GST pull-down and co-immunoprecipitation assays of wild-type and mutant constructs, it has been shown that, following phosphorylation of Tyr281, G6F interacts with the SH2 domains of the adaptor signalling proteins Grb2 and Grb7.[31] These interactions and increased phosphorylation of p42/44 mitogen-activated protein kinase (MAPK) after antibody cross-linking of G6F indicate that G6F could be a novel co-stimulatory molecule involved in cellular activation, *via* its association with downstream signal transduction pathways involving Grb2 and Grb7, including the Ras-MAPK pathway. G6F is expressed on the surface of platelets, where it has been found to interact with Grb2 and is thought to potentiate platelet activation and aggregation.[32]

15.2.3 Lymphocyte Antigen 6 Superfamily Members LY6G6C, LY6G6D, LY6G6E, LY6G5B and LY6G5C

Members of the lymphocyte antigen 6 (LY6) superfamily typically contain 70–80 amino acid domains that have 8–10 Cys residues. Most LY6 proteins are attached to the cell surface by a glycosylphosphatidylinositol (GPI) anchor that is directly involved in signal transduction, although some LY6 family members are secreted. Little is known about the function of LY6 proteins, except for CD59 and uPAR. CD59 protects host cells from the activation of the complement cascade by inhibiting formation of the membrane attack complex,[33] while uPAR plays an important role in the proteolysis of extracellular matrix proteins.[34] LY6G6C, LY6G6D and mouse Ly6g6e (the human equivalent is a pseudogene) are glycosylated, GPI-anchored, cell surface molecules located on filopodia, which could act as cell surface receptors with potential roles in signalling or cell–cell interactions.[35] LY6G5C and LY6G5B, however, are glycosylated, potentially secreted proteins that could act as ligands for other cell surface receptors and, thus, might participate in cell signalling.[35]

An interesting feature in relation to the regulation of expression of LY6G5B and LY6G6D involves an intron-retention event.[36] The intron retained is the first in the open reading frame and interrupts the protein just after the signal peptide by introducing a premature stop codon. The presence of this premature block to transcription should cause the intron-retaining transcript to undergo degradation by nonsense mediated decay (NMD). However, this transcript is present and is generally more abundant than the correctly spliced partner in all cell lines and tissues analyzed, as it is able to escape NMD.[36,37] Another interesting feature of these genes is that exonic sequences from LY6G5B and LY6G6D are found in transcripts derived from their upstream genes.[37] This phenomenon is known as transcription-induced chimerism (TIC), or tandem chimerism, and is being promoted as a novel way to increase combinatorial complexity of the proteome. LY6G6D forms a chimera with G6F to yield the megakaryocyte enhanced gene transcript 1 (MEGT1) protein while LY6G5B forms a chimera with CSK2B.

15.2.4 Activating Natural Killer Receptor

The natural cytotoxicity receptor (NCR) family are natural killer (NK)-activating receptors whose members are NCR1 (NKp46), NCR2 (NKp44) and NCR3 (NKp30). By carrying out NK cell-expression screening, combined with functional analysis, Pende *et al.*[38] identified a monoclonal antibody that reacted with a 30 kDa integral membrane protein, NCR3, on NK cells. Treatment of NK cells with this anti-NCR3 inhibited natural cytotoxicity against normal target cells and most tumour cells. Pende *et al.*[38] isolated a cDNA-encoding NCR3 which was found to be identical to a splice variant of the 1C7 gene described by Neville and Campbell.[39] NCR3 encodes six alternatively spliced isoforms. The expressed protein has a signal peptide followed by an extracellular region that forms a V-type or C-type (depending on isoform) Ig-like domain, a hydrophobic transmembrane region with a positively charged Arg residue and a 33, 25 or 12 (depending on isoform) amino acid cytoplasmic tail lacking any obvious signalling motif. However, NCR3 is found associated with the immunoreceptor tyrosine-based activation motif (ITAM)-containing CD3Z polypeptide and can thus transduce activating signals *via* CD3Z. Interestingly, NCR3 is a pseudogene in mouse.[40]

15.2.5 Butyrophilin-like Protein 2

Butyrophilin-like protein 2 (BTNL2) is a member of the Ig superfamily with homology to butyrophilin genes (*e.g.* BTN1A1) and to the B7 family of co-stimulatory molecules. BTNL2 is a cell surface glycoprotein, which contains two pairs of Ig-like domains separated by a seven-amino acid heptad repeat, but lacks the C-terminal B30-2 domain found in other butyrophilins.[41] Instead, the C-terminal cytoplasmic tail comprises just nine amino acids. In functional assays, a soluble BTNL2-Fc fusion protein was found to inhibit the proliferation of murine $CD4^+$ T cells and to reduce proliferation and cytokine production from T cells activated by anti-CD3 and B7-related protein 1.[42,43] These data suggest a role for BTNL2 as a negative co-stimulatory molecule. Despite its sequence similarity to the B7 family, BTNL2 does not bind any of the known B7 family receptors.[42] Characterization of the human BTNL2 receptor on immature monocyte-derived dendritic cells (DCs) revealed that it is the cell surface molecule DC-SIGN,[44] where binding of DC-SIGN was dependent on the tumour- and/or tissue-specific glycosylation status of BTNL2.

15.2.6 Advanced Glycosylation End-products Receptor

The advanced glycosylation end products receptor (AGER), also called RAGE, is a member of the Ig superfamily that transduces the biological impact of discrete families of ligands [including advanced glycation end products, several members of the S100/calgranulin family, high mobility group box-1 (HMGB1), the β2 integrin Mac-1 and amyloid-β peptide and β-sheet fibrils] in homeostasis, development and inflammation.[45] AGER is a type I transmembrane glycoprotein, which contains a leader peptide, an extracellular region with a single V-type Ig-like and two C-type Ig-like domains, a transmembrane segment and a short cytoplasmic tail.[46] AGER is expressed by multiple, distinct cell types, including endothelial cells, monocytes and macrophages, neutrophils and neurons. AGER adheres strongly to human neutrophils and can act as an endothelial adhesion receptor which mediates direct interaction with Mac-1.[47] The interplay between AGER and Mac-1 in neutrophil recruitment and activation is enhanced by HMGB1.[48] Extensive studies have also highlighted roles for AGER signalling in monocyte and macrophage migration and activation.[45] Interaction of S100A12 and related members of the S100/calgranulin superfamily with AGER triggered cellular activation, with the generation of key proinflammatory mediators.[49] As AGER is expressed in many distinct cell types ligand binding impacts diverse signal transduction pathways; for example, multiple members of the MAPK family are activated by AGER, including p44/p42 (ERK) MAPK, p38 MAPK, and JNK MAPK.[50] In monocytes and macrophages, activation of nuclear factor of kappa light chain gene enhancer in B cells 1 (NF-κB) is a central function of AGER.

15.2.7 Leukocyte-specific Transcript 1

The leukocyte-specific transcript 1 (LST1) gene is expressed as at least nine protein isoforms because of a complex pattern of alternative splicing that involves at least nine exons.[51,52] Both membrane-bound and soluble isoforms of LST1 are expressed, depending on the usage of two possible open reading frames. Treatment of monocytes with interferon-γ (IFN-γ) has been shown to result in a shift from expression of both soluble and membrane-bound LST1 to expression of the soluble isoform alone.[52] One of the membrane-bound isoforms, LST1/c, has been shown to have a profound inhibitory effect on lymphocyte proliferation.[53] The membrane-bound isoform of LST1 (molecular weight 11 kDa) reported by Raghunathan *et al.*,[54] when overexpressed in the HeLa cell line, induces the formation of filopodia and microspikes at the cell surface. An alternative splice variant, LST1/f, that lacks the transmembrane domain, does not have this activity. This latter splice variant has been used in protein interaction studies and two specific interaction partners for LST1/f were identified: a putative E3 ubiquitin ligase (KIAA1333, containing a RING finger and a HECTc domain) and NY-REN-24, a protein similar to *Drosophila* cactin, which is involved in Inhibitor of kappa light chain gene enhancer in B cells (IκB) signalling.[55] These interactions suggest that LST1/f might form a complex with human cactin and an ubiquitin ligase, which could regulate the human equivalents of cactus *via* the ubiquitination–proteasome pathway.

15.3 Genes that Encode Cytokines

15.3.1 Tumour Necrosis Factor

Tumour necrosis factor (TNF) is one of the most prominent inflammatory mediators and plays a central role in triggering the inflammatory reactions of the innate immune system, including induction of cytokine production, activation and expression of adhesion molecules and growth stimulation.[56] TNF is initially produced as a biologically active 26 kDa membrane-anchored precursor protein,[57] which is subsequently cleaved, principally by TNF-converting enzyme,[58] to release a 17 kDa free protein. These proteins form biologically active homotrimers[59] that act on the ubiquitously expressed TNF receptors 1 and 2 (TNFR1 andTNFR2).[60] The presence of homotrimeric TNF is essential for the juxtaposition of the intracellular domains (ICDs) of the TNF receptor, where dimerization or trimerization of these domains is required for signal transduction. The TNF–ligand interaction causes intracellular signalling leading to phosphorylation of NF-κB and thus activation of the p50-p65 subunit, which then increases transcription of pro-inflammatory genes, such as IL-8, IL-6 and TNF itself.[61]

Although TNF is produced by numerous cell types, monocytes and tissue macrophages are the primary cell sources of TNF synthesis, at least during the inflammatory response. TNF gene expression is induced by various stimuli that include viruses, bacterial and parasitic products, tumour cells, IL-1β, IL-2, IFN-γ and TNF itself.[62] TNF is trafficked from the Golgi to the recycling endosome, where vesicle-associated membrane protein-3 (VAMP3) mediates its delivery to the cell surface at the site of phagocytic cup formation.[63] Fusion of the recycling endosome with the cup simultaneously allows rapid release of TNF and expands the membrane for phagocytosis. Synthesis of TNF is tightly controlled at several levels to ensure no inappropriate expression of the gene. However, TNF is known to play a causative role in inflammatory diseases such as RA.[64]

15.3.2 Lymphotoxin-α and -β

Lymphotoxin-α (LTA) was first characterized as a biological factor, produced by mitogen-stimulated lymphocytes, with a cytotoxic effect on neoplastic cell lines.[65] The LTA gene was subsequently mapped to the MHC Class III region[66] together with the lymphotoxin-β (LTB) gene,[67] where these flank the TNF gene, with which they share significant sequence similarity. LTA and LTB occur in three distinct forms, a secreted homotrimer of LTA (LTA3), and two membrane-anchored heterotrimers, LTA1B2 (predominant form) and LTA2B1, with the LTB subunit providing the membrane anchor.[67,68] The homotrimer LTA, like TNF, binds to the TNFR1 and TNFR2 receptors, while LTA1B2 signals *via* the LTB receptor (LTBR).[69] Although LTA and TNF are structurally and functionally related they play distinct roles in the immune system.[68]

Studies on $Lta^{-/-}$, $Ltb^{-/-}$ and $Ltr^{-/-}$ knockout mice revealed that LTA is instrumental in the development of lymph nodes and Peyer's patches, tissues in which primary immune responses are initiated.[70] Activation of the p52/RelB pathway by LTBR signalling results in the translocation of the NF-κB dimers to the nucleus, which leads to the transcription of chemokines, such as CXCL13, CCL19 and CCL21, and other genes involved in the development of lymphoid organs and maintenance of architecture in secondary lymphoid organs. Thus, the spleens of $Lta^{-/-}$ mice lack organized B and T cell areas, marginal zones and germinal centres. These mice generally have poor humoral immune responses and have defective isotype switching, affinity maturation, generation of B cell memory and antibody production.

Recently, Lo *et al.*[71] have identified LTA and another TNF superfamily member LIGHT as critical regulators of key enzymes that control lipid metabolism. In low-density lipoprotein (LDL) receptor-deficient mice, which lack the ability to control lipid levels in the blood, inhibition of Lta and Light signalling with a soluble LtbR decoy protein was found to attenuate the dyslipidaemia.

15.3.3 Allograft Inflammatory Factor 1

The allograft inflammatory factor 1 (AIF1) gene encodes a cytokine-inducible, tissue-specific protein that is transiently expressed in response to vascular trauma and is thought to play a fundamental role in chronic immunological inflammatory processes.[72] AIF1 contains a consensus EF-hand helix loop domain that is a conserved feature of calcium-binding proteins, together with a leucine zipper motif, a hydrophobic region and a nuclear localization sequence.[73] The expression of AIF1 is largely restricted to the monocyte–macrophage lineage, although it has also been detected in spleen, lymph node and thymus. AIF1 can be induced by IFN-γ, which in turn upregulates the expression of several cytokines, including IL-6, IL-10 and IL-12 in murine macrophages.[74] Stimulation of macrophages with oxidized LDL has also been found to increase AIF1 expression significantly. siRNA studies revealed that inhibition of AIF1 protein expression leads to reduced macrophage proliferation and migration, which suggests a tight association between AIF1 expression and macrophage activation.[75] Furthermore, AIF1 expression was upregulated when T lymphocytes were activated, while overexpression of AIF1 led to increased T lymphocyte migration and proliferation.[76]

Acute and transient expression of AIF1 has been observed in vascular smooth muscle cells (VSMCs) in several models of arterial injury.[77] In unstimulated VSMCs, AIF1 was found to co-localize with F-actin, but translocated to lamellipodia upon platelet-derived growth factor (PDGF) stimulation. AIF1 also co-localized with RAC1 and RAC2, while AIF1-transduced VSMCs showed a constitutive and enhanced activation of RAC1 and RAC2,[78] which highlights an important role for these proteins in the inflammation-driven VSMC response to injury.

15.4 Genes that Encode Proteins Involved in Response to Stress

15.4.1 70 kDa Heat Shock Proteins HSPA1A, HSPA1B and HSPA1L

70 kDa heat shock proteins (HSP70s) have been implicated in the synthesis, folding, unfolding, assembly and degradation of proteins and protein complexes during normal cellular processes and in response to stress, where the prevention of protein aggregation by the inducible HSP70s is key to cell recovery.[79,80] These functions rely on the ability of the HSP70s to act as chaperones that can recognize and bind unfolded proteins and peptides. Peptide–protein binding is mediated by the ~28 kDa C-terminal domain of the HSP70s, while the N-terminal region of 44 kDa is an adenosine triphosphatase (ATPase) domain.

A cluster of three HSP70 genes, HSPA1A, HSPA1B and HSPA1L, is located within the MHC Class III region.[81] This finding led to the suggestion that these proteins might participate in antigen processing and/or presentation. In response to heat shock (42 °C) both protein and mRNA levels are substantially elevated for HSPA1A and -B (which encode an identical protein product) and there is translocation of the protein from the cytoplasm to the nucleus, particularly the nucleoli.[81,82] In contrast, HSPA1L mRNA and protein levels are unaltered by heat shock, but the intracellular localization of the protein changes from cytoplasmic to nuclear.[81,82] IFN-γ, a known inducer of MHC Class I molecules and components of the antigen processing pathway [*e.g.* transporter associated with antigen processing (TAP) and proteasome subunits], has been shown to upregulate HSPA1L expression, but causes a small reduction in HSPA1A and -B protein levels in HeLa cells. However, when human subjects were injected systemically with bacterial lipopolysaccharide (LPS; a potent inducer of the inflammatory response) HSPA1A and -B mRNA levels in white blood cells were elevated seven-fold after three hours, while HSPA1L and the co-chaperone Hdj2 mRNAs were induced four-fold and nine-fold, respectively (other HSPs and DnaJ homologues showed little or no change in expression).[82] These data led to the suggestion that the MHC-encoded HSP70s and the co-chaperone Hdj2 might have specific functions in the inflammatory response.

HSP70s have been implicated in cross-priming, where HSP70–peptide complexes released following the lysis of infected or cancerous cells can be taken up by antigen presenting cells (APCs) and the peptides presented by MHC Class I molecules, resulting in the initiation of a cytotoxic T cell (CTL) response.[83] The effects of the complexes are thought to involve initiation of both innate and adaptive immune responses.[83,84] Not only does the endocytosis of HSP70–peptide complexes by APCs allow peptides to be channelled into the MHC Class I-associated antigen-processing pathway, but HSP70 itself has been implicated in the induction of pro-inflammatory cytokines, the activation of NK cells and the maturation of DCs. All these activities require the interaction of HSP70 with target cells and a number of HSP receptors have been described.[85] Relative affinity measurements of HSP70 for various proposed receptors, following their over-expression on non-APCs, found that only the lectin-like scavenger receptor, LOX-1, showed significant binding to HSP70.[86] It has also been shown that the exosomes of heat-stressed B cells contain high levels of HSPs, including HSP70, but that these are not surface exposed.[87] This raises the possibility that the uptake and intracellular processing of exosomes by DCs provides an alternative route for HSP70–peptide delivery to DCs. Although the mechanism(s) by which HSP70 participates in cross-priming are not fully defined this activity provides exciting possibilities for tumour immunotherapy.[88]

15.4.2 Activating Transcription Factor 6β

Stress-induced changes within the endoplasmic reticulum (ER) result in the misfolding of newly synthesized proteins. Cells react to this by initiating the unfolded protein response (UPR), whereby the expression of genes that encode ER chaperones, or proteins involved in ubiquitin-dependent proteasome degradation, is enhanced.[89,90] This response requires the activation of transcription factors that bind to ER stress-response elements (ERSEs) in the promoters of relevant genes. Activating transcription factor 6α (ATF6α) and ATF6β, which are resident in the ER membrane, have been implicated in regulating the expression of stress-response genes, where ATF6β is encoded in the MHC Class III region.[91,92] Both ATF6α and β are activated by stress-induced proteolysis, releasing N-terminal cytoplasmic fragments that are able to translocate to the nucleus. Studies using siRNAs to inhibit ATF6β expression suggest that this protein is a transcriptional repressor that regulates the ATF6α-mediated expression of ER stress-response genes,[93] while gel shift assays have indicated that these two proteins can compete with each other for binding to the ERSE or the ER chaperone BiP.[94] However, a recent study using knockout mice revealed an unaltered stress responsiveness in ATF6$\beta^{-/-}$ embryonic fibroblasts compared to controls, which indicates that ATF6β is not a negative regulator of ATF6α, while the embryonic lethality of ATF6α and β double knockout was taken as evidence that both proteins are positive regulators of ER chaperone induction.[95]

15.4.3 FK-506 Binding Protein-like

FK-506 binding protein-like (FKBPL) has been independently described as an immunophilin-like gene DIR1[96] and WISp39.[97] DIR1 was reported to be transiently repressed in response to the treatment of cells with low doses of radiation,[96] while repression of DIR1 expression resulted in increased DNA repair and cell survival in radio-resistant, but not radio-sensitive, cell lines.[98] The latter suggests that DIR1 might be involved in the phenomenon known as induced radio-resistance, whereby radio-protective mechanisms are upregulated in cells exposed to small doses of ionizing radiation (IR) or other DNA-damaging agents. Consistent with this, the suppression of DIR1 inhibits cell-cycle arrest in response to IR, where IR is known to induce tumour suppressor p53-dependent expression of the cyclin-dependent kinase inhibitor p21, an important regulator of cell-cycle progression that causes G1 phase arrest in response to various stress stimuli. DIR1 (WISp39) has been shown to bind to newly synthesized p21 and prevents its degradation by the proteasome,[97] where this relies on the recruitment of HSP90 by WISp39, which allows the formation of a trimeric WISp39–HSP90–p21 complex.

15.5 Genes that Encode Proteins Involved in Protein Ubiquitination

15.5.1 E3 Ubiquitin-protein Ligase RNF5

Selective ubiquitination, which is well-known as a way to mark proteins as targets for degradation by the proteasome, also controls the intracellular localization of many proteins and, thus, is important in the regulation of cell differentiation, growth and transformation. Ubiquitination is a multi-step process, involving three classes of enzymes – E1 (ubiquitin-activating enzymes), E2 (ubiquitin-conjugating enzymes) and E3 (ubiquitin protein ligases). RNF5 (or RMA1) was identified as a novel RING-finger gene[99] which encodes a membrane-bound protein with E3 ligase activity.[100,101]

RNF5 in the plasma membrane has been shown to interact with and mediate the ubiquitination of paxillin – a key component of the focal adhesions that link the actin cytoskeleton to the extracellular matrix and provide docking sites for many signalling and structural molecules.[101] RNF5 does not appear to affect the stability of paxillin, but it does cause its relocation from focal adhesions to the cytoplasm; this activity requires the RING domain of RNF5, which suggests that the altered cellular distribution results from paxillin ubiquitination. Furthermore, RNF5-mediated loss of paxillin from the focal adhesions results in reduced cell motility.[101] More generally, RNF5 could be responsible for regulating the localization of proteins involved in cytoskeleton organization and thereby act as a regulator of cell migration. RNF5 has also been shown to occur (together with the quality-control protein Derlin-1 and the E2 enzyme Ubc6e) as part of an ER membrane-associated ubiquitin ligase complex that cooperates with the cytosolic HSC70–CHIP E3 complex[102] in the clearance of misfolded proteins.

RNF5 has been implicated in protection against *Salmonella*.[103] Following host-cell infection, SopA, a key effector protein of *Salmonella*, is recognized and ubiquitinated by RNF5. While mono-ubiquitination appears to promote the release of *Salmonella* from vacuoles into the cytoplasm, poly-ubiquinated SopA is rapidly degraded by the host proteasome, thus providing protection against enteropathogenicity.

15.5.2 Scythe (BAT3)

Scythe is a highly conserved protein that was originally isolated from *Xenopus* oocytes and shown to be a Reaper-binding protein and a negative regulator of apoptosis.[104] Scythe has features in common with BAG1, a bcl-2-binding anti-apoptotic protein that acts as a negative regulator of HSP70 function by associating with its ATPase domain (to form a complex with HSP70, the co-chaperone Hdj1-1 and partially folded substrate) and preventing substrate release.[105] In particular, both BAG1 and Scythe have N-terminal ubiquitin-like domains and C-terminal BAG domains. Moreover, Scythe has been shown to bind, *via* its BAG domain, to the ATPase domain of HSP70 or HSC70, where this interaction inhibits HSP70-mediated protein refolding *in vitro*.[105] Reaper was found to abrogate the association of Scythe (but not BAG1) with HSP70 and, thereby, its inhibition of HSP70 activity.[105] This led to the suggestion that Scythe can sequester cytochrome *c* releasing factor(s) and retain these in a soluble, partially folded state through the formation of complexes with HSP70. Reaper-mediated release of the unfolded substrate from HSP70 could subsequently trigger mitochondrial cytochrome *c* release and caspase activation. Reaper, Scythe and HSP70 or HSC70 might represent a regulatory network able to modulate the activity of signalling molecules involved in cell proliferation, apoptosis and stress responses.[105,106]

Scythe is a nuclear protein whose localization is not altered in response to apoptotic stimuli.[107] Since HSP70 translocates from the cytoplasm to the nucleus in response to stress, this is consistent with Scythe acting as a regulator of the HSP70 chaperone function. Knockdown of Scythe in an osteosarcoma cell line was found to substantially reduce apoptosis in response to DNA damage, concomitant with almost complete abolition of acetylation of the tumour suppressor protein p53, which indicates an essential role for Scythe as a positive regulator of p53-mediated apoptosis following genotoxic stress.[108] Scythe$^{-/-}$ mice are characterized by embryonic or perinatal lethality, with abnormal development of the brain or lung and kidney, respectively.[109] These abnormalities were associated with extensive defects in apoptosis and cell proliferation.

15.6 Genes that Encode Proteins Involved in Transcriptional Control

15.6.1 Serine/Threonine Protein Kinase 19

The serine/threonine protein kinase 19 (STK19) gene is predicted to encode a 364 amino acid ubiquitously expressed protein with an extremely hydrophilic N-terminal region and a C-terminal region that comprises alternating hydrophilic and hydrophobic sequences, a putative nuclear localization signal and multiple potential phosphorylation sites.[110] Sargent *et al.*[111] isolated a cDNA that encoded a 258 amino acid STK19 protein (termed G11-Y), which lacks the N-terminal 110 amino acids present in STK19, and identified an inphase STK19 splice variant of 254 amino acids (G11-X). STK19 is expressed as 41 kDa (G11-Z) and 30 kDa (G11-Y) intracellular proteins, with the former localized primarily in the nucleus.[112] Functional analysis demonstrated that STK19 has a manganese-dependent protein kinase activity that phosphorylates α-casein at Ser–Thr residues and histone at Ser residues, although the target of STK19 phosphorylation *in vivo* was not determined. However, in yeast 2 hybrid analysis STK19 interacted with four nuclear proteins known to be involved in DNA replication or transcription, including the transcription factor Sp3 and the splicing factor SF3bsu4, which may be either substrates or cofactors for this kinase.[55]

15.6.2 Negative Elongation Factor Polypeptide E

The RD gene[113] [or negative elongation factor polypeptide E (NELF-E)], which is ubiquitously expressed, was so named because it contains a 58 amino acid central segment that consists almost entirely of an Arg (R)–Asp (D) dipeptide repeat. RD binds to various RNA elements as part of a protein complex originally identified as being required for 5,6-dichloro-1-beta-D-ribofuranosyl benzimidazole (DRB, a nucleoside analogue)-sensitive transcription.[114] This complex, which is designated NELF, is composed of five polypeptides – NELF-E, which is identical to RD, NELF-A [Wolf–Hirschhorn syndrome candidate 2 (WHSC2)], NELF-B and either the alternatively spliced NELF-C or

NELF-D (TH1L). NELF acts with DRB sensitivity-inducing factor (DSIF), a heterodimer of SPT4 [suppressor of Ty 4 homologue 1 (SUPT4H1)] and SPT5 [suppressor of Ty 5 homologue (SUPT5H)], to cause transcriptional pausing of RNA polymerase II.[115] This repression is reversed by positive transcription elongation factor B (PTEFB)-dependent phosphorylation of the RNA polymerase II C-terminal domain. Recently, it has been shown that NELF interacts with the nuclear cap-binding complex (CBC),[116] a multifunctional factor that plays important roles in several mRNA processing steps. Thus, NELF is a new factor that coordinates different mRNA processing steps during transcription.

15.6.3 Nuclear Factor of Kappa Light Chain Gene Enhancer in B Cells Inhibitor-like 1 (NFKBIL1)

Nuclear factor of kappa light chain gene enhancer in B cells inhibitor-like 1 (NFKBIL1) contains 2–3 ankyrin repeats that most closely resemble the second and third ankyrin repeats of NF-κB, and three PEST motifs (a sequence that is rich in Pro, Ser, Asp and Thr residues).[117] These features led to the suggestion that NFKBIL1 might be a member of the IκB family of proteins,[118] where these contain multiple ankyrin repeats that serve as protein–protein interaction domains and are necessary for binding to the transcription factor NF-κB.[61] However, in contrast to other IκB proteins the NFKBIL1 protein is targeted to the nucleus, where it accumulates in nuclear speckles.[119] Recently, it has been shown that NFKBIL1 does not bind NF-κB proteins or downregulate inflammatory signalling, but instead binds mRNA, which suggests a role in mRNA processing.[120]

15.6.4 Euchromatic Histone-lysine N-methyltransferase 2

The product of the euchromatic histone-lysine N-methyltransferase 2 (EHMT2) gene has been identified as a nuclear protein,[121,122] which has histone methyltransferase activity specific for Lys9 and Lys27 of histone H3 (H3-K9 and H3-K27).[123] Histone lysine methylation plays a central epigenetic role in the organization of chromatin domains and the regulation of gene expression.[124] Histone H3 Lys9 (H3-K9) methylation is a key element in the transcriptional silencing of genes and EHMT2 is the major H3-K9 methyltransferase that targets euchromatic regions,[125] having been implicated in the silencing of developmentally regulated genes.[126] The closely related G9a-like protein (GLP), or EHMT1, also methylates H3-K9 and it has been shown that EHMT1 and EHMT2 form homo- and heteromeric complexes *via* their SET (suvar3-9, enhancer-of-zeste, trithorax) domains.[127] Although both proteins can independently methylate H3-K9 *in vitro*, it seems that the formation of heteromers is essential for this activity *in vivo*.[127] Recently, the purification of EHMT2 complexes from mouse embryonic stem cells led to the identification of Wiz, a zinc finger protein that interacts with both EHMT1 and EHMT2 *via* their SET domains.[128] Knockdown of Wiz with siRNA revealed that Wiz contributes to

EHMT2 stability by promoting the formation of EHMT1–EHMT2 and also links these complexes to the C-terminal binding protein (CtBP) co-repressor machinery,[128] where CtBP interacts with a variety of transcriptional silencing molecules that include histone deacetylase and lysine-specific histone demethylase 1 (LSD1). In addition, Wiz can directly bind DNA and might be involved in targeting EHMT1–EHMT2 to specific genetic loci.

15.7 Genes that Encoding Proteins Involved in mRNA Processing

The LSM2, UAP56, DOM3Z and SKIV2L genes encode proteins that are orthologues of yeast proteins involved in mRNA processing (LSM2, UAP56, Rai1p and SKI2). In addition, BAT2, STK19, PBX2 and BAT5 have been found to interact with proteins previously implicated in RNA processing.

LSM2 is known to be a component of two doughnut-shaped heptameric complexes of LSM proteins. The two complexes have LSM proteins 2 to 7 in common, differing only in the seventh subunit (LSM8 or LSM1).[129] In yeast two-hybrid experiments, LSM2 has been shown to interact with LSM3 and either LSM1 or LSM8 depending upon the complex in question.[130] It is likely that the LSM2–8 complex is assembled in the cytoplasm and migrates to the nucleus, where it interacts with the 3′ end of the U6 snRNA, stabilizing the U6 snRNP and the U4/U6 snRNA interaction.[131] The LSM1–7 complex, however, has been shown to accumulate in cytoplasmic foci together with other components of the mRNA decapping and degradation machinery.[132]

DOM3Z is homologous to the yeast protein Rai1p, which interacts with, and stabilizes, the yeast nuclear Rat1p 5′–3′ exoribonuclease.[133] Rai1p appears to play a critical role in enhancing the efficiency of pre-rRNA processing and degradation carried out by Rat1p and the nuclear exosome component Rrp6p.[134]

SKIV2L has a typical helicase domain with seven conserved boxes that consist of structural motifs for ATP binding and hydrolysis, and RNA binding and unwinding activities.[135] SKI2VL is localized to the nucleoli and cytoplasm, where it is associated with the 40S subunit of the ribosomes.[136] The yeast homologue of SKIV2L, SKI2, encodes a putative RNA helicase which is required to repress translation of poly(A) mRNA and for normal 3′ mRNA degradation,[137] which suggests that the human protein performs a similar function.

The BAT2 gene encodes a large proline-rich protein. In yeast two-hybrid experiments, BAT2 was found to interact with heterogeneous nuclear ribonucleoprotein (hnRNP) A1 and hnRNP M, which are components of the spliceosome, Gemin3, which is a component of the survival motor neuron (SMN) complex, and complement C1q binding protein (C1QBP), which interacts with the ASF/SF2 splicing factor.[55] These data suggest that BAT2 may play a role in the regulation of pre-mRNA splicing.

UAP56 belongs to the DEXD/H-box helicase family of proteins. UAP56 is an essential factor in pre-mRNA splicing as well as in the export of mature mRNA from the nucleus to the cytoplasm.[138] Its role in splicing was first identified by virtue of its ability to bind the mammalian splicing factor $U2AF_{65}$ as a component of the 3′ splice site complex.[139] During the splicing reaction, UAP56 appears to facilitate the removal of $U2AF_{65}$ from the polypyrimidine sequence downstream of the splice branchpoint, thereby allowing U2 snRNP to interact with the branchpoint sequence.[139] Subsequently, Luo *et al.*[140] demonstrated that UAP56 interacts with Aly, leading to the recruitment of the mRNA export machinery. In addition, both UAP56 and Aly are part of the transcription–export (Trex) complex that is thought to couple transcription and mRNA export.[141]

15.8 Genes that Encode Proteins Involved in Cell Signalling

15.8.1 Neurogenic Locus Notch Homologue Protein 4

The neurogenic locus notch homologue protein 4 (NOTCH4) gene encodes a type I membrane glycoprotein whose extracellular region contains 29 EGF-like domains and 3 Lin/Notch repeats. The intracellular portion contains six Ankyrin repeats and a PEST domain.[142] Although synthesized as a single-pass transmembrane precursor, the NOTCH4 protein is cleaved into two subunits within the trans-Golgi network, to yield a heterodimeric protein expressed at the cell surface. The core NOTCH4 signalling pathway involves ligand-induced activation of the receptor *via* binding of Jagged1, Jagged2 and Delta-like 4;[143] this results in a series of proteolytic cleavages within the NOTCH4 transmembrane domain, culminating in the release of the ICD. After translocation to the nucleus, the ICD of NOTCH4 associates with the DNA-binding protein RBP-J (also called CSL), recruits the co-activators PCAF and GCN5, and triggers transcription of NOTCH4 target genes.[144] The most extensively characterized function of NOTCH4 is its involvement in vascular development and remodelling.[145,146]

15.8.2 Casein Kinase II β Subunit

The casein kinase II β subunit (CSK2B) gene encodes the β subunit of the ubiquitously expressed serine/threonine casein kinase II, a tetrameric complex which is composed of catalytic α and α′ subunits and two regulatory β subunits. The β subunit plays a complex role in modulating the ability of the α subunits to interact with and phosphorylate substrates. Even though the CSK2 α and α′ subunits are highly related, there are indications that the two isozymes exhibit functional specificity. The identification of its substrates and binding partners has revealed roles for CSK2 in a diversity of cellular processes, ranging from transcription and translation to the regulation of the cell cycle, the actin cytoskeleton, circadian rhythms, apoptosis, transformation and tumorigenesis.[147,148]

When using the CSK2 β subunit as a bait in yeast two-hybrid screens, in addition to its self-association and its interactions with the α and α′ subunits, an interaction with the tyrosine kinase Lyn was also identified.[55] CSK2 and Lyn phosphorylate some of the same proteins and it is possible that they could be part of a larger regulatory complex.

15.9 Genes that Encode Enzymes

15.9.1 Sialidase

The neuraminidase-1 (Neu1) gene encodes an intralysosomal sialidase[149,150] that occurs as a complex with protective protein–cathepsin A (PPCA) and β-galactosidase, which are required for its localization to the lysosomes.[151] Sialidase (or neuraminidase) enzymes, of which there are four in humans, have essential roles in the modulation of cellular functions by regulating the sialic acid content of glycoproteins and glycolipids. Deficiency in lysosomal sialidase activity, arising from point mutations in Neu1 itself or in PPCA, gives rise to sialidaosis[152] or galactosialidosis, respectively. Sialidosis patients suffer frequent infections,[152] which is likely to be because of the involvement of Neu1 in the immune response.

The characterization of mice with different Neu1 allotypes indicates that this sialidase is upregulated in activated T cells with a concomitant reduction in cell surface sialic acid content, where the latter is a requirement for T cell responsiveness to alloreactive B cells.[153] More recently, reduced sialylation of cell surface molecules on activated human $CD4^+$ and $CD8^+$ T cells has been shown to correlate with increased Neu1 activity at the cell surface, where Neu1 was also found to be important in the production of IFN-γ by T cells.[154] Neu1 has also been implicated in the regulation of IL-4 expression during T cell activation, which might be important in determining the balance between the initiation of helper T cell Th2- versus Th1-type responses.[155] In monocytes, the upregulation of Neu1 in response to LPS or TNF has been implicated in the switching of CD44 from an inactive to an active hyaluronan (HA)-binding form.[156,157] The differentiation of monocytes into macrophages is accompanied by increased Neu1 expression and its relocalization from lysosomes to the cell surface *via* MHC Class II compartments.[158] The effects of Neu1 siRNA in this system indicated a role for Neu1 in antigen uptake and cytokine production by macrophages and suggest that it might be involved in antigen presentation.

15.9.2 G2 Subunit of Vacuolar H^+-ATPase

The ATP6G2 gene encodes the G2 subunit of the vacuolar H^+-ATPase (V-ATPase).[39,159] ATP-dependent proton pumps are found within the membranes of most organelles (including endosomes, lysosomes and secretory vesicles), where they play a variety of crucial roles that include protein glycosylation in the Golgi, lysosomal degradation of cellular debris, the processing of receptor–ligand complexes following endocytosis, and synaptic transmission in neuronal cells. The V-ATPases are highly conserved in evolution and contain at least 13 subunits, organized into two domains; a membrane-anchored V_o domain, which forms the pore that mediates proton translocation, and a peripheral domain (V_1) that is responsible for ATP hydrolysis.[160] The V_1 domain contains eight subunits, A–H, with three copies of the A and B subunits and two copies of subunit G, while V_o contains six different subunits, a, d, c, c′, c″ and e, with multiple copies of the c subunits. Mammalian V-ATPases contain tissue-specific isoforms of some of these subunits, for example three isoforms of the G subunit, G1, G2 and G3, have been described.[161] According to the current model of V-ATPase, subunit G2 is thought to interact with subunit E, and this has been confirmed by yeast two-hybrid analysis.[55] Northern blot analysis of mRNA from mouse tissues revealed that ATP6G2 expression is specific to the neurons of the central nervous system.[159] In particular, V-ATPases that contain subunit G2 localize to cell bodies and dendrites as well as axons,[159] where the latter suggests a role for the G2 isoform in synaptic vesicles. Neville and Campbell[39] also observed expression of ATP6G2 mRNA in cell lines that represent B cells, T cells, monocytes and macrophages, consistent with a possible role for V-ATPases that contain this subunit in the immune response.

15.10 Conclusions

As can be seen from the above overview, the MHC Class III region genes encode proteins with a plethora of different functions. Although these proteins participate in diverse biological pathways, certain unifying themes are beginning to develop. For example, at least one-third of the intracellular proteins encoded by genes in the Class III region are likely to have roles in mRNA processing, which is significantly greater than would be expected if the genes were randomly distributed in the human genome. This clustering of mRNA-processing genes is consistent with the notion that the MHC contains clusters of loci that are both functionally and evolutionary related, thereby supporting the concept of a higher order organization of functionally related genes within eukaryotic genomes.[162–164] In addition, one-third of the genes in the Class III region encode proteins with known or suspected roles in the immune and inflammatory responses. These include cytokines, cell surface receptors, complement system proteins, signal transduction proteins, transcription factors and HSPs. Some, such as TNF, LTA, LTB, AIF1, Factor B, C4, AGER and BTNL2, have already been implicated in the pathological mechanisms that underlie certain autoimmune diseases, either through over or under expression, inappropriate expression or expression of variant gene products. However, given that it is proving very difficult to identify which genes contribute to a given disease susceptibility, it is highly likely that many pathologies are influenced by the products of multiple Class III region genes.

With the new generation single nucleotide polymorphism (SNP) maps and genotyping technologies now available, coupled with access to large patient cohorts, it will soon be possible to pinpoint much more accurately those genes involved in autoimmune disease susceptibility. Meaningful interpretation of this genetic data will be greatly enhanced as the molecular functions of the proteins encoded by each gene become much better understood. However, there is still a significant amount of work to be done as 25% of the genes in the MHC Class III region encode proteins with no function so far identified. In addition, the functional complexity of this genomic region is increased markedly by the number of multiple transcripts generated by each gene, which results in different protein isoforms, different proteins and even transcript chimeras. Changes in the regulation of expression of these transcripts could be crucial in relation to their involvement in disease.

References

1. S. Beck and J. Trowsdale, *Annu. Rev. Genomics Hum. Genet.*, 2000, **1**, 117.
2. R. Horton, L. Wilming, V. Rand, R. Lovering, E. A. Bruford, V. K. Khodiyar, M. J. Lush, S. Povey, C. C. Talbot Jr., M. W. Wright, H. M. Wain, J. Trowsdale, A. Ziegler and S. Beck, *Nature Rev. Genet.*, 2004, **5**, 889.
3. J. Trowsdale, *Curr. Opin. Immunol.*, 2005, **17**, 498.
4. A. M. Little and P. Parham, *Rev. Immunogenet.*, 1999, **1**, 105.
5. The MHC Sequencing Consortium, *Nature*, 1999, **401**, 921.
6. F. Pociot and M. F. McDermott, *Genes Immun.*, 2002, **3**, 235.
7. B. A. Lie and E. Thorsby, *Curr. Opin. Immunol.*, 2005, **17**, 526.
8. A. Silman and J. E. Pearson, *Arthritis Res.*, 2002, **4**(Suppl. 3), S265.
9. A. V. S. Hill, *Annu. Rev. Genomics Hum. Genet.*, 2001, **2**, 373.
10. P. D. Burrows and M. D. Cooper, *Adv. Immunol.*, 1997, **65**, 245.
11. M. C. Carroll, R. D. Campbell, D. R. Bentley and R. R. Porter, *Nature*, 1984, **307**, 237.
12. M. C. Carroll, R. D. Campbell and R. R. Porter, *Proc. Natl. Acad. Sci. USA*, 1985, **82**, 521.
13. P. C. White, D. Grossberger, B. J. Onufer, D. D. Chaplin, M. I. New, B. Dupont and J. L. Strominger, *Proc. Natl. Acad Sci. USA*, 1985, **82**, 1089.
14. J. Bristow, W. Carey, D. Egging and J. Schalkwijk, *Am. J. Med. Genet*, 2005, **139C**, 24.
15. S. Nimkarn and M. I. New, *Nature Clin. Pract.*, 2007, **3**, 405.
16. M. Ibba and D. Soll, *Annu. Rev. Biochem.*, 2000, **69**, 617.
17. A. Lynn, R. Soucek and G. V. Bomer, *Chrom. Res.*, 2007, **15**, 591.
18. R. H. Ashley, *Mol. Membr. Biol.*, 2003, **20**, 1.
19. F. Palm, M. L. Onozato, Z. Luo and C. S. Wilcox, *Am. J. Physiol. Heart Circ. Physiol.*, 2007, **293**, H3227.
20. D. W. Leung, *Front. Biosci.*, 2001, **6**, D944.
21. M. E. Linder and R. J. Deschenes, *Biochemistry*, 2003, **43**, 4311.
22. B. Dahlback and L. B. Nielsen, *Curr. Opin. Lipidol.*, 2006, **17**, 291.

23. T. D. Capellini, G. Di Giacomo, V. Salsi, A. Brendolan, E. Ferretti, D. Srivastava, V. Zappavigna and L. Selleri, *Development*, 2006, **133**, 2263.
24. J. B. Blumer, A. V. Smrcka and S. M. Lanier, *Pharmacol. Therap.*, 2007, **113**, 488.
25. G. Ribas, M. Neville, J. L. Wixon, J. Cheng and R. D. Campbell, *J. Immunol.*, 1999, **163**, 278.
26. E. C. de Vet, B. Aguado and R. D. Campbell, *J. Biol. Chem.*, 2001, **276**, 42070.
27. E. C. de Vet, S. A. Newland, P. A. Lyons, B. Aguado and R. D. Campbell, *FEBS Lett.*, 2005, **579**, 2355.
28. S. A. Newland, I. C. Macaulay, A. R. Floto, E. C. de Vet, W. H. Ouwehand, N. A. Watkins, P. A. Lyons and R. D. Campbell, *Blood*, 2007, **109**, 4806.
29. Y. A. Senis, M. G. Tomlinson, A. García, S. Dumon, V. L. Heath, J. Herbert, S. P. Cobbold, J. C. Spalton, S. Ayman, R. Antrobus, N. Zitzmann, R. Bicknell, J. Frampton, K. Authi, A. Martin, M. J. Wakelam and S. P. Watson, *Mol. Cell. Proteomics*, 2007, **6**, 548.
30. J. Li, M. Cadeiras, M. P. Von Bayern, L. Zhang, A. I. Colovai, R. Dedrick, E. A. Jaffe, N. Suciu-Foca and M. C. Deng, *Hum Immunol.*, 2007, **68**, 708.
31. E. C. de Vet, B. Aguado and R. D. Campbell, *Biochem. J.*, 2003, **375**, 207.
32. A. García, Y. A. Senis, R. Antrobus, C. E. Hughes, R. A. Dwek, S. P. Watson and N. Zitzmann, *Proteomics*, 2006, **19**, 5332.
33. A. Davies, D. L. Simmons, G. Hale, R. A. Harrison, H. Tighe, P. J. Lachmann and H. Waldmann, *J. Exp. Med.*, 1989, **170**, 637.
34. T. Tarui, A. P. Mazar, D. B. Cines and Y. Takada, *J. Biol. Chem.*, 2001, **276**, 3983.
35. M. Mallya, R. D. Campbell and B. Aguado, *Protein Science*, 2006, **15**, 2244.
36. M. Mallya, R. D. Campbell and B. Aguado, *Genomics*, 2002, **80**, 113.
37. V. Calvanese, M. Mallya, R. D. Campbell and B. Aguado, *BMC Mol. Biol.*, 2008, Submitted.
38. D. Pende, S. Parolini, A. Pessino, S. Sivori, R. Augugliaro, L. Morelli, E. Marcenaro, L. Accame, A. Malaspina, R. Biassoni, C. Bottino, L. Moretta and A. Moretta, *J. Exp. Med.*, 1999, **190**, 1505.
39. M. J. Neville and R. D. Campbell, *J. Immunol.*, 1999, **162**, 4745.
40. M. Hollyoake, R. D. Campbell and B. Aguado, *Mol. Biol. Evol.*, 2005, **22**, 1661.
41. M. Stammers, L. Rowen, D. Rhodes, J. Trowsdale and S. Beck, *Immunogenetics*, 2000, **51**, 373.
42. H. A. Arnett, S. S. Escobar, E. Gonzalez-Suarez, A. L. Budelsky, L. A. Steffen, N. Boiani, M. Zhang, G. Siu, A. W. Brewer and J. L. Viney, *J. Immunol.*, 2007, **178**, 1523.
43. T. Nyugen, X. K. Liu, Y. Zhang and C. Dong, *J. Immunol.*, 2006, **176**, 7354.

44. G. Malcherek, L. Mayr, P. Roda-Navarro, D. Rhodes, N. Miller and J. Trowsdale, *J. Immunol.*, 2007, **179**, 3804.
45. K. Herold, B. Moser, Y. Chen, S. Zeng, S. F. Yan, R. Ramasamy, J. Emond, R. Clynes and A. M. Schmidt, *J. Leukoc. Biol.*, 2007, **82**, 204.
46. M. Neeper, A. M. Schmidt, J. Brett, S. D. Yan, F. Wang, Y.-C. E. Pan, K. Elliston, D. Stern and A. Shaw, *J. Biol. Chem.*, 1992, **267**, 14998.
47. T. Chavakis, A. Bierhaus, N. Al-Fakhri, D. Schneider, S. Witte, T. Linn, M. Nagashima, J. Morser, B. Arnold, K. T. Preissner and P. P. Nawroth, *J. Exp. Med.*, 2003, **198**, 1507.
48. V. V. Orlova, E. Y. Choi, C. Xie, E. Chavakis, A. Bierhaus, E. Ihanus, C. M. Ballantyne, C. G. Gahmberg, M. E. Bianchi, P. P. Nawroth and T. Chavakis, *EMBO J.*, 2007, **26**, 1129.
49. M. A. Hofmann, S. Drury, C. Fu, W. Qu, A. Taguchi, Y. Lu, C. Avila, N. Kambham, A. Bierhaus, P. Nawroth, M. F. Neurath, T. Slattery, D. Beach, J. McClary, M. Nagashima, J. Morser, D. Stern and A. M. Schmidt, *Cell*, 1999, **97**, 889.
50. R. Ramasamy, S. J. Vannucci, S. S. Yan, K. Herold, S. F. Yan and A. M. Schmidt, *Glycobiology*, 2005, **15**, 16R.
51. M. J. Neville and R. D. Campbell, *DNA Seq.*, 1997, **8**, 155.
52. A. de Baey, B. Fellerhoff, S. Maier, S. Martinozzi, U. Weidle and E. H. Weiss, *Genomics*, 1997, **45**, 591.
53. B. Rollinger-Holzinger, B. Eibl, M. Pauly, U. Griesser, F. Hentges, B. Auer, G. Pall, P. Schratzberger, D. Niederwieser, E. H. Weiss and H. Zwierzina, *J. Immunol.*, 2000, **164**, 3169.
54. A. Raghunathan, R. Sivakamasundari, J. Wolenski, R. Poddar and S. M. Weissman, *Exp. Cell Res.*, 2001, **268**, 230.
55. B. Lerner, J. I. Semple, S. E. Brown, D. Counsell, R. D. Campbell and C. M. Sanderson, *Genomics*, 2004, **83**, 153.
56. T. Hehlgans and K. Pfeffer, *Immunology*, 2005, **115**, 1.
57. M. Kriegler, C. Perez, K. DeFay, I. Albert and S. D. Lu, *Cell*, 1988, **53**, 45.
58. Y. Zheng, P. Saftig, D. Hartmann and C. Blobel, *J. Biol. Chem.*, 2004, **279**, 42898.
59. R. A. Smith and C. Baglioni, *J. Biol. Chem.*, 1987, **262**, 6951.
60. M. Brockhaus, H. J. Schoenfeld, E. J. Schlaeger, W. Hunziker, W. Lesslauer and H. Loetscher, *Proc. Natl. Acad. Sci. USA*, 1990, **87**, 3127.
61. P. A. Baeuerle and D. Baltimore, *Cell*, 1996, **87**, 13.
62. A. J. Schottelius, L. L. Moldawer, C. A. Dinarello, K. Asadullah, W. Sterry and C. K. Edwards, *Exp. Dermatol.*, 2004, **13**, 193.
63. R. Z. Murray, J. G. Kay, D. G. Sangermani and J. L. Stow, *Science*, 2005, **310**, 1492.
64. F. M. Brennan, D. Chantry, A. Jackson, R. Maini and M. Feldmann, *Lancet*, 1989, **2**, 244.
65. P. W. Gray, B. B. Aggarwal, C. V. Benton, T. S. Bringman, W. J. Henzel, J. A. Jarrett, D. W. Leung, B. Moffat, P. Ng, L. P. Svedersky, M. A. Palladino and G. E. Nedwin, *Nature*, 1984, **312**, 721.

66. U. Muller, C. V. Jongeneel, S. A. Nedospasov, K. F. Lindahl and M. Steinmetz, *Nature*, 1987, **325**, 265.
67. J. L. Browning, A. Ngam-ek, P. Lawton, J. De Marinis, R. Tizard, E. P. Chow, C. Hession, B. O'Brine-Greco, S. F. Foley and C. F. Ware, *Cell*, 1993, **72**, 847.
68. K. Schneider, K. G. Potter and C. F. Ware, *Immunol. Rev.*, 2004, **202**, 49.
69. P. D. Crowe, T. L. van Arsdale, B. N. Walter, C. F. Ware, C. Hession, B. Ehrenfels, J. L. Browning, W. S. Din, R. G. Goodwin and C. A. Smith, *Science*, 1994, **264**, 707.
70. J. Rangel-Moreno, J. E. Moyron-Quiroz and T. D. Randall, *Mod. Asp. Immunobiol.*, 2005, **15**, 41.
71. J. C. Lo, Y. Wang, A. V. Tumanov, M. Bamji, Z. Yao, C. A. Reardon, G. S. Getz and Y.-X. Fu, *Science*, 2007, **316**, 285.
72. G. Liu, H. Ma, L. Jiang and Y. Zhao, *Autoimmunity*, 2007, **40**, 95.
73. U. Utans, W. C. Ouist, B. M. McManus, J. E. Wilson, R. J. Arceci, A. F. Wallace and M. E. Russell, *Transplantation*, 1996, **61**, 1387.
74. K. Watano, K. Iwabuchi, S. Fujii, N. Ishimori, S. Mitsuhashi, M. Ato, A. Kitabatake and K. Onoe, *Immunology*, 2001, **104**, 307.
75. Y. Tian, S. E. Kelemen and M. V. Autieri, *Am. J. Physiol. Cell Physiol.*, 2006, **290**, 1083.
76. S. E. Kelemen and M. V. Autieri, *Am. J. Path.*, 2005, **167**, 619.
77. M. V. Autieri, S. E. Kelemen and K. W. Wendt, *Circ. Res.*, 2003, **92**, 1107.
78. Y. Tian and M. V. Autieri, *Am. J. Physiol. Cell Physiol.*, 2007, **292**, C841.
79. F. U. Hartl, *Nature*, 1996, **381**, 571.
80. J. L. Brodsky and G. Chiosis G, *Curr. Top. Med. Chem.*, 2006, **6**, 1215.
81. C. M. Milner and R. D. Campbell, *Immunogenetics*, 1990, **32**, 242.
82. A. M. Fourie, P. Peterson and Y. Yang, *Cell Stress Chap.*, 2001, **6**, 282.
83. P. Srivastava, *Ann. Rev. Immunol.*, 2002, **20**, 395.
84. B. Javid, P. A. MacAry and P. J. Lehner, *J. Immunol.*, 2007, **179**, 2035.
85. R. J. Binder, R. Vatner and P. Srivastava, *Tissue Antigens*, 2004, **64**, 442.
86. J. R. Thériault, S. S. Mambula, T. Sawamura, M. A. Stevenson and S. K. Calderwood, *FEBS Letts.*, 2005, **579**, 1951.
87. A. Clayton, A. Turkes, H. Navabi, M. D. Mason and Z. Tabi, *J. Cell Sci.*, 2005, **118**, 3631.
88. S. K. Calderwood, J. R. Thériault and J. Gong, *Eur. J. Immunol.*, 2005, **35**, 2518.
89. K. Mori, *Cell*, 2000, **101**, 451.
90. B. Bukau, J. Weissman and A. Horwich, *Cell*, 2006, **125**, 443.
91. J. Min, H. Shukla, H. Kozono, S. K. Bronson, S. M. Weissman and D. D. Chaplin, *Genomics*, 1995, **30**, 149.
92. A. Khanna and R. D. Campbell, *Biochem. J.*, 1996, **319**, 81.
93. D. J. Thuerauf, L. Morrison and C. C. Glembotski, *J. Biol. Chem.*, 2004, **279**, 21078.

94. D. J. Thuerauf, M. Marcinko, P. J. Belmont and C. C. Glembotski, *J. Biol. Chem.*, 2007, **282**, 22865.
95. K. Yamamoto, T. Sato, T. Matsui, M. Sato, T. Okada, H. Yoshida, A. Harada and K. Mori, *Dev. Cell*, 2007, **13**, 365.
96. T. Robson, M. C. Joiner, G. D. Wilson, W. McCullough, M. E. Price, I. Logan, H. Jones, S. R. McKeown and D. G. Hirst, *Radiation Res.*, 1999, **152**, 451.
97. T. Jascur, H. Brickner, I. Salles-Passador, V. Barbier, A. El Khissiin, B. Smith, R. Fotedar and A. Fotedar, *Mol. Cell.*, 2005, **17**, 237.
98. T. Robson, M. E. Price, M. L. Moore, M. C. Joiner, V. J. McKelvey-Martin, S. R. McKeown and D. G. Hirst, *Int. J. Rad. Biol.*, 2000, **76**, 617.
99. H. Kyushiki, Y. Kuga, M. Suzuki, E. Takahashi and M. Horie, *Cytogenet. Cell Genet.*, 1997, **79**, 114.
100. N. Matsuda, T. Suzuki, K. Tanaka and A. Nakano, *J. Cell Sci.*, 2001, **114**, 1949.
101. C. Didier, L. Broday, A. Bhoumik, S. Israeli, S. Takahashi, K. Nakayama, S. M. Thomas, C. E. Turner, S. Henderson, H. Sabe and Z. Ronai, *Mol. Cell. Biol.*, 2003, **23**, 5331.
102. M. J. Younger, L. Chen, H. Y. Ren, M. F. Rosser, E. L. Turnbull, C. Y. Fan, C. Patterson and D. M. Cyr, *Cell*, 2006, **126**, 571.
103. Y. Zhang, W. Higashide, S. Dai, D. M. Sherman and D. Zhou, *J. Biol. Chem.*, 2005, **280**, 38682.
104. K. Thress, W. Henzel, W. Shillinglaw and S. Kornbluth, *EMBO J.*, 1998, **17**, 6135.
105. K. Thress, J. Song, R. I. Morimoto and S. Kornbluth, *EMBO J.*, 2001, **20**, 1033.
106. S. Takayama and J. C. Reid, *Nature Cell Biol.*, 2001, **3**, E237.
107. S. T. Manchen and A. V. Hubbersty, *Biochem. Biophys. Res. Comm.*, 2001, **287**, 1075.
108. T. Sasaki, E. C. Gan, A. Wakeham, S. Kornbluth, T. W. Mak and H. Okada, *Genes Dev.*, 2007, **21**, 848.
109. F. Desmots, H. R. Russell, Y. Lee, K. Boyd and P. J. McKinnon, *Mol. Cell. Biol.*, 2005, **25**, 10329.
110. L. Shen, L. Wu, S. Sanlioglu, R. Chen, A. R. Mendoza, A. W. Dangel, M. C. Carroll, W. Zipf and C. Y. Yu, *J. Biol. Chem.*, 1994, **269**, 8466.
111. C. A. Sargent, M. J. Anderson, S.-L. Hsieh, E. Kendall, N. Gomez-Escobar and R. D. Campbell, *Hum. Mol. Genet.*, 1994, **3**, 481.
112. N. Gomez-Escobar, C.-F. Chou, W.-W. Lin, A.-L. Hsieh and R. D. Campbell, *J. Biol. Chem.*, 1998, **273**, 30954.
113. M. Levi-Strauss, M. C. Carroll, M. Steinmetz and T. Meo, *Science*, 1988, **240**, 201.
114. Y. Yamaguchi, T. Takagi, T. Wada, K. Yano, A. Furuya, S. Sugimoto, J. Hasegawa and H. Handa, *Cell*, 1999, **97**, 41.
115. T. Narita, Y. Yamaguchi, K. Yano, S. Sugimoto, S. Chanarat, T. Wada, D. Kim, J. Hasegawa, M. Omori, N. Inukai, M. Endoh, T. Yamada and H. Handa, *Mol. Cell. Biol.*, 2003, **23**, 1863.

116. T. Narita, T. M. Yung, J. Yamamoto, Y. Tsuboi, H. Tanabe, K. Tanaka, Y. Yamaguchi and H. Handa, *Mol. Cell*, 2007, **26**, 349.
117. M. R. Albertella and R. D. Campbell, *Hum. Mol. Genet.*, 1994, **3**, 793.
118. N. Perkins, *Nature Rev. Mol. Cell Biol.*, 2007, **8**, 49.
119. J. I. Semple, S. E. Brown, C. M. Sanderson and R. D. Campbell, *Biochem. J.*, 2002, **361**, 489.
120. D. Greetham, C. D. Ellis, D. Mewar, U. Fearon, S. N. Ultaigh, D. J. Veale, F. Guesdon and A. G. Wilson, *Hum. Mol. Genet.*, 2007, **16**, 3027.
121. C. M. Milner and R. D. Campbell, *Biochem. J.*, 1993, **290**, 811.
122. S. E. Brown, R. D. Campbell and C. M. Sanderson, *Mammalian Genome*, 2001, **12**, 916.
123. M. Tachibana, K. Sugimoto, T. Fukushima and Y. Shinkai, *J. Biol. Chem.*, 2001, **276**, 25309.
124. T. Jenuwein, *FEBS J.*, 2006, **273**, 3121.
125. J. C. Rice, S. D. Briggs, B. Ueberheide, C. M. Barber, J. Shabanowitz, D. F. Hunt, Y. Shinkai and C. D. Allis, *Mol. Cell*, 2003, **12**, 1591.
126. I. Gyory, J. Wu, G. Fejer, E. Seto and K. L. Wright, *Nat. Immunol.*, 2004, **5**, 299.
127. M. Tachibana, J. Ueda, M. Fakuda, N. Takeda, T. Ohta, H. Iwanari, T. Skaihama, T. Kdama, T. Hamakubo and Y. Shinkai, *Genes Dev.*, 2005, **19**, 815.
128. J. Ueda, M. Tachibana, T. Ikura and Y. Shinkai, *J. Biol. Chem.*, 2006, **281**, 20120.
129. C. Kambach, S. Walke, R. Young, J. M. Avis, E. de la Fortelle, V. A. Raker, R. Luhrmann, J. Li and K. Nagai, *Cell*, 1999, **96**, 375.
130. B. Lerner and C. M. Sanderson, *Genome Res.*, 2004, **14**, 1315.
131. T. Achsel, H. Brahms, B. Kastner, A. Bachi, M. Wilm and R. Luhrmann, *EMBO J.*, 1999, **18**, 5789.
132. D. Ingelfinger, D. J. Arndt-Jovin, R. Luhrmann and T. Achsel, *RNA*, 2002, **8**, 1489.
133. Y. Xue, X. Bai, I. Lee, G. Kallstrom, J. Ho, J. Brown, A. Stevens and A. W. Johnson, *Mol. Cell. Biol.*, 2000, **20**, 4006.
134. F. Fang, S. Phillips and J. S. Butler, *RNA*, 2005, **11**, 1571.
135. A. W. Dangel, L. Shen, A. R. Mendoza, L. C. Wu and C. Y. Yu, *Nucleic Acids Res.*, 1995, **23**, 2120.
136. X. Qu, Z. Yang, S. Zhang, L. Shen, A. W. Dangel, J. H. Hughes, K. L. Redman, L.-C. Wu and C. Y. Yu, *Nucleic Acids Res.*, 1998, **26**, 4068.
137. W. R. Widner and R. B. Wickner, *Mol. Cell. Biol.*, 1993, **13**, 4331.
138. R. Reed and E. Hurt, *Cell*, 2002, **108**, 523.
139. J. Fleckner, M. Zhang, J. Valcarcel and M. R. Green, *Genes Dev.*, 1997, **11**, 1864.
140. M. L. Luo, Z. Zhou, K. Magni, C. Christoforides, J. Rappsilber, M. Mann and R. Reed, *Nature*, 2001, **413**, 644.
141. K. Strasser, S. Masuda, P. Mason, J. Pfannstiel, M. Oppizzi, S. Rodrigues-Navarro, A. G. Rondon, A. Aquilera, K. Struhl, R. Reed and E. Hurt, *Nature*, 2002, **417**, 304.

142. L. Li, G. M. Huang, A. B. Banta, Y. Deng, T. Smith, P. Dong, C. Friedman, L. Chen, B. J. Trask, T. Spies, L. Rowen and L. Hood, *Genomics*, 1998, **51**, 45.
143. A. de la Coste and A. A. Freitas, *Immunol. Lett.*, 2006, **102**, 1.
144. K. Tanigaki and T. Honjo, *Nature Immunol.*, 2007, **8**, 451.
145. K. G. Leong, X. Hu, L. Li, M. Noseda, B. Larrivee, C. Hull, L. Hood, F. Wong and A. Karsan, *Mol. Cell. Biol.*, 2002, **22**, 2830.
146. F. MacKenzie, P. Duriez, B. Larrivee, L. Chang, I. Pollet, F. Wong, C. Yip and A. Karsan, *Blood*, 2004, **104**, 1760.
147. B. Guerra and O. G. Issinger, *Electrophoresis*, 1999, **20**, 391.
148. D. A. Canton and D. W. Litchfield, *Cell. Sig.*, 2006, **18**, 267.
149. E. Bonten, A. van der Spoel, M. Fornerod, G. Grosveld and A. d'Azzo, *Genes Dev.*, 1996, **10**, 3156.
150. C. M. Milner, S. V. Smith, M. B. Carrillo, G. L. Taylor, M. Hollinshead and R. D. Campbell, *J. Biol. Chem.*, 1997, **272**, 4549.
151. A. van der Spoel, E. Bonten and A. d'Azzo, *EMBO J.*, 1998, **17**, 1588.
152. V. Seyrantepe, H. Poupetova, R. Froissart, M. T. Zabot, I. Maire and A. V. Pshezhetsky, *Hum. Mut.*, 2003, **22**, 343.
153. N. F. Landolfi, J. Leone, J. E. Womack and R. G. Cook, *Immunogenetics*, 1985, **22**, 159.
154. X. Nan, I. Carubelli and N. M. Stamatos, *J. Leuk. Biol.*, 2007, **81**, 284.
155. X. P. Chen, E. Y. Enioutina and R. A. Daynes, *J. Immunol.*, 1997, **158**, 3070.
156. S. Katoh, T. Miyagi, H. Taniguchi, Y. Matsubara, J. Kadota, A. Tominaga, P. W. Kincade, S. Matsukura and S. Kohno, *J. Immunol.*, 1999, **162**, 5058.
157. K. Gee, M. Kozlowski and A. Kumar, *J. Biol. Chem.*, 2003, **278**, 37275.
158. F. Liang, V. Seyrantepe, K. Landry, R. Ahad, A. Ahmad, N. M. Stamatos and A. V. Pshezhetsky, *J. Biol. Chem.*, 2006, **281**, 37526.
159. Y. Murata, G. H. Sun-Wada, T. Yoshimizu, T. Yamamoto, Y. Wada and M. Futai, *J. Biol. Chem.*, 2002, **277**, 36296.
160. E. E. Norgett, K. J. Borthwick, R. S. Al-Lamki, Y. Su, A. N. Smith and F. E. Karet, *J. Biol. Chem.*, 2007, **282**, 14421.
161. B. P. Crider, P. Andersen, A. E. White, Z. Zhou, X. Li, J. P. Mattsson, L. Lundberg, D. J. Keeling, X. S. Xie, D. K. Stone and S. B. Peng, *J. Biol. Chem.*, 1997, **272**, 10721.
162. H. Caron, B. van Schaik, M. van der Mee, F. Baas, G. Riggins, P. van Sluis, M. C. Hermus, R. van Asperen, K. Boon, P. A. Voute, S. Heisterkamp, A. van Kampen and R. Versteeg, *Science*, 2001, **291**, 1289.
163. M. J. Lercher, A. O. Urrutia and L. D. Hurst, *Nature Genet.*, 2002, **31**, 180.
164. J. Trowsdale, *Genome Biol.*, 2002, **3**, 2002.1.

Section 6
Hyaluronan-Binding Proteins in Inflammation

CHAPTER 16

Hyaluronan-Binding Proteins in Inflammation

ANTHONY J. DAY,[a] CHARLES D. BLUNDELL,[b] DAVID J. MAHONEY,[c] MARILYN S. RUGG[d] AND CAROLINE M. MILNER[a]

[a] Wellcome Trust Centre for Cell-Matrix Research, Faculty of Life Sciences, University of Manchester, Manchester M13 9PT, UK; [b] Manchester Interdisciplinary Biocentre, University of Manchester, Manchester M1 7DN, UK; [c] Botnar Research Centre, Nuffield Department of Orthopaedic Surgery, University of Oxford, Oxford OX3 7LD, UK; [d] MRC Immunochemistry Unit, Department of Biochemistry, University of Oxford, Oxford OX1 3QU, UK

16.1 About the Authors

Tony Day joined the Immunochemistry Unit in September 1984 to do a 10-month research project on factor H, under the supervision of Bob Sim, for the final year of his chemistry degree at the University of Oxford. He continued working on factor H for his DPhil (1985–1987) – Tony was Bob's second PhD student – and remained in the Unit as a postdoctoral fellow, being involved for part of this time in the 'Amylin project' with Garth Cooper and Ken Reid. In September 1991 Tony moved to the ground floor of the Rex Richards Building to take up an Arthritis Research Campaign (ARC) Fellowship within Iain Campbell's group. It was during this time that Tony began his work on hyaluronan-binding proteins after Bob brought to his attention a paper on the newly discovered inflammation-associated protein tumour necrosis factor (TNF)-stimulated gene 6 (TSG-6) that contained a complement–Uegf–BMP-1

Molecular Aspects of Innate and Adaptive Immunity
Edited By Kenneth BM Reid and Robert B Sim

Published by the Royal Society of Chemistry, www.rsc.org

(CUB) module. Tony rejoined the Immunochemistry Unit in June 1998 (moving upstairs again!) as a member of the MRC Senior Scientific Staff, and continued his research on hyaluronan–protein interactions and the functional characterization of TSG-6, and also renewed his interest in factor H through a fruitful collaboration with Bob and the co-supervision of a joint DPhil student, Simon Clark. In October 2005 Tony took up a Chair in Biochemistry at the University of Manchester and he and his group physically relocated from the Unit in September 2006.

Charlie Blundell worked in the Immunochemistry Unit as a summer student ahead of doing his fourth year undergraduate project with Tony Day in the autumn of 1998. Between 1999 and 2002 he did his DPhil, which was co-supervised by Tony and Iain Campbell, to determine the nuclear magnetic resonance (NMR) structure of the TSG-6 Link module in its ligand-bound conformation. After completing his DPhil he worked as a research assistant with Tony and Iain for one year before joining Andrew Almond's group in 2003 and relocating with Andy to Manchester in 2005.

Dave Mahoney did his undergraduate research project with Tony Day (1996) followed by a DPhil in Tony's laboratory (1997–2000) – mapping the hyaluronan-binding site on TSG-6 by site-directed mutagenesis – moving with Tony into the Immunochemistry Unit in 1998. Dave then spent the next 6 years as a postdoctoral scientist in the Unit, working jointly with Tony and Caroline Milner. In October 2006 he moved to the Botnar Centre (Oxford) to work with Afsie Sabokbar on a collaborative project with Tony and Caroline to investigate the role of TSG-6 in bone remodelling.

Marilyn Rugg joined the Immunochemistry Unit in August 1968, working first as a technician with Rodney Porter and then with Alan Williams from 1970 to 1978, at which point she had a career break to raise her family. In August 1998 she returned to the Unit as a technician with Tony Day, where she characterized the role of TSG-6 in 'heavy chain transfer' as well as keeping laboratory F1 running smoothly. Marilyn remained in the Unit after Tony's departure, where she is still currently engaged in TSG-6-related projects.

Caroline Milner began her 10-month chemistry research project within the Immunochemistry Unit in September 1987, followed by a DPhil (1988–1991) on the characterization of novel genes within the human major histocompatibility complex (MHC) – both supervised by Duncan Campbell. Caroline then continued in Duncan's group as a MRC Training Fellow (1991–1994) and ARC Fellow (1994–2000), working on the functional characterization of MHC-encoded proteins. Caroline stayed in the Unit on Duncan's relocation to Cambridge in 1998 and shared laboratory facilities with Tony Day's group. From 2000 onwards Caroline and Tony have worked together on the role of TSG-6 in inflammatory processes and in September 2006 they relocated to Manchester. In November 1991 Caroline and Tony were married, after meeting in the Unit, falling in love, *etc.*, and they have two children. They recently acquired two goldfish.

16.2 Introduction and Historical Perspective

Hyaluronan (HA) is a linear, high molecular mass (up to ~10^7 Da) polysaccharide composed entirely of a repeating disaccharide of glucuronic acid (GlcA) and *N*-acetyl glucosamine (GlcNAc). It has diverse functional roles in vertebrates, including acting as a key structural component of the extracellular matrix (ECM) and as an important mediator of leukocyte adhesion and migration.[1–3] HA is also central to a wide variety of physiological and pathological processes, such as ovulation and inflammatory disease. The diversity of HA function is thought to result from its interaction with specific HA-binding proteins (hyaladherins), with the hypothesis that they stabilize different conformations of the polysaccharide, which gives rise to a range of multi-molecular complexes with different overall architectures that have different functional properties, for example, with respect to leukocyte binding.[2,4–6] Over the past 15 years or so Tony Day's group (in conjunction with a large number of collaborators) has been determining the molecular basis and regulation of HA–protein interactions as well as studying the ligand-binding properties and functions of a particular hyaladherin, TSG-6, associated with a wide range of inflammatory processes. These structural and functional studies, many of which were carried out within the MRC Immunochemistry Unit, form the basis of this review with reference to the particular Unit members involved.

16.3 Structural and Molecular Studies on Hyaluronan-Binding Proteins

Most hyaladherins interact with HA through a common structural domain of approximately 90 amino acids, termed the Link module, and thus belong to the Link module superfamily.[7,8] These include CD44, the major cell-surface receptor for HA, and the inflammation-associated protein, TSG-6, in addition to the lectican and Link protein families. Determination of the tertiary structure of the Link module from human TSG-6 by NMR spectroscopy[9] revealed that this domain has a similar α/β fold to that found in C-type lectins and, based on structural and sequence similarities, that these domains are likely to be evolutionarily related.[10,11] However, the Link module lacks the long Ca^{2+}-binding loop present in classical C-type lectins (*e.g.*, as found in the collectins) and interacts with HA in a metal ion-independent fashion.

As illustrated in Figure 16.1, the Link module superfamily can be divided into three subgroups (Types A, B and C) according to the size of their HA-binding domains (HABDs).[7,8] Type A HABD, the smallest of these, comprises a single independently folded Link module, typified by TSG-6 – a protein composed mainly of a Link and a CUB module arranged contiguously – in which all of its HA-binding activity is present within the Link domain. In this regard, the Link module fold is composed of two triple-stranded antiparallel β-sheets and two α-helices.[9,11] Determination of the structure of the HABD from human CD44 identified that, in this case, the β-sheet structure of

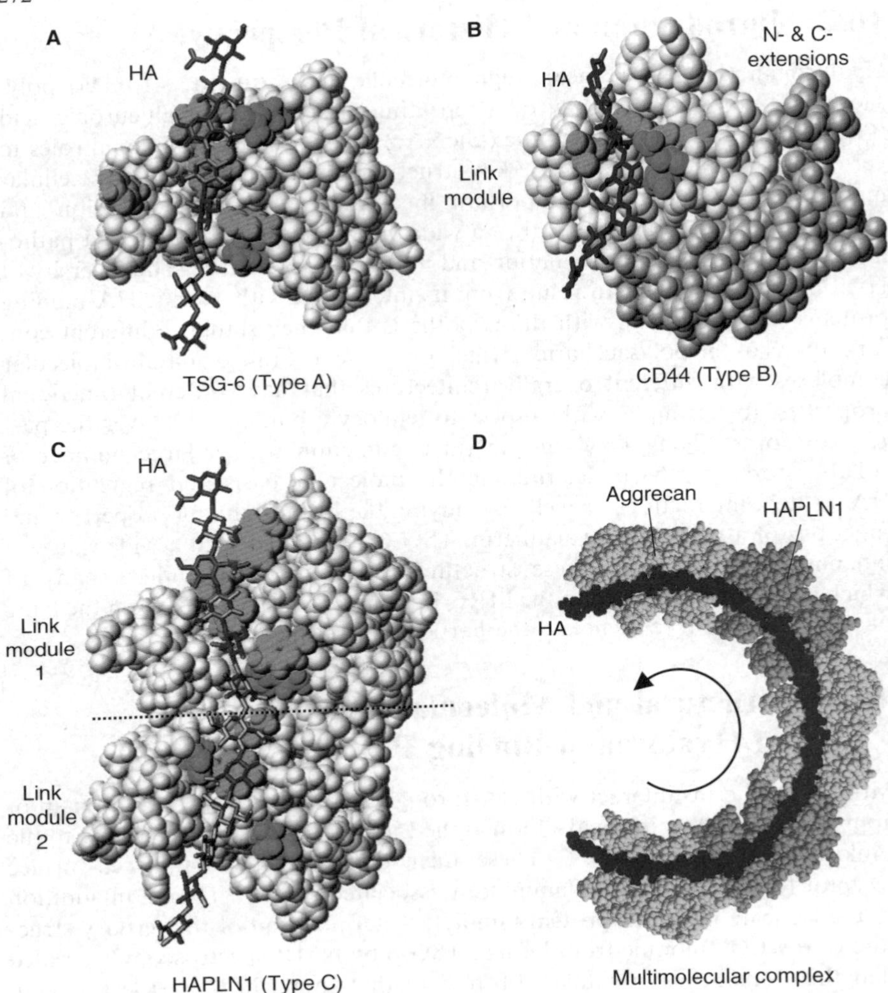

Figure 16.1 Structural models of HA-binding domains. (A) NMR structure of the Link module from human TSG-6 (a Type A HABD) in its HA-bound conformation with a HA octasaccharide (HA_8; blue) modelled into the binding groove.[15] Functionally important basic (Lys11 and Arg81) and aromatic residues (Tyr12, Tyr59, Phe70, Tyr78) are coloured green and red, respectively. (B) X-ray crystal structure of murine CD44 HABD co-complexed with HA_8;[28] the HABD of CD44 (Type B), which is composed of a Link module domain (white) with additional N- and C-terminal extensions (grey), is shown in what is likely to be its high-affinity conformation. Residues that make critical contacts with the HA are indicated in green (Arg45 and Arg82) and red (Tyr46, Tyr83, Ile100). (C) Model of the tandem Link modules from HAPLN1 (cartilage link protein), which constitutes a Type C HABD, in complex with a HA dodecamer (HA_{12}).[15] Predicted HA-binding residues are shown in green (Arg/Lys) and red (Trp/Tyr/Val). (D) Hypothetical organization of a multimolecular complex comprising alternating aggrecan (yellow) and HAPLN1 (pink) tandem Link module pairs on a superhelical HA.[15]

the Link module is extended by N- and C-terminal flanking sequences, which contribute 1 and 3 β-strands, respectively.[12] Thus, this Type B HABD, which is ~150 amino acids in length, represents a novel elaboration of Link module fold. Type C HABD – found in four related link proteins [hyaluronan and proteoglycan-binding link (HAPLN) 1–4] and in the G1 domains of the chondroitin sulfate (CS) proteoglycans aggrecan, versican, neurocan and brevican that constitute the lectican family[8] – is composed of two tandem Link modules, but at present no tertiary structures are available for this, the largest of the HABD subtypes.

16.3.1 TSG-6 and Type A HABD

The initial solution structure determined for the TSG-6 Link module,[9] performed while AJD was a research fellow with Iain Campbell, provided the basis for further structural and functional analysis. For example, Jan Kahmann used an HA octasaccharide (HA_8) to make a preliminary map of the HA-binding site by NMR chemical shift mapping; isothermal titration calorimetry (ITC) had shown that an HA 8-mer was the minimum length of oligomer that bound maximally.[13] In parallel with these studies, Dave Mahoney identified functional residues by site-directed mutagenesis in combination with ITC and other HA-binding assays.[14] Charlie Blundell continued the NMR investigations initiated by Jan, and determined high-resolution NMR structures for the TSG-6 Link module in both its free and HA_8-bound conformations.[11] The solution structure of the free protein corrected some inaccuracies in our original NMR model,[9] in particular with regard to the orientation of the α2 helix. Comparison of the free and ligand-bound structures revealed that, upon interaction with HA, conformational changes and subtle side-chain rearrangements result in the opening of a shallow groove on the protein surface, where this groove contains the key HA-binding residues (*i.e.*, Lys11, Tyr12, Tyr59, Phe70, Tyr78)[14] identified by site-directed mutagenesis. The opening of the ligand-binding groove on interaction with HA largely results from a change in conformation of the β4/β5 loop (this loop is indicated on Figure 16.2) caused by an alteration in the geometry of the disulfide bridge between Cys47 and Cys68.[11] Charlie found that the polarity (and approximate register) of HA within the binding groove could be inferred from the analysis of distinct chemical shift changes in the Link module caused by its interaction with different lengths of HA oligosaccharide.[11] This information, along with the observation that the groove contains two adjacent tyrosine residues (Tyr59 and Tyr78) that are likely to form CH–π stacking interactions with sequential rings in the sugar, were used to construct a model of a Link module–HA complex in collaboration with Andrew Almond.[15] The model, which is shown in Figure 16.1A, was then tested against various experimental information, including NMR data sets derived from isotopically-labelled HA (made in collaboration with Paul DeAngelis, Oklahoma, USA) in complex with the protein. A major (and unexpected) finding of this analysis was that acetamido side-chains of two

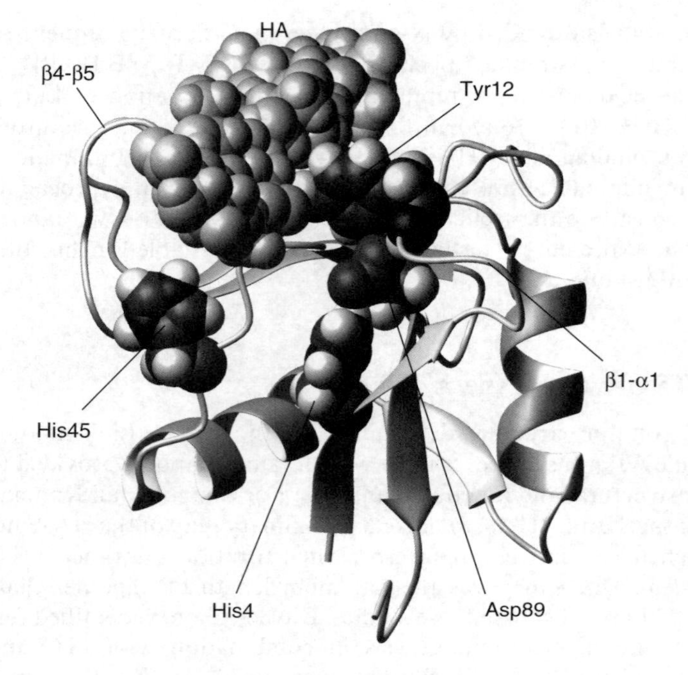

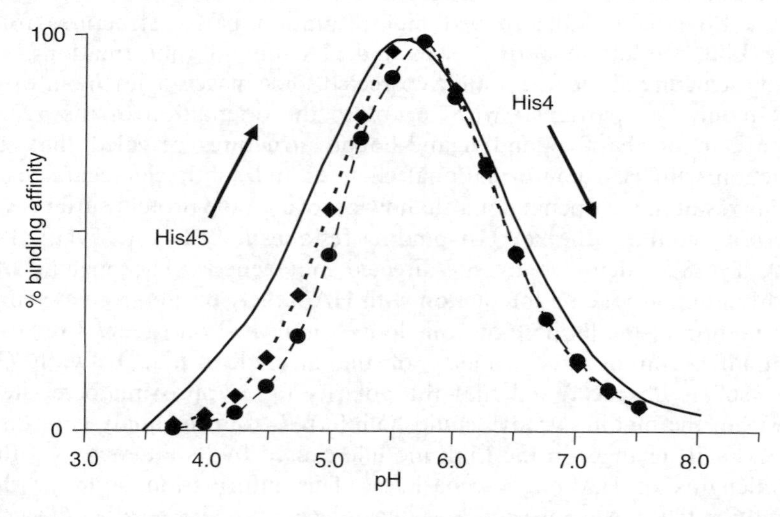

Figure 16.2 Molecular basis for the pH-dependent binding of TSG-6 with HA. The affinity of HA binding to the Link module of TSG-6 is pH dependent as shown in the bottom panel (solid line). The top panel shows the molecular network of pH-dependent interactions that connect HA binding at the hydroxyl group of Tyr12 to His4 *via* Asp89. The lower panel illustrates that the experimentally observed pH-dependency profile (solid line) can be simulated on the basis of the pK_a values for His4, His45 and Asp89 (diamonds), where the inclusion of pH-dependent protein folding (circles) causes an improvement to this model curve.[21]

GlcNAc rings fit into hydrophobic pockets present at the bottom of the binding groove. Furthermore, two basic amino acids (Lys11 and Arg81 shown in green in Figure 16.1A) have a separation that matches that of GlcA residues in the sugar, consistent with the formation of ionic interactions. This is supported by ITC experiments that indicated the TSG-6–HA interaction involves the formation of 1–2 salt bridges.[16] We believe that the combination of aromatic ring stacking interactions, salt bridges and hydrophobic pockets provide the exquisite specificity for the interaction of TSG-6 (and other hyaladherins – see below) with HA.[15]

More recent NMR studies conducted by Vicky Higman have shown that, in the absence of ligand, the β4/β5 loop is highly dynamic [on the nanosecond (ns) to picosecond (ps) timescale], but that on HA_8 binding this dynamic behaviour becomes dampened.[17] However, slower timescale motions still occur in the Link module–HA complex, which suggests a degree of dynamic matching between the protein and sugar that may decrease the entropic penalty of complex formation – HA is a highly dynamic molecule.[18] Interestingly, in the free state the β4/β5 loop also undergoes slow timescale motions such that, while the binding groove is likely to be essentially closed, it is likely to be sampling a variety of conformations (mediated by movement of the Cys47–Cys68 disulfide bond). This may include the 'open' HA-bound conformation, thus, facilitating the capture of HA by exposing the functional residues within the groove, including the tyrosines that are thought to stack against the sugar rings. Determination of the crystal structure for the free Link module of TSG-6 (by Vicky in conjunction with Martin Noble)[17] showed that the Cys47–Cys68 disulfide bond can adopt several conformations in addition to those seen in the solution structures, and confirmed that the conformation of the β4/β5 loop is highly plastic, which may be important given that this region of TSG-6 also mediates binding to other ligands, such as the bikunin chain of inter-α-inhibitor (IαI).[19]

Around the time that the initial NMR structure for the TSG-6 Link module was being determined (about 1995), it was discovered by Ash Parkar that its interaction with HA has an unusual pH dependency. As can be seen from Figure 16.2 (lower panel) the interaction is maximum at about pH 6 with a dramatic loss of binding activity on either side of this optimum.[20] We believe that pH gradients could have an important role in differentially modulating TSG-6 function in different tissue microenvironments. Investigations into the molecular basis for this pH dependency have been an ongoing saga that many members of the Day group have been involved in for more than 10-years (including major contributions from Dave Mahoney, Charlie Blundell and Martin Cordell) – these studies have finally yielded what we consider to be a plausible explanation.[21] Experimental evidence based on mutagenesis, ITC and NMR spectroscopy has revealed that the increase in HA binding between pH 3.5 and 6.0 results from the change in protonation state of His45. This histidine has a low pK_a, where its loss of positive charge is believed to lead to stabilization of the β4/β5 loop, which forms one side of the HA-binding groove (top panel of Figure 16.2). The decrease in ligand binding activity between pH 6.0 and pH 8.0 is also caused by a change in a histidine charge state. At pH 6.0,

His4 makes a salt bridge to one side-chain oxygen atom of a buried aspartic acid (Asp89), while its other oxygen is simultaneously hydrogen bonded to a key HA-binding residue (Tyr12). The resulting molecular network transmits the change in ionization state of His4 to the HA-binding site, through loss of the salt bridge that connects His4 and Asp89, with increasing pH. This is likely to lead to a local structural perturbation of the $\alpha1/\beta1$ loop, which forms the other side of the HA-binding groove and contains the functionally important residues Lys11 and Tyr12.

Overall, we have made much progress on understanding the molecular basis of HA binding to TSG-6 and how this could be regulated. Clearly, determination of a high-resolution structure for the TSG-6 Link module in complex with HA will provide further insights and reveal the accuracy of our proposed model.

16.3.2 Type C HABD in the Link Proteins and Lecticans

Type C HABDs are composed of a tandem pair of Link modules and are found in the G1-domains of CS proteoglycans that form Link protein-stabilized aggregates with HA.[8] For example, multimolecular complexes formed between aggrecan, HAPLN1 (cartilage link protein) and HA are crucial in providing the load-bearing properties of cartilage. Comparison of the Link module sequences for these proteins with that from TSG-6 indicate that they are likely to use a similar mechanism of HA recognition. That is, there is a high degree of conservation of functional residues, including those that could support aromatic ring stacking interactions and contribute to the formation of the specificity pockets, as well as basic amino acids that may form ionic interactions with the HA carboxylate groups.[15] The determination of the TSG-6 Link module structure in its HA-bound form (see above) has allowed the modelling of the Link modules from Type C HABDs in their active conformations and the identification of potential HA-binding residues;[15] as noted above no structures are yet determined for this hyaladherin subgroup. As illustrated for HAPLN1 in Figure 16.1C, its two Link modules were docked together (using bound HA as a guide), which led to the identification of hydrophobic residues likely to form an intra-Link module interface.[15] In addition, amino acids that could be involved in supporting intermolecular interactions between link proteins and CS proteoglycans were also identified, from which a mechanism was proposed for the generation of multimolecular complexes likely to have a higher order helical structure (Figure 16.1D).

To obtain structural information on Type C HABDs, Gillian McVey expressed the tandem Link modules from HAPLN1, and the G1 domains from aggrecan and versican in *Escherichia coli*. Although high yields were obtained for all three human proteins, only HAPLN1 could be refolded to give functionally active material.[22] This is consistent with the notion that the N-terminal immunoglobulin fold present in these proteins may be required to stabilize the Link modules, at least in the case of the lectican proteoglycans.[23] Production of functionally active HAPLN1 and G1 domains from aggrecan and versican in *Drosophila* S2 cells (which was initiated by Gillian and then continued by Nick Seyfried) allowed their HA-binding activities to be characterized and link protein-stabilized 'ternary' complexes to be formed.[22] These recombinant proteins were found to have properties similar to those described for the native proteins; *e.g.* HAPLN1 was able to form dimers and HA 10-mers were the minimum size oligomers that could compete effectively for their binding to polymeric HA. Nick showed [by gel filtration chromatography and a combination of protein cross-linking and matrix assisted laser desorption/ionization–time of flight (MALDI-TOF) peptide fingerprinting] that HAPLN1 and the G1 domain of aggrecan interact in both the absence and presence of HA. Conversely, HAPLN1 and the versican G1 domain do not bind directly to each other in solution, and yet form ternary complexes in the presence of HA_{24}. Interestingly, the length of HA oligosaccharides necessary to accommodate two aggrecan G1 domains (*i.e.*, aggrecan's footprint on HA) was significantly larger than we had anticipated. The most likely explanation for this is that aggrecan stabilizes an HA conformation that inhibits binding of another aggrecan close by or, in other words, there is negative co-operativity.[22]

Although the proteins produced in S2 cells have been useful in investigating the functional characteristics of the Type C hyaladherins, they have not proved suitable to determine high-resolution three-dimensional structures because of the relatively low expression levels and heterogeneity caused by N-linked glycosylation. Given that removal of N-linked carbohydrate from the G1 domains of aggrecan and versican was demonstrated by Nick Seyfried to have no effect on either HA binding or ternary complex formation[22] we revisited their production in a bacterial system. In this regard, Helen Fielder has expressed the versican G1 domain in *E. coli* and developed a refolding method that generates functionally active material; it binds HA in addition to the protein ligands thrombospondin-1 (TSP1) and TSG-6.[24] While we haven't, as yet, obtained a tertiary structure for this G1 domain, Helen has been investigating the formation of multimolecular complexes between polymeric HA of different sizes and the protein (using multi-angle laser light scattering) that have given novel insights into the interaction of versican with HA (H. L. Fielder and A. J. Day, unpublished).

16.3.3 Type B HABD of CD44

Three-dimensional structures of the HABD from human CD44 have been determined by both NMR spectroscopy (by Peter Teriete) and X-ray crystallography, where these studies were in collaboration with Iain Campbell, David Jackson and Martin Noble.[12] This revealed that the N-terminal and C-terminal extensions to the Link module, known to be necessary for protein folding,[25] come together in space to form an extra lobe of structure in intimate contact with the Link module (see Figure 16.1B).

Mapping of 'functional' residues previously implicated by mutagenesis[26,27] onto the CD44 HABD structure indicated that four key amino acids (including

Arg41) form a central patch on one face of the Link module (at a similar position to the HA-binding surface determined for TSG-6). Although some of the other 'important' residues were in close proximity to this site, others (*e.g.*, those on the C-terminal extension),[26] were widely spaced, being difficult to reconcile with the recognition of a single HA molecule.[12] Furthermore, Peter Teriete found in his NMR studies that HA binding causes widespread chemical shift alterations likely to result from a ligand-induced conformational change, which is consistent with the observed perturbation of the hydrogen bond network. In this regard, the structure of HABD from mouse CD44 has recently been determined in its free state and in complex with a HA_8 oligosaccharide.[28] Two different bound conformations were identified that are likely to correspond to low- and high-affinity states of the receptor (the latter complex is illustrated in Figure 16.1B). The interchange between these two conformations might represent the regulatory switch between what has been considered previously as the 'on' and 'off' states of CD44 (*e.g.*, on the leukocyte surface); *i.e.*, involved in switching CD44 into a pro-adhesive state in the context of inflammation. Interestingly, while HA is bound in a position on the CD44 Link module equivalent to that predicted in our TSG-6 model,[15] where the polarity of the HA within the binding groove is the same, the network of stabilizing interactions is different.[28] For example, there are no CH–π stacking interactions or salt bridges involved in stabilizing the CD44–HA complex (in either the low- or high-affinity forms), but instead there is an unexpectedly large contribution from aliphatic side-chain residues. The lack of aromatic ring stacking is not particularly surprising given that the residues in TSG-6 implicated in this interaction are not conserved in CD44.[15] However, the *N*-acetyl group from a GlcNAc sugar is accommodated within a hydrophobic pocket in a similar position to one of the specificity pockets seen in our Link module model.[28] The structure of the CD44–HA complex clearly indicates that most, if not all, atoms in close contact with the bound sugar come from residues in the Link module domain, which is inconsistent with previous mutagenesis data that implicated four basic amino acids in the C-terminal extension of human CD44 as being important in HA binding.[26] To address this issue, Alan Wright made equivalent mutations within the context of the human CD44 HABD (*i.e.*, R150A, R154A, K158A and R162A) and the interactions of these mutants with HA was assessed by NMR and surface plasmon resonance (SPR). All four mutants gave NMR spectra essentially identical to wild-type protein in the presence of HA oligosaccharides and were found to have similar binding affinities, which shows that these residues do not make a significant contribution to HA binding.[28] However, Alan's mutagenesis studies confirmed the importance of Arg41 (in human CD44) in HA binding, where its mutation to alanine causes a large reduction in the affinity. Interestingly, the two crystal forms determined for the murine CD44–HA complex have different conformations in a loop that connects the β1 strand and α1 helix, which results in a side-chain rearrangement that brings Arg45 (equivalent to Arg41 in human CD44) into contact with HA. This conformational alteration is largely consistent with the HA-induced chemical shift perturbations seen previously.[12] Alan showed that in the absence of the Arg41 side chain (*i.e.*, in the human R41A mutant) no conformational change occurs, which indicates that this basic amino acid has a pivotal role in transducing the conformational change that switches an initial CD44–HA recognition complex to a high affinity 'on' state.[28]

16.4 Investigating TSG-6 Biology

TSG-6 was originally identified by Jan Vilcek's lab as a TNF-inducible protein secreted by human fibroblasts,[29,30] and its expression has been reported in a diversity of cell and tissue types after treatment with inflammatory mediators and some growth factors (reviewed in Milner and Day[31]). Not surprisingly, TSG-6 is expressed in the context of inflammatory pathologies, including arthritis[32,33] and asthma,[34] but there is growing evidence that it is produced in some healthy adult tissues.[34,35] In addition, TSG-6 expression is upregulated in ovarian follicles during ovulation.[36]

TSG-6 is a 35-kDa protein composed almost entirely of contiguous Link and CUB modules (reviewed in Milner and Day[31] and Milner *et al.*[37]). Tony's original work on TSG-6 was aimed at determining the three-dimensional structure of the CUB module (which is also found in the complement proteins C1r and C1s). However, attempts to refold this isolated domain when expressed in *E. coli* were unsuccessful and the laboratory's focus switched to the Link module. Later, Hilke Nentwich developed a protocol for the production of the CUB domain together with the 27-amino acid C-terminal segment of TSG-6 (termed CUB_C_TSG6) in *E. coli*, although we were unable to obtain a preparation that was folded homogeneously, and this work was again put to one side. More recently, using methodologies developed for refolding proteins for crystallization,[17,38] Dave Mahoney and Tariq Ali succeeded in producing fully folded and functionally active CUB_C protein.[39] Work by Tariq and David Briggs in Manchester has now finally led to the determination of an X-ray crystal structure for the CUB module (D. C. Briggs, T. Ali, D. J. Mahoney, C. M. Milner and A. J. Day, unpublished data). Collaborative studies with David Roberts (Bethesda, USA) have identified what is currently the only known ligand for TSG-6 that binds exclusively to the CUB_C domain of TSG-6, namely fibronectin – an important component of the ECM and a major plasma protein.[39] TSG-6 binds to type III repeats 9–14 of fibronectin, a region of the protein that contains the Arg–Gly–Asp motif responsible for its interaction with $\alpha5\beta1$ integrin as well as a heparin-binding site. The TSG-6–fibronectin interaction has been shown to enhance fibronectin-mediated fibril formation by human fibroblasts,[39] where this might involve the production of ternary complexes with TSG-6 acting as a bridge between fibronectin bound to its CUB_C domain and TSP-1 associated with the Link module.

The Link module of TSG-6, expressed in *E. coli* for structural studies (termed Link_TSG6), has been used in extensive characterization of the ligand-binding properties of this domain, initially by Ash Parkar and then, over many years, by Dave Mahoney. In addition, the production of full-length recombinant human (rh)TSG-6 in *Drosophila* S2 cells by Hilke Nentwich and Marilyn Rugg[40] and, more recently, the CUB_C domain[39] have been invaluable in investigating TSG-6 function. These studies revealed that the TSG-6 Link module supports the binding not only to HA (see above), but also to the glycosaminoglycans heparin, heparan sulfate (HS), chondroitin-4-sulfate (C4S) and dermatan sulfate (DS) and to the proteins aggrecan, versican, pentraxin-3 (PTX3), TSP-1 and bikunin.[19,20,24,41–44] In a recent collaboration with Afsie Sabokbar we have confirmed the observation of Tsukahara and colleagues[45] that TSG-6 binds to the bone morphogenetic protein-2 (BMP-2) and have also showed that it interacts with the receptor activator of nuclear factor (NF) kappaB ligand (RANKL), the major mediator of osteoclast activation, where these two interactions are likely to involve both the Link and CUB modules of TSG-6.[35] As described above, the HA-binding surface on Link_TSG6 has been mapped in detail and the interaction site for bikunin has been shown to be similar, involving many of the same residues (*i.e.*, Lys11, Tyr12, Tyr59 and Tyr78).[19] However, heparin binds to a distinct surface on the Link module that involves lysine residues at positions 20, 34, 41 and 54. Like HA binding, heparin binding to TSG-6 is pH sensitive[21] and association with heparin has been shown to inhibit subsequent HA binding,[19] which is consistent with the hypothesis that TSG-6 function is regulated by tissue microenvironment. In addition to binding non-covalently to the ligands described above, TSG-6 also forms covalent interactions with the heavy chains (HC) of IαI[46] (see Figure 16.3 and Section 16.4.1).

16.4.1 TSG-6 in Ovulation

Ovulation is an acute inflammation-like process that involves extensive tissue remodelling, where the cumulus oocyte complex (COC; an oocyte surrounded by adherent cumulus cells) expands in volume 20- to 40-fold *via* the production of a HA-rich matrix that is essential for fertilization *in vivo* (reviewed in Espey and Richards[47]). This process is initiated by a luteinizing hormone surge that stimulates the expression of specific gene products at defined time points within the ovarian follicle. We have generated a TSG-6-specific antiserum, RAH-1,[48] which has been used in collaborative studies to reveal that TSG-6 is expressed in response to ovulatory stimulus in mice, where it is associated with cumulus cells and the granulosa cells that line the follicle wall and it appears to co-localize with HA.[49,50]

Work by Csaba Fülöp and Vince Hascall (Cleveland, USA) revealed the importance of TSG-6 in ovulation, where the most obvious phenotype of the tsg-6$^{-/-}$ mouse (made by Kati Mikecz in Chicago) is severe female infertility caused by a defect in COC expansion.[51] Csaba showed that our rhTSG-6 could rescue expansion *ex vivo*. JoAnne Richards' laboratory (Houston, USA) has also found that Cox-2$^{-/-}$ and EP2$^{-/-}$ mice are female infertile, and a collaborative study demonstrated that this is because of a lack of TSG-6 protein synthesis in the ovarian follicle.[52] Early work on TSG-6 indicated its presence in arthritic synovial fluids as a complex with IαI,[32,53] an abundant serum proteoglycan that consists of

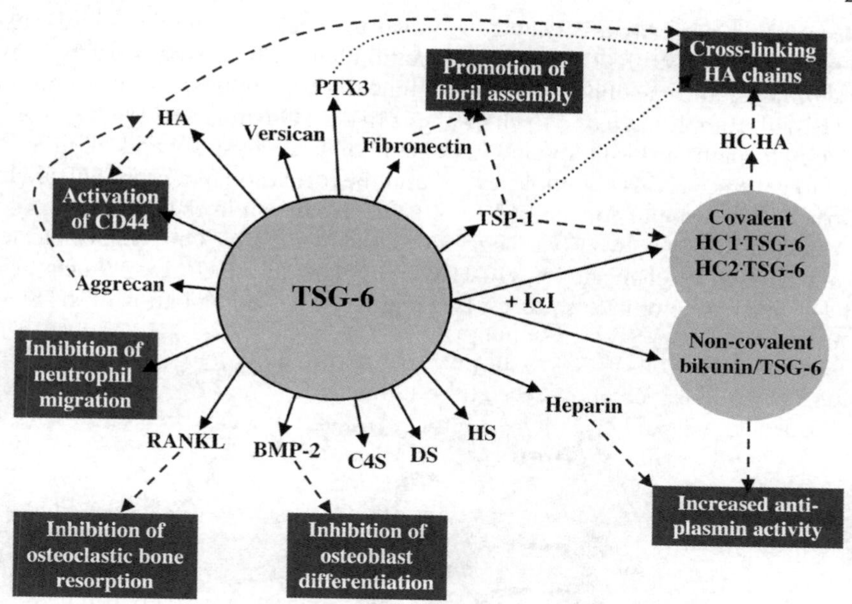

Figure 16.3 TSG-6: ligands and functional activities. Major functions of TSG-6 are shown in grey boxes, with dashed arrows indicating where these (or a particular binding process) are affected and/or mediated through the interactions with one of more of its ligands; the dotted arrows indicate functional outcomes that are, at present, hypothetical. Detailed discussion on the ligand-binding properties and functional activities of TSG-6 can be found in the text and in Day and de la Motte,[2] Milner and Day,[31] Mahoney *et al.*,[35] Milner *et al.*,[37] Kuznetsova *et al.*[39] and Inoue *et al.*[45]

two heavy chains (HC1 and HC2), connected covalently to a CS moiety that is attached to the serine protease inhibitor bikunin.[54] Analysis of the ECM from murine-ovulated COCs showed these to contain covalent HC1·TSG-6 and HC2·TSG-6 complexes.[50] The biosynthesis of IαI and the related pre-α-inhibitor (PαI, which comprises bikunin and a single heavy chain, HC3[55]) are dependent on the expression of bikunin. In this regard, female bikunin$^{-/-}$ mice, like tsg-6$^{-/-}$ animals, exhibit severe sub-fertility. In both cases this arises from a failure in the formation and structural integrity of the HA-rich cumulus matrix, such that cumulus cells are shed from the COC in the follicle.[51,56,57] This is because of a lack of formation of covalent complexes between HC and HA (termed HC·HA or SHAP-HA), which are necessary for HA cross-linking.[2,58] The resulting naked oocytes do not fertilize *in vivo,* which indicates that the cumulus matrix is required for sperm docking.

There was good evidence that co-localization of TSG-6 and IαI in the ovulating follicle results in the rapid incorporation of TSG-6 into TSG-6·HC1 and TSG-6·HC2 complexes,[50,52] and we had shown that this is also the case when rhTSG-6 and purified IαI are combined *in vitro.*[40] Furthermore, a monoclonal antibody, A38, specific for the TSG-6 Link module, which was made in collaboration with Jayne Lesley and shown to block HA binding,[59] was also found to inhibit both TSG-6·HC formation and COC expansion.[60] These observations, and the findings that ovarian extracts from tsg-6$^{-/-}$ mice contain no HC·HA complexes, led to the hypothesis that TSG-6 might have a key role in HC·HA formation during ovulation. The inability of tsg-6$^{-/-}$ and bikunin$^{-/-}$ mice to generate HC·HA complexes[51,57] and the formation of a cross-linked HA matrix would account for the failure of COC expansion in these animals.

A detailed study by Marilyn Rugg, aided and abetted by Tony Willis, has led to a proposed mechanism for HC·HA formation, in which TSG-6 acts as a catalyst and co-factor in this process (see Figure 16.4). Co-incubation of rhTSG-6

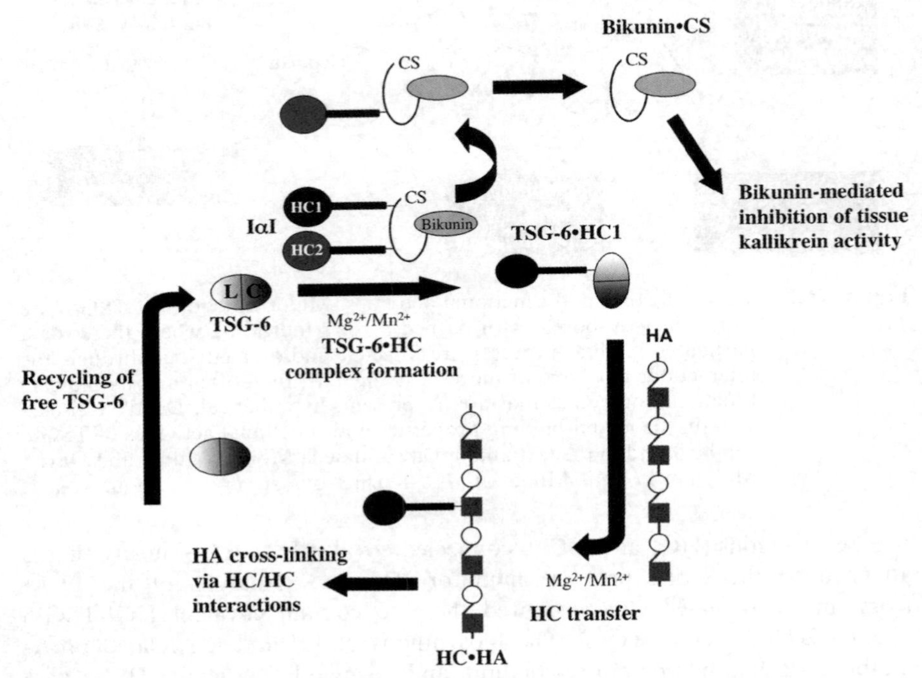

Figure 16.4 TSG-6-mediated formation of HC·HA complexes and release of bikunin·CS. TSG-6, composed mainly of Link (L) and CUB (C) modules, reacts with IαI to form covalent TSG-6·HC complexes that act as intermediates in the formation of HC·HA;[46] this is illustrated for HC1, but a similar mechanism occurs for HC2. HC transfer from IαI (where the HC is attached to its CS chain *via* ester bonds) onto HA is thought to occur *via* two transesterification reactions, where TSG-6 is both an essential cofactor and true catalyst: the first transfers the HC onto TSG-6 and the second transfers the HC onto HA. HA modified by the covalent attachment of HC is believed to be transiently cross-linked *via* non-covalent interactions between HCs, which leads to altered HA function (*e.g.*, enhanced cell-binding properties[6]). Another functional consequence is that the HC·bikunin by-products of the TSG-6·HC complex formation breakdown over time to release bikunin·CS that, unlike the intact IαI or HC·bikunin, is a potent inhibitor of the serine protease tissue kallikrein.[34]

and IαI [purified from human serum, provided by Erik Fries (Uppsala, Sweden)] gave rise to the formation of a ~130 kDa species and a ~120 kDa doublet, detectable on a Coomassie blue-stained sodium dodecyl sulfate polyacrylamide gel electrophoresis (SDS-PAGE) gel.[46] Characterization of these species revealed that the 130 kDa band represents a mixture of bikunin linked, *via* CS, to either HC1 or HC2, while the upper and lower bands of the 120 kDa doublet correspond to covalent HC2•TSG-6 and HC1•TSG-6 complexes, respectively,[46] consistent with the results of a similar study by others.[61] The inclusion of HA in the TSG-6–IαI reaction mixture gave rise to HC•HA complexes, indicating that TSG-6 can mediate the formation of HC•HA,[46] which is likely to occur *via* two sequential transesterification reactions; where HC1 or HC2 are understood to be transferred from GalNAc sugars in the CS chain of IαI, *via* an unknown site(s) on TSG-6, onto GlcNAc moieties of HA.[46,62] In this regard, Link_TSG6 cannot form complexes with HCs or catalyze the transfer of HCs onto HA,[63] which suggests that residues outside the Link module (*e.g.*, in the CUB_C region) are required for these processes. The addition of limiting quantities of rhTSG-6 to the IαI–HA reaction mixture revealed that TSG-6 is released after HC•HA formation and can interact with a new IαI molecule, thus acting as a true catalyst, as well as a co-factor, in HC•HA formation.[46]

Both TSG-6•HC and HC•HA formation are dependent on metal ions – Mg^{2+} or Mn^{2+} have been shown to be necessary for these reactions.[46] The CUB module of TSG-6 contains at least one predicted divalent metal ion-binding site,[37,46] where Wannarat Tongsoongnoen has generated single-site mutants of rhTSG-6 that are being used to test the importance of cation binding in the catalysis of HC transfer. Marilyn has also produced mutants of rhTSG-6 to investigate the role of the Link module–HA interaction in TSG-6•HC and HC•HA formation, and revealed, for example, that mutant Y94F substantially reduces HA-binding capability (as expected from the corresponding Link_TSG6 mutant),[15] but has HC transfer activity similar to that of wild-type rhTSG-6.[64,65] These data indicate that TSG-6HC might have a HA-binding site that differs from that of free TSG-6.

In a collaborative study with Alberto Mantovani (Milan, Italy), we showed that Link_TSG6 binds to the long PTX3, *via* a site distinct from its HA-binding surface.[44] This led to the suggestion that TSG-6 might contribute to the stability of the cumulus matrix through its interaction with multimers of PTX3, and thereby form nodes to which several HA molecules can bind,[2,36,44] which is consistent with the severe sub-fertility of female Ptx3$^{-/-}$ mice.[44] In these PTX3-deficient animals, the cumulus cells around the oocytes are rapidly dispersed after ovulation, which indicates that the structural integrity of the cumulus matrix is compromised, resulting in the failure of fertilization. Recent work from Antonietta Salustri's laboratory[66] has shown that PTX3 interacts directly with the HCs of IαI. This suggests that PTX3 is likely to mediate HA cross-linking in the cumulus matrix through direct interaction with HC•HA complexes, while TSG-6 contributes to this indirectly *via* its catalysis of HC transfer onto HA. This is consistent with the observation that TSG-6 within ovulating follicles is predominantly in the form TSG-6•HC, with very little free protein available.[52] Furthermore, HA_6, which binds to the Link module of TSG-6 and can effectively inhibit TSG-6 binding to polymeric HA,[59] has no effect on COC expansion *in vitro* and does not form HC•HA_6.[36,67] This suggests that the contribution of TSG-6 to cumulus matrix expansion is restricted to its catalysis of HC transfer; *i.e.* interactions of TSG-6–PTX-3 or TSG-6•HC complexes with HA do not appear to be directly involved in HA cross-linking, at least in the context of COC expansion. Work is currently in progress by Antonio Inforzato in Manchester to further characterize the ligand-binding properties of PTX3 and investigate its oligomeric structure.

Work by Dave Mahoney, as part of a collaboration with David Roberts, has revealed that the TSG-6 Link module interacts with the N-terminal domain of TSP-1, an ECM protein that is induced by inflammatory mediators and has been detected in the ovarian follicles of rats.[43] TSP-1$^{-/-}$ mice have mildly impaired fertility of unknown cause (see Kuznetsova *et al.*[43]) and, in this regard, the interaction of TSG-6 with TSP-1 promotes the formation of TSG-6•HC complexes and thereby augments the transfer of HCs onto HA.[43] It has also been suggested that TSP-1 might cross-link HA *via* its association with TSG-6,[2] but as yet this has not been investigated.

16.4.2 TSG-6 in Arthritis and Inflammatory Disease

In collaboration with Mike Bayliss we showed that TSG-6 is expressed within the inflamed synovial tissue of arthritis patients,[33] which is consistent with its upregulation by inflammatory mediators. A number of studies, by others, have shown that TSG-6 is a potent inhibitor of inflammation and joint damage in mouse models of arthritis[68–72] (reviewed in Milner and Day[31]). A number of mechanisms might contribute to these activities of TSG-6. For example, work by Dave Mahoney has shown that TSG-6 can potentiate the anti-plasmin activity of IαI through the non-covalent interaction with bikunin *via* its Link module.[19,63] This effect is significantly augmented when TSG-6 is associated with certain sulfated glycosaminoglycans, *e.g.* heparin.[19] We have suggested that TSG-6 dimers, which form as a result of heparin binding, could associate with the two kunitz-type domains of bikunin, thereby changing their relative orientation, which makes the protease-binding sites more accessible and gives rise to enhanced anti-plasmin activity.[17,19] In this regard, plasmin is a key activator of matrix metalloproteinases (MMPs), which degrade cartilage proteoglycans in inflamed joints, and the treatment of arthritic mice with recombinant TSG-6 has been shown to inhibit the formation of aggrecan fragments characteristic of MMP activity.[70] Therefore, the co-localization of TSG-6, IαI and heparin or HS proteoglycans in inflamed joints could downregulate the protease network.[19] In this regard, tsg-6$^{-/-}$ mice with proteoglycan-induced arthritis (PGIA) exhibit more severe symptoms than wild-type animals and have elevated levels of plasmin activity in their inflamed paw joints.[72]

In a collaborative study with Mauro Perretti, we found that TSG-6 is a potent inhibitor of acute inflammation *in vivo,* causing significant inhibition of neutrophil migration, where the Link module was as active as the full-length protein (*e.g.*, in mouse air-pouch models).[63] The use of intravital microscopy has indicated that TSG-6 might influence multiple aspects of neutrophil adhesion and transendothelial migration.[73] The neutrophil inhibitory activity of TSG-6 is consistent with the observation that the progression and severity of PGIA is significantly increased in tsg-6$^{-/-}$ mice compared to wild-type controls, where this is associated with early and more extensive infiltration of neutrophils into the synovia of the knock-out mice.[72]

In addition to its effects on cartilage protection in arthritis, recent data from a collaborative study with Afsie Sabokbar indicate that TSG-6 might directly inhibit bone erosion in inflamed joints. Dave Mahoney found that TSG-6 binds directly to RANKL, the key mediator of osteoclast activation, where this interaction might be responsible for the inhibition of bone resorption that we have observed when rhTSG-6 is added to osteoclast precursors *in vitro*.[35] Furthermore, we have confirmed the findings of Tsukahara and colleagues[45] that TSG-6 can interact with BMP-2 and inhibit osteoblast differentiation, which leads us to the hypothesis that, in the absence of inflammation, TSG-6 might play a key role in bone homeostasis.[35]

Working with Rosanna Forteza (Miami, USA), we have shown that TSG-6 expression is upregulated in the lungs of asthmatics and smokers, where it is released into the airway secretions from submucosal glands and epithelial cells.[34] We also detected TSG-6·HC and HC·HA complexes in the bronchiolar lavage fluids of asthmatics, which indicates that IαI ingresses into lung tissues in this condition. Marilyn generated a bikunin-specific antiserum and used this to demonstrate that bikunin·CS is released as a by-product of TSG-6-mediated HC transfer (Figure 16.4) and Rosanna's group showed that it can inhibit the activity of tissue kallikrein (TK), a serine protease that has been implicated in airway inflammation.[34] This represents a novel means by which TSG-6 might modulate protease activity, in addition to its potentiation of the anti-plasmin activity of IαI (see above), and suggests that it could be important in the protection of the bronchial epithelium. This is supported by our observation, using antisera specific for TSG-6, bikunin and HC3, that TSG-6 and PαI are expressed locally by surface epithelial cells in the airways, both constitutively and at elevated levels in inflammatory conditions.[34]

16.5 Summary and Final Thoughts

Our recent work on the HA-binding properties of TSG-6, CD44 and Type C hyaladherins indicates that there is likely to be considerable variability in the way in which HA is recognized by proteins. This is consistent with the hypothesis that the diversity of HA function arises through the formation of HA–protein complexes with very different molecular architectures. Clearly, although much progress has been made in understanding the molecular basis of HA–protein interactions, we still have many unanswered questions. We hope these will be addressed by our future structural studies.

Over the past 10 years or so considerable insights have been made into the role of TSG-6 in inflammatory processes and in characterizing the underpinning molecular mechanisms. It has been an exciting time to work on this protein, with the discoveries that it has both a crucial role in mammalian reproductive biology and serves to protect joint (and other) tissues from damage during inflammation, as well as the identification of numerous new ligands for TSG-6; it is so much more than just a HA-binding protein! There remains much to do and our interest in TSG-6 is still unabated, especially with the recent finding that it may play an important role in the modulation of bone metabolism. Where it will take us to next, only time will tell.

Acknowledgements

We take this opportunity to thank all of the members of the Immunochemistry Unit for making it such a special place to work, including the PIs, technical staff, research fellows, postdoctorates, PhD students, undergraduate project students and visitors who have contributed to its success. In particular, much credit must go to Ken Reid for his excellent stewardship of the Unit for more than 20 years and for its continued high-quality output. Tony would like to say a special 'thank you' to Bob Sim for getting him started in research (and for his good advice on matters too numerous to mention) and to Tony Willis for friendship – all the help with mass spectrometry, protein sequencing, amino acid analysis, HPLC, *etc*., go without saying! Tony and Caroline would also like to acknowledge their wonderful team of co-workers in Oxford and Manchester (both past and present) without whom none of the research could happen. In addition, we thank our many collaborators in Oxford and across the globe for their crucial contributions to the work described above.

References

1. M. Tammi, A. J. Day and E. A. Turely, *J. Biol. Chem.*, 2002, **277**, 4581.
2. A. J. Day and C. A. de la Motte, *Trends Immunol.*, 2005, **26**, 637.
3. S. P. Evanko, M. I. Tammi, R. H. Tammi and T. N. Wight, *Adv. Drug Delivery Rev.*, 2007, **59**, 1351.
4. A. J. Day and J. K. Sheehan, *Curr. Opin. Struct. Biol.*, 2001, **11**, 617.
5. J. Lesley, I. Gál, D. J. Mahoney, M. R. Cordell, M. S. Rugg, R. Hyman, A. J. Day and K. Mikecz, *J. Biol. Chem.*, 2004, **279**, 25745.
6. L. Zhuo, A. Kanamori, R. Kannagi, N. Itano, J. Wu, M. Hamaguchi, N. Ishiguroand and K. Kimata, *J. Biol. Chem.*, 2006, **281**, 20303.
7. A. J. Day and G. D. Prestwich, *J. Biol. Chem.*, 2002, **277**, 4585.
8. C. D. Blundell, N. T. Seyfried and A. J. Day, in *Chemistry and Biology of Hyaluronan*, eds. H.G. Garg and C. A. Hales, Elsevier, Amsterdam, 2004, p. 189.

9. D. Kohda, C. J. Morton, A. A. Parkar, H. Hatanaka, F. M. Inagakai, I. D. Campbell and A. J. Day, *Cell*, 1996, **86**, 767.
10. A. J. Day, at http://www.glycoforum.gr.jp/science/hyaluronan/HA16/HA16E.html, 2001.
11. C. D. Blundell, D. J. Mahoney, A. Almond, P. L. DeAngelis, J. D. Kahmann, P. Teriete, A. R. Pickford, I. D. Campbell and A. J. Day, *J. Biol. Chem.*, 2003, **278**, 49261.
12. P. Teriete, S. Banerji, M. Noble, C. D. Blundell, A. J. Wright, A. R. Pickford, E. Lowe, D. J. Mahoney, M. I. Tammi, J. D. Kahmann, I. D. Campbell, A. J. Day and D. G. Jackson, *Mol. Cell*, 2004, **13**, 483.
13. J. D. Kahmann, R. O'Brien, J. Werner, D. Heinegård, J. E. Ladbury, I. D. Campbell and A. J. Day, *Structure*, 2000, **8**, 763.
14. D. J. Mahoney, C. D. Blundell and A. J. Day, *J. Biol. Chem.*, 2001, **276**, 22764.
15. C. D. Blundell, A. Almond, D. J. Mahoney, P. L. DeAngelis, R. D. Campbell and A. J. Day, *J. Biol. Chem.*, 2005, **280**, 18189.
16. C. D. Blundell, J. D. Kahmann, A. Perczel, D. J. Mahoney, M. R. Cordell, P. Teriete, I. D. Campbell and A. J. Day, in *Hyaluronan, volume 1*, eds. J. F. Kennedy, G. O. Phillips, P. A. Williams and V. C. Hascall, Woodhead Publishing Ltd, Abington, Cambridge, UK, 2002, p. 161.
17. V. A. Higman, C. D. Blundell, D. J. Mahoney, C. Redfield, M. E. M. Noble and A. J. Day, *J. Mol. Biol.*, 2007, **371**, 669.
18. C. D. Blundell, P. L. DeAngelis, A. J. Day and A. Almond, *Glycobiology*, 2004, **14**, 999.
19. D. J. Mahoney, B. Mulloy, M. J. Forster, C. D. Blundell, E. Fries, C. M. Milner and A. J. Day, *J. Biol. Chem.*, 2005, **280**, 27044.
20. A. A. Parkar, J. D. Kahmann, S. L. T. Howat, M. T. Bayliss and A. J. Day, *FEBS Lett.*, 1998, **428**, 171.
21. C. D. Blundell, D. J. Mahoney, M. R. Cordell, A. Almond, J. D. Kahmann, A. Perczel, J. D. Taylor, I. D. Campbell and A. J. Day, *J. Biol. Chem.*, 2007, **282**, 12976.
22. N. T. Seyfried, G. F. McVey, A. Almond, D. J. Mahoney, J. Dudhia and A. J. Day, *J. Biol. Chem.*, 2005, **280**, 5435.
23. H. Watanabe, S. C. Cheung, N. Itano, K. Kimata and Y. Yamada, *J. Biol. Chem.*, 1997, **272**, 28057.
24. S. A. Kuznetsova, P. Issa, E. M. Perruccio, B. Zeng, J. M. Sipes, Y. Ward, N. T. Seyfried, H. L. Fielder, A. J. Day, T. N. Wight and D. D. Roberts, *J. Cell Sci.*, 2006, **119**, 4499.
25. S. Banerji, A. J. Day, J. D. Kahmann and D. G. Jackson, *Protein Expres. Purif.*, 1998, **14**, 371.
26. R. J. Peach, D. Hollenbaugh, I. Stamenkovic and A. Aruffo, *J. Cell Biol.*, 1993, **122**, 257.
27. J. Bajorath, B. Greenfield, S. B. Munro, A. J. Day and A. Aruffo, *J. Biol. Chem.*, 1998, **273**, 338.
28. S. Banerji, A. J. Wright, M. Noble, D. J. Mahoney, I. D. Campbell, A. J. Day and D. G. Jackson, *Nat. Struct. Mol. Biol.*, 2007, **14**, 234.

29. T. H. Lee, G. W. Lee, E. B. Ziff and J. Vilcek, *Mol. Cell Biol.*, 1990, **10**, 1982.
30. T. H. Lee, H. G. Wisniewski and J. Vilcek, *J. Cell Biol.*, 1992, **116**, 545.
31. C. M. Milner and A. J. Day, *J. Cell Sci.*, 2003, **116**, 1863.
32. H. G. Wisniewski, R. Maier, M. Lotz, S. Lee, L. Klampfer, T. H. Lee and J. Vilcek, *J. Immunol.*, 1993, **151**, 6593.
33. M. T. Bayliss, S. L. T. Howat, J. Dudhia, J. M. Murphy, F. P. Barry, J. C. W. Edwards and A. J. Day, *Osteoarthritis Cartilage*, 2001, **9**, 42.
34. R. Forteza, S. Casalino-Matsuda, M. E. Monzon-Medina, M. S. Rugg, C. M. Milner and A. J. Day, *Am. J. Respir. Cell Mol. Biol.*, 2007, **36**, 20.
35. D. J. Mahoney, K. Mikecz, T. Ali, D. Benayahu, A. Plaas, C. M. Milner, A. J. Day and A. Sabokbar, manuscript submitted.
36. A. J. Day, M. S. Rugg, D. J. Mahoney and C. M. Milner, in *Hyaluronan: structure, metabolism, biological activities, therapeutic applications*, eds. E. A. Balasz and V. C. Hascall, Vol. II. Matrix Biology Institute, Edgewater, NJ, 2005, p. 675.
37. C. M. Milner, V. A. Higman and A. J. Day, *Biochem. Soc. Trans.*, 2006, **34**, 446.
38. B. E. Prosser, S. Johnson, P. Roversi, A. P. Herbert, B. S. Blaum, J. Tyrrell, T. A. Jowitt, S. J. Clark, E. Tarelli, D. Uhrin, P. N. Barlow, R. B. Sim, A. J. Day and S. M. Lea, *J. Exp. Med.*, 2007, **204**, 2277.
39. S. A. Kuznetsova, D. J. Mahoney, G. Martin-Manso, T. Ali, H. A. Nentwich, J. M. Sipes, T. Vogel, A. J. Day and D. D. Roberts, *Matrix Biology*, 2008, **27**, 201.
40. H. A. Nentwich, Z. Mustafa, M. S. Rugg, B. D. Marsden, M. R. Cordell, D. J. Mahoney, S. C. Jenkins, B. Dowling, E. Fries, C. M. Milner, J. Loughlin and A. J. Day, *J. Biol. Chem.*, 2002, **277**, 15354.
41. D. J. Mahoney, J. D. Whittle, C. M. Milner, S. J. Clark, B. Mulloy, D. J. Buttle, G. C. Jones, A. J. Day and R. D. Short, *Anal. Biochem.*, 2004, **330**, 123.
42. A. A. Parkar and A. J. Day, *FEBS Lett.*, 1997, **410**, 413.
43. S. A. Kuznetsova, A. J. Day, D. J. Mahoney, D. F. Mosher and D. D. Roberts, *J. Biol. Chem.*, 2005, **280**, 30899.
44. A. Salustri, C. Garanda, E. Hirsch, M. De Acetis, A. Maccagno, B. Bottazzi, A. Doni, A. Bastone, G. Mantovani, P. Beck Peccoz, G. Salvatori, D. J. Mahoney, A. J. Day, G. Siracusa, L. Romani and A. Mantovani, *Development*, 2004, **131**, 1577.
45. I. Inoue, R. Ikeda and S. Tsukahara, *J. Pharmacol. Sci.*, 2006, **100**, 205.
46. M. S. Rugg, A. C. Willis, D. Mukohpadhyay, V. C. Hascall, E. Fries, C. Fülöp, C. M. Milner and A. J. Day, *J. Biol. Chem.*, 2005, **280**, 25674.
47. L. L. Espey and J. S. Richards, in *Physiology of Reproduction*, ed. J. D. Neill, Elsevier, Amsterdam, Vol. 1. p. 42.
48. T. Fujimoto, R. C. Savani, M. Watari, A. J. Day and J. F. Strauss, *Am. J. Pathol.*, 2002, **160**, 1495.
49. O. Carrette, R. V. Nemade, A. J. Day, A. Brickner and W. J. Larsen, *Biol. Reprod.*, 2001, **65**, 301.

50. D. Mukhopadhyay, V. C. Hascall, A. J. Day, A. Salustri and C. Fülöp, *Arch. Biochem. Biophys.*, 2001, **394**, 173.
51. C. Fülöp, S. Szántó, D. Mukhopadhyay, T. Bárdos, R. V. Kamath, M. S. Rugg, A. J. Day, A. Salustri, V. C. Hascall, T. T. Glant and K. Mikecz, *Development*, 2003, **130**, 2253.
52. S. A. Ochsner, D. L. Russell, A. J. Day, R. M. Breyer and J. S. Richards, *Endocrinology*, 2003, **144**, 1008.
53. H. G. Wisniewski, W. H. Burgess, J. D. Oppenheim and J. Vilcek, *Biochemistry*, 1994, **33**, 7423.
54. J. J. Enghild, I. B. Thogersen, F. Cheng, L. A. Fransson, P. Roepstorff and H. Rahbek-Nielsen, *Biochemistry*, 1999, **38**, 11804.
55. J. J. Enghild, I. B. Thogersen, S. V. Pizzo and G. Salvesen, *J. Biol. Chem.*, 1989, **264**, 15975.
56. H. Sato, S. Kajikawa, S. Kuroda, Y. Horisawa, N. Nakamura, N. Kaga, C. Kakinuma, K. Kato, H. Morishita and H. Niwa, *et al.*, *Biochem. Biophys. Res. Commun.*, 2001, **281**, 1154.
57. L. Zhuo, M. Yoneda, M. Zhao, W. Yingsung, N. Yoshida, Y. Kitagawa, K. Kawamura, T. Suzuki and K. Kimata, *J. Biol. Chem.*, 2001, **276**, 7693.
58. W. Yingsung, L. Zhuo, M. Mörgelin, M. Yoneda, D. Kida, H. Watanabe, N. Ishiguro, H. I. Iwata and K. Kimata, *J. Biol. Chem.*, 2003, **278**, 32710.
59. J. Lesley, N. M. English, I. Gál, K. Mikecz, A. J. Day and R. Hyman, *J. Biol. Chem.*, 2002, **277**, 26600.
60. S. A. Ochsner, A. J. Day, M. S. Rugg, R. M. Breyer, R. H. Gomer and J. S. Richards, *Endocrinology*, 2003, **144**, 4376.
61. K. W. Sanggaard, H. Karring, Z. Valnickova, I. B. Thogersen and J. J. Enghild, *J. Biol. Chem.*, 2005, **280**, 11936.
62. K. W. Sanggaard, C. S. Sonne-Schmidt, C. Jabobsen, I. B. Thøgersen, Z. Valnickova, H. G. Wisniewski and J. J. Enghild, *Biochemistry*, 2006, **45**, 7661.
63. S. J. Getting, D. J. Mahoney, T. Cao, M. S. Rugg, E. Fries, C. M. Milner, M. Perretti and A. J. Day, *J. Biol. Chem.*, 2002, **277**, 51068.
64. W. Selbi, A. J. Day, M. S. Rugg, C. Fulop, C. A. de la Motte, T. Bowen, V. C. Hascall and A. Phillips, *J. Am. Soc. Nephrol.*, 2006, **17**, 1553.
65. C. M. Milner, W. Tongsoongnoen, M. S. Rugg and A. J. Day, *Biochem. Soc. Trans.*, 2007, **35**, 672.
66. L. Scarchilli, A. Camaioni, B. Bottazzi, V. Negri, A. Doni, L. Deban, A. Bastone, G. Salvatori, A. Mantovani, G. Siracusa and A. Salustri, *J. Biol. Chem.*, 2007, **282**, 30161.
67. D. Mukhopadhyay, A. Asari, M. S. Rugg, A. J. Day and C. Fülöp, *J. Biol. Chem.*, 2004, **279**, 1119.
68. C. Mindrescu, G. J. Thorbecke, M. J. Klein, J. Vilcek and H. G. Wisniewski, *Arthritis Rheum.*, 2000, **43**, 2668.
69. C. Mindrescu, A. A. M. Dias, R. J. Olszewski, M. J. Klein, L. F. L. Reis and H. G. Wisniewski, *Arthritis Rheum.*, 2002, **46**, 2453.
70. T. Bárdos, R. V. Kamath, K. Mikecz and T. T. Glant, *Am. J. Pathol.*, 2001, **159**, 1711.

71. T. T. Glant, R. V. Kamath, T. Bárdos, I. Gal, S. Szanto, Y. M. Murad, J. D. Sandy, J. S. Mort, P. J. Roughley and K. Mikecz, *Arthritis Rheum.*, 2002, **46**, 2207.
72. S. Szántó, T. Bárdos, I. Gál, T. T. Glant and K. Mikecz, *Arthritis Rheum.*, 2004, **50**, 3012.
73. T. V. Cao, M. La, S. J. Getting, A. J. Day and M. Perretti, *Microcirculation*, 2004, **11**, 615.

Subject Index

Page references to *figures* and *tables* are shown in *italics*.